QUANTUM EFFECTS IN BIOLOGY

QUANTUM EFFECTS IN BIOLOGY

Sisir Roy
National Institute of Advanced Studies, India

NEW JERSEY • LONDON • SINGAPORE • BEIJING • SHANGHAI • HONG KONG • TAIPEI • CHENNAI • TOKYO

Published by

World Scientific Publishing Co. Pte. Ltd.
5 Toh Tuck Link, Singapore 596224
USA office: 27 Warren Street, Suite 401-402, Hackensack, NJ 07601
UK office: 57 Shelton Street, Covent Garden, London WC2H 9HE

British Library Cataloguing-in-Publication Data
A catalogue record for this book is available from the British Library.

QUANTUM EFFECTS IN BIOLOGY

ISBN 978-981-310-982-7 (hardcover)
ISBN 978-981-310-983-4 (ebook for institutions)
ISBN 978-981-310-984-1 (ebook for individuals)

For any available supplementary material, please visit
https://www.worldscientific.com/worldscibooks/10.1142/10038#t=suppl

Desk Editor: Rhaimie B Wahap

Typeset by Stallion Press
Email: enquiries@stallionpress.com

This work is dedicated to my beloved wife Malabika — without whom all that is said here would not have come about.

Preface

Quantum theory is considered to be the most successful theory in explaining the behavior of the most of the physical phenomena in the universe to an unprecedented level of accuracy. So people thought that the functional process in a living organism may be also explained by the laws of quantum physics. We are not much concerned with this fact in quantum biology. One of the biggest questions in biology is whether the processes of life are able to exploit quantum effects to improve their lot. If nature has not found a way to exploit quantum mechanics, an equally important question is : why not? Is it merely an oversight on the part of evolution or is there some other deeper reason why evolution can not exploit quantum mechanics? Indeed this is a very deep question. We do not know the answer yet or even I am not going to give an answer to this question in this book. The main focus of this book to identify some of the biological processes which can not be explained using Newtonian dynamics. Someone may call it as non-trivial applications of quantum theory in contrast to the idea of universe made of elementary entities like electrons, protons, etc, which are governed by the laws unveiled in microscopic domain and hence known as trivial applications of quantum theory.

In 2013, Lambert made an overview of the two main candidates for biological systems which may harness such functional quantum effects: photosynthesis and magnetoreception. In this book two chapters have been devoted to discuss how quantum principles are used to understand the photosynthesis. Here, quantum coherence plays an important role. In Chapter 2, the Role of Quantum Mechanics in Photosynthesis has been discussed in details with comprehensive historical perspective. We start with a definition of Photosynthesis that is a process in which light energy is captured and stored by an organism, and the stored energy is used to drive cellular processes. Infact, molecular mechanisms of energy transport in photosynthetic light-harvesting complexes have been a subject of study for decades. The plants, algae, and phototropic bacteria use solar energy to synthesize high-energy molecular species that power life. The physical chemist Theodor Förster[1,2] opined, out of his pioneering research and observations that observed electronic excitation are transferable by means of a physical process, i.e., incoherent, radiation-less scattering process. The theoretical framework developed by Förster, followed by others, explained in details about the process of related electronic excitation migration in the photosynthetic apparatus of plants, algae, and bacteria which starting from the light absorbing pigments follows up to the so-called reaction centers (RC). In this chapter we elaborated the seminal contributions of Förster and his followers.

It is worth mentioning that the researchers reported evidence that photosynthesis — the process with the help of which, green plants and some other bacteria turn sunlight into chemical energy — gains light-harvesting efficiency through the exploitation of the phenomenon, called "quantum coherence". That is, even though the biological molecules are constructed of atoms, the associated forces, being of a 'quantum origin', bound them together. But, still the role of quantum coherence (and more recently, entanglement), being dependent on time or energy scales, have a direct and dictating role to play in biological function, even if keeping their roles still controversial. The recent experimental results on photosynthetic 'Light Harvesting Complexes'(LHC) together with their constituents, e.g., the Fenna–Matthews–Olson (FMO) pigment-protein complex in green sulfur bacteria, have put on the strong evidences which suggest, especially, the presence of quantum coherence playing an unavoidable and a crucial role, not only, in one of the very fundamental but also an important biological processes, i.e., the energy transport together with energy conversion, this way, persuading the proper investigation further on. In this experiment, critical experimental observations have been performed in order to measure coherent oscillations or "quantum beats", observed via femtosecond laser spectroscopy. This comprise the observation of coherent oscillations, observable in many different light harvesting systems. Consequently, substantial number of important experimental and theoretical studies have been followed recently, with the aim of observing the way which could have been followed during ultrafast energy transfer in light harvesting, particularly, with the aim of investigating the role of coherence. We have discussed Quantum coherence and photosynthesis in Chapter 3.

Olfaction process, is accepted as one of the most ancient and again at the same time, one of the most intriguing characteristics of living organisms. It is the oldest and most fundamental aspect of chemical sensing which are being applied by almost all kind of lifeforms in interpreting their surroundings. The process of smelling is caused by certain kind of small molecules, neutral, volatile in nature, known as odorant. For example, in human beings, the odorant molecules, binding to specific sites on olfactory receptors in nasal cavity, causes finally the olfaction process. This process has certain typically interesting and fascinating characteristics, thus attracting the science community greatly, deliveing a great number of unique theories each of which tried to define and explain henceforth the mechanism behind such process. Even though there have been considerable knowledge of structure of ORs (Olfactory Receptors), detailed knowledge about the molecular mechanisms needed in discriminating different odorants has not yet been fully understood except the already accepted as most obvious characteristic of an odorant molecule, is its shape (Lock and Key concept). The works of Luca Turin regarding the different smell even from same shape raise lot of debates among the community. The quantum model of olfaction, firstly proposed by Dyson and refined by Wright, is based on the idea that the signature of scent is caused due to the odorant's unique vibrational spectrum not its structure. An unique scent is attributed to its unique spectrum in the same way a colour is associated to its unique frequency of light. At present, it is an well known fact that the Quantum effects in biology, a continually growing field of interest, includes, for example, coherent energy transfer within photosynthetic bacteria proteins, mechanism of the avian magnetic compass and the possibility of inelastic electronic transfer (IET) occurring in olfactory receptors to name a few. This is critically discussed in Chapter 4.

"How migratory birds know which way to fly" — one of the most intriguing scientific puzzles which baffled the scientists since 19th century. Bird navigation is a complex enterprise, requiring birds to make repeated and varying orientation decisions based on directional and positional information. Birds are aided by multiple physiological compass systems, among them a physiological magnetic compass. Since long, scientists have speculated that certain animals are making use of the planet's magnetic fields to find their way, but biologists are mystified as to how they might do it. Now some answers might be coming from one of the most perplexing interactions in physics, related to the planet's magnetic fields. There are currently two leading hypotheses to explain this remarkable ability: (a) the magnetite-based mechanism, and (b) the radical pair mechanism. In certain species, including certain birds, fruit flies and even plants, the evidence supports a so-called Radical Pair (RP) mechanism. This process involves the quantum evolution of a spatially-separated pair of electron spins and such a model is supported by several results from the field of spin chemistry. An artificial chemical compass operating according to this principle has been demonstrated experimentally, and a very recent theoretical studies examines the presence of entanglement within such a system. Chapter 5 unfolds this story of applications of quantum principles in navigation of birds.

Ion channel plays very important role in information processing of brain. The opening and closing of gate of the channel is a random process characterized by Markovian one. The noise associated with this random process may also affect the cognitive domain. The relevance of quantum theory and the concept of coherence in understanding the dynamics of the movement of ions in ion channel are critically analyzed in Chapter 6.

Quantum theory is not only a successful theory explaining most of the physical phenomena in this universe but also helps to understand functionality of the some of the living organisms. In recent decades the idea of quantum coherence, quantum entanglement and other nonclassical effects attracted large attention to clarify some of the fundamental issues towards systems of increasing complexity. On the otherhand, more and more refined explanations of macroscopic phenomena for living organisms are put forward based on the improved understanding of molecular structures and mechanisms. These developments in animate and inanimate objects help us to study some of the phenomena related to living organisms using the concepts of quantum theory. Recent experimental findings clearly suggest that the classical probability theory is still not successful to explain modalities in human cognition, especially in connection to decision making. The major problem seems to be the presence of epistemic uncertainty and its effects on cognition at any time point. Moreover, the stochasticity in the model arises due to the unknown path or trajectory (the definite state of mind at each point in time) a person follows. The consideration of "black box model of human mental functions" produces much ambiguity. A generalized version of probability theory, borrowing the idea from the quantum paradigm, may be a plausible approach. Quantum theory enables a person to be in an indefinite state (superposition state) in the context of neurobiology, especially in relation to central nervous system and allows all these states to be potentially (of course, with prior probability amplitudes) expressible at each moment. Thus, a superposition state seems to provide a better representation of the conflict, ambiguity or uncertainty that a person experiences at each moment. However, the framework of quantum probability considering the superposition of mental states is an abstract framework devoid of material content like the concept of elementary particles and various fundamental constants in nature. This framework can be applied to any branch of science dealing with decision making, such as biology and the social sciences. Very few attempts have been made so far in the context of neuroscience and higher order cognitive activities. The quantum principles may shed new light in decision making and the various mental states as envisaged in ancient Indian wisdom.

References

[1] Förster T. (1946); Naturwissenschaften.; 6:166–175.
[2] Förster T. (1948); Zwischenmolekulare Energiewanderung und Fluoreszenz; Ann Phys (Leipzig). 1948; 2:55–75.

Sisir Roy, Bengaluru, 2024

Acknowledgements

The author is indebted to National Institute of Advanced Studies for giving a wonderful opportunity and environment to write such a comprehensive book. The author is grateful to his collaborator Dr. Sudharsana Iyenger for her help in compilation of the text and figures. I am indebted to my another collaborator Mr. Sarangam Majumdar for preparing the figures with proper resolutions and beautiful cover design. Finally I must say that without the help of team of World Scientific Publishers under the guidance of the Desk editor Rhaimie B Wahap it would not be possible to publish the book in its present form.

Contents

Chapter 1

Introduction

Quantum mechanics is certainly the best theory of nature with the exception of gravity, explaining all known phenomena from atoms to condensed matter properties, and, furthermore, to an unprecedented level of accuracy. So the functional process in a living organism may be explained by the laws of quantum physics. We are not much concerned with this fact in quantum biology. One of the biggest question in biology is whether the processes of life are able to exploit quantum effects to improve their lot. If nature has not found a way to exploit quantum mechanics, an equally important question is: why not? Is it merely an oversight on the part of evolution or is there some other deeper reason why evolution can not exploit quantum mechanics?

There are good reasons why the possibility of quantum biology could have, until now, seemed remote. Basic quantum phenomena (the wave particle duality, the Heisenberg uncertainty relations, quantum entanglement, the peculiar nature of quantum statistics, to name a few), until recently, have been thought to occur only at the level of one or a few subatomic particles and to fade away in the presence of larger physical aggregates (like the ones studied in molecular biology). Equally, the need for an observer and/or his highly sophisticated instruments was an impediment. Biological phenomena seem to have no need of human observers. None seem to have been around when evolution began. In spite of such barriers, and others that might be mentioned, some of the founders of quantum physics believed that in the end quantum physics would provide the real explanation for life. Among these were Erwin Schrödinger,[1] Niels Bohr.[2] It is now beginning to be conceivable that they were right. Recently Lambert *et al.*[3] made an overview of the two main candidates for biological systems which may harness such functional quantum effects: photosynthesis and magnetoreception. Photosynthesis is arguably the fundamental process of life, since it enables energy from the Sun to enter the food chain on the Earth. It is a remarkable non-equilibrium process in which photons are converted to many body excitations, which traverse a complex biomolecular membrane, where they are captured and fuel chemical reactions within a reaction center (RC) in order to produce nutrients. The precise nature of these dynamical processes—which lie at the interface between quantum and classical behavior and involve both noise and coordination is still being explored. Nowadays there is an increasing amount of controversies and implications of a contemporary scientific scenarios trying to establish the fundamentals of several biologists' research activities i.e. how quantum theory is needed to understand such biological activities. Before going into the discussions of the existing controversies, let us elaborate briefly the "quantum weirdness" — for example, superposition principle, concept-of non-locality, quantum coherence and entanglement.

- Superpositon: one of the intriguing aspects of physics is related to the fact that quantum states are well-defined even when they refer to situations, which we would describe as a coexistence of mutually excluding possibilities in classical physics. We usually adhere to Aristotle's statement of non contradiction that "the same attribute cannot at the same time belong and not belong to the same subject." Quantum physics teaches us that we either have to give up this wisdom or otherwise renounce other well-established intellectual concepts.
- Concept of Non-locality: Einstein-Podolsky-Rosen(EPR) suggested an experiment which gives rise to the debate on the "incompleteness" of quantum mechanics (QM). The concept of non-locality was introduced in QM. This is a metaphysical concept. Non-locality suggests that universe is in fact profoundly different from our habitual understanding of it, and that the "separate" parts of the universe are actually potentially connected in an intimate and immediate way. In fact, Einstein was so upset by the conclusions on non-locality at one point that he declared that the whole of quantum theory must be wrong, and he never accepted the idea of non-locality till his dying day.
- Quantum Entanglement: Quantum entanglement is a strange and non-intuitive aspect of the quantum theory of matter, which has puzzled and intrigued physicists since the earliest days of the quantum theory. Quantum entanglement represents the extent to which measurement of one part of a system affects the state of another; for example,

measurement of one electron influences the state of another that may be far away.

- Quantum Coherence: one of the hallmarks of quantum mechanics, as opposed to to classical mechanics, is the existence of coherence in particle mechanics, caused by interference of probability amplitudes. (The Schrodinger equation is, after all, a wave equation.) The classic example of this is the "2-slit problem" as is usually discussed in the introductory lectures of a quantum mechanics course: a particle has two possible paths from source to detector (e.g., a screen) by going through hole 1 or hole 2 in a barrier; there is a (complex) amplitude associated with each path, one adds these amplitudes, and the square (modulus) of the net amplitude gives the probability distribution at the screen. There is an oscillatory structure due to the cross term when squaring the sum of the two amplitudes. (If there were an infinite number of "slits", e.g., as for a crystal, rather than just two, then the peaks in the oscillatory pattern would narrow up to be delta functions at the Bragg diffraction angles.) If one neglects this cross term, then one obtains the result from classical mechanics, i.e., the probability distribution at the screen is the sum of the probability distributions for the particle going through hole 1 or hole 2, and there is no oscillatory diffraction pattern.
- Classical Coherence: However, there can also be coherence in classical mechanics. Consider, for example, the position-position time correlation function, $x(0)x(t)$, for a classical (or quantum) harmonic oscillator; an elementary calculation gives a result that is proportional to $cos(\omega t)$, where ω is the vibrational frequency, i.e., one has "coherent" vibrational motion. In classical wave optics, one can get interference of waves and define coherence too. How does one distinguish between coherence phenomena that are of a classical origin and those that are intrinsically quantum mechanical? Coherence effects may be of quantum or classical origin and it is not always so obvious which it is. To make matters even more ambiguous, sometimes, whether the observed coherence stems from a quantum mechanical or classical origin, depends upon what is being observed!
- Quantum weirdness in Biology: why do we so often ignore quantum dynamics in studying biomolecular functions? When this question is asked, many of those who do computer simulations give the answer that most of the atoms in biomolecules are heavy and so they have a small wavelength; thus they act merely as points in their motion. The simulation experts might also say quantum effects such as tunneling are

seen but only in low-temperature cryogenic experiments, where owing to the lower velocity of atomic motion the thermal de Broglie wavelengths are larger. When pressed further, this glib answer might be amplified by acknowledging that, yes, proton transfer reactions and electron transfer reactions are important at physiological temperatures throughout the cell and that these processes involve quantum effects but, again, these random transfer events only occur once the ponderous motions of the heavy atoms of the proteins have tuned the energetics of the reacting molecules to allow such intrinsic quantum processes to occur near a resonance between two alternatives.

1.1 Quantum Coherence in Biology

We will concentrate our discussions on the issue of quantum coherence in the following areas of biology.

- Photosynthesis: The evidence for quantum coherence is considered to be strong.
- Sense of smell: The conventional postulate for the mechanism of smell is a lock and key model, in which different types of odorant molecules bind to different types of olfactory sensors. However, the nose actually behaves like a vibrational spectrometer. The only known mechanism that can explain this vibrational sensitivity of the sense of smell is quantum mechanical.
- The avian compass: some birds, such as homing pigeons, possess a small piece of magnetite in their beaks which functions as a compass, allowing them to tell North from South.
- Ion Channel:Ions propagate through the channels which are responsible for the transmission of information in the brain. Noise assisted ionic movement in the channel and its coherence indicate the application of quantum theory in the brain.

1.1.1 *Quantum Entanglement and Photosynthesis*

The Fenna-Matthews-Olson (FMO) complex (a light harvesting complex) can be seen as a multi-arm interferometer where each arm of the interferometer corresponds to a site of FMO complex. The propagation of an exciton through the FMO complex is analogous to propagation of a photon through an interferometer. A single photon propagating through the interferometer produces entanglement between the arms and so the propagation of excitation may also produce entanglement between the sites of the FMO complex. However, there are some open issues which need to be analyzed carefully before making any conclusive remark. First, the entanglement between the arms of the interferometer occur depending on the initial state concerning photon. In the case of photosysnthesis in which the entire light harvesting complex (FMO complex) is illuminated by weak classical light, not by a single photon, the entanglement may therefore occur in the FMO complex.

1.1.2 *Quantum Physics and Olfaction*

Smell is one of our five basic senses, but the key steps in the mechanism of smell (olfaction) remain unknown. We know that particles move from the source of smell through the air to our nasal membranes by the process of diffusion, and we are very familiar with the neurological pathways that happen after the odor has been detected, but the happenings in between are still in debate. It is usually suggested that molecules and receptors function as simple 'lock and keys' — that a molecule has a categorical shape that fits into a corresponding receptor in our noses. This is known as the Shape Theory — smell influenced by the shape of odorant and receptor. However, it is to be mentioned that similarly-shaped molecules sometimes have different smells, and very different-shaped molecules sometimes have similar smells. This led to the proposal of a new theory — Vibration Theory. In order to explain the mechanisms of vibration theory, we have to turn to quantum mechanics — specifically electron tunneling. If correct, this is truly exciting science: every repulsive or irresistible smell has an explanation directly grounded in quantum physics.

1.1.3 *Navigating Birds and Magnetorecptor*

Bird navigation is a complex enterprise, requiring birds to make repeated and varying orientation decisions based on directional and positional information. Birds are aided by multiple physiological compass systems, among them a physiological magnetic compass. The existence of a magnetic compass was discovered in orientation experiments with birds in cages by Wolfgang Wiltschko *et al.* in late 1960s.[4] In the 1970s, studies[5] indicated that weak magnetic fields can influence chemical processes involving photoactivated radical pair intermediates, i.e., a transient pair of molecules with an un-paired electron spin each. The underlying mechanism was shown to be based on the effects of magnetic fields on the electron spin evolution in each of the radical pairs and investigation of such effects opened the now mature field of spin chemistry. It was Klaus Schulten[6] who first suggested that this radical pair mechanism might operate in the compass of migratory birds.

1.1.4 *Ion Channel and Information Propagation in Brain*

On short time scales, the brain represents, transmits, and processes information through the electrical activity of its neurons. On long time scales, the brain stores information in the strength of the synaptic connections between its neurons. Electrical activity in neurons is mediated by many small membrane proteins called ion channels. Although single ion channels are known to open and close stochastically, the macroscopic behavior of populations of ion channels are often approximated as deterministic. This is based on the assumption that the intrinsic noise introduced by stochastic ion channel gating is so weak as to be negligible. Both computers and brains work electrically. But their charge carriers respectively are different electrons in a solid ion lattice and ions in a polar liquid. It is an intellectual and technological challenge to join these different systems directly on the level of electronic and ionic signals. Already in the 18th century, Luigi Galvani established the electrical coupling of inorganic solids and excitable living tissue. Today, after fifty years of dramatic developments in semiconductor microtechnology and cellular neurobiology, we may envisage such an integration by far more complex interactions, right on the level of individual nerve cells and microelectronic devices. Conceptually three significant functional domains of all ion channels are:

- Ion conducting pore: An aqueous pathway for ions with a narrow selectivity filter that distinguishes among the ions that do go through and the ions that do not.
- Gates: a part of the channel that can open and close the conducting pore.
- Sensors: detectors of stimuli that respond to electrical potential changes or chemical signals. The sensors couple to the channel gates to control the probability that they open or close.

Scientists try to explain the dynamics of ion channel using the two approaches within classical physics:

- Molecular dynamics
- Brownian dynamics

However, the present author along with his collaborators have shown[7] that quantum mechanical description is needed to explain the dynamics of $K+$ ion channel and it raises a lot of interesting issues such as decoherence time, etc.

1.2 Implications and Controversies

Nowadays there is an increasing amount of controversies and implications of a contemporary scientific scenarios trying to establish the fundamentals of several biologists' research activities and the role of quantum mechanics. Pivotal to this story is the dispute between the largely accepted fact that quantum mechanics lies at the heart of all micro and nano-world phenomena, and its role played in biology. Just to start with as an example, the entangled strands of DNA, the famous double helix of the molecule of life, are considered to be held together by a quantum phenomenon known as hydrogen bonding. The way in which those strands untwist and build new double helixes during the process of reproduction is at heart a quantum phenomenon, closely related to the way in which quantum entities such as electrons can be both wave and particle at the same time.

It has been argued for some time that a hot and wet biological environment limits or destroys subtle, non-obvious quantum effects (coherence, tunneling, entanglement, etc.). Recent evidence, however, suggests that a variety of organisms may harness some of the unique, non-trivial features of quantum mechanics thus gaining a biological advantage over classical equivalents. Results for non-trivial quantum effects in photosynthetic light harvesting, avian magneto reception and several other candidates for functional quantum biology are discussed by Lambert *et al.*[3] Regarding light harvesting in photosynthesis, Sarovar and collaborators[8] presented strong evidence for entanglement observation in a protein structure that is central to photosynthesis by green anoxygenic bacteria. This constitutes the first rigorous quantification of entanglement in a biological system.

Recently, that the role, quantum coherence may play for efficient light harvesting, has caught some notoriety (discussed also in a prior Perspective,[9] particularly due to experimental results by Fleming, Engel and Scholes. In time-resolved two-dimensional spectroscopy it is possible to see oscillations of exciton state populations, special initial states prepared by carefully chosen laser pulses. The oscillations, lasting up to a few hundred femto seconds, are attributed to quantum coherence emerging as a result of the initially prepared coherent quantum state and decay rapidly (compared to the typical lifetime of excitation in photosynthetic systems of one nanosecond). However, Panitchayangkoon *et al.* investigated the possibility of long lived quantum coherence in photosynthesis at physiological temperature[10] in a series of recent papers. It is interesting to note that Scholes argues that "nature's solar cells have been improved through billions of years of evolution. It is therefore not so surprising that plants, algae, and other photosynthetic organisms have developed tricks that lie behind their success. It has emerged recently that one of these strategies, at least for some organisms, is to use quantum coherence to direct energy flow from molecule to molecule through antenna proteins. Natural light-harvesting antenna systems are a prominent component in the photosynthetic machinery".

So the following broad, interdisciplinary questions for future works and queries arise in our investigating minds. For example, how can chemists use quantum coherence in synthetic systems (perhaps in organic photovoltaics)? Why did certain photosynthetic organisms evolve to use quantum coherence in light harvesting? Are these electronic excitations entangled? More trivially, there is a plethora of evidence showing that tunneling occurs in enzymes and that, most importantly, tunneling can speed up a reaction by a thousand times or more. As pointed out elsewhere, some enzymes seem to adjust the amount of tunneling to suit their operating temperature.[11]

It is difficult to explain many biological processes at molecular level in terms of classical principles. This shortcoming suggests that the structure of living matter, as described by most of the approaches employed in biology, is inadequate. In fact, life is a final and sophisticated product elaborated by the quantum nature of inanimate matter. Actually, the 'new orthodoxy' in the physics foundations presumes all nature quantum (i.e., that all natural phenomena are most completely and adequately described quantum-mechanically), whereas classical physics provides only a rough approximation. But, by no means is biology simply applied quantum mechanics.

We expect that our humble and sincere trial of writing this book will enhance the reasonable extrapolation from the known into the unknown, and plausible speculation to give an accessible overview of a revolutionary transformation in our understanding of the living world, might be looking at robins or migratory birds, flying high in the blue or black, dark sky towards north or South pole with more respect in future.

In this book, we discuss the details of various light harvesting mechanisms and the challenges for modelling in Chapter 2. The recent availability of an sufficient and increasing number of atomic resolution structures, as well as availing the ultrafast spectroscopy data for different photosynthetic systems made it possible for such detail analysis. It was Förster who proposed in his pioneering work as well as applied quantum physics for the description of fundamental biological process. Here, we mainly discuss Förster's theory in the context of photosynthesis.

In recent times, due to the availability of an sufficient and increasing number of atomic resolution structures, as well as availing the ultrafast spectroscopy data for different photosynthetic systems, the details of various light harvesting mechanisms have been possible, for comparison. However, further modeling challenges are provided by multi-subunit light harvesting systems. Because, both these kind of theoretical and experimental facts, i.e., multiple protein-pigment complexes, interact systematically, resulting to an intense as well as detailed studies which have taken the pieces together regarding the the evolutionary history of photosynthesis. Quantum coherence plays a key role understanding the applicability of quantum theory in biological systems especially photosynthesis. Quantum coherence and quantum entanglement are discussed in the context of photosynthesis in Chapter 3. Navigation of birds and the role of quantum theory become one of the fascinating areas of quantum biology. This is discussed in Chapter 4. Theory of smell is usually based on the use of "Lock and Key concept". But the recent discovery of Luca Turin indicated the failure of "lock and Key Concept" and the proposition of quantum mechanical ionic vibration attract a large attention. We elaborate this in Chapter 5. Ion channel plays an important role in iformation processing of the brain. The noise assisted movement of ions and the issue of coherence become a fascinating area of research. This is discussed in Chapter 6. Finally, the challenges and future directions of research are envisaged in Chapter 7.

References

[1] Schrödinger, E; (1944); What Is Life?; Cambridge University Press.

[2] Niels Bohr; (1929); Collected Works Volume 6; 1985; Pg; 219–221, 223–253.

[3] Lambert N, Chen YN, Cheng YC, Li CM, Chen GY *et al.*, (2013) Quantum Biology. Nature Physics 9:10–18.

[4] Wolfgang Wiltschko and Friedrich W. Merkel (1966). Orientierung zugunruhiger Rotkehlchen im statische Magnetfeld. Verhandlungen der Deutschen Zoologischen Gesellschaft, 59:362–367.
[5] Roswitha Wiltschko and Wolfgang Wiltschko (2006) Magnetoreception. BioEssays, 28:157–168.
[6] K. Schulten (1982) Magnetic field effects in chemistry and biology J. Treusch (Ed.), Festkörperprobleme, 22, Vieweg, Braunschweig, pp. 61–83.
[7] Sisir Roy (2014) Quantum effects in Biological systems, Nature's Longest Threads, World Scientific Publishers, pp. 27–36.
[8] Sarovar M, Ishizaki A, Fleming GR, Whaley KB (2010) Quantum entanglement in photosynthetic light-harvesting complexes, Nature Physics 6: 462–467.
[9] Scholes G. (2010) Quantum-coherent electronic energy transfer: Did Nature think of it first? J. Phys. Chem. Lett.; 1:2–8
[10] Panitchayangkoon G, Hayes D, Fransted KA, Caram JR, Harel E, Wen J, Blankenship RE, Engel GS.; (2010); Proc Natl Acad Sci USA.;107(29): 12766–12770.
[11] Ball P (2004) Enzymes: by chance, or by design? Nature 431: 396–397.

Chapter 2

Photosynthesis: Förster's Theory & The Role of Quantum Mechanics in Photosynthesis

"Whoever in the pursue of Science, seeks after immediate practical utility, may rest assured that he seeks in vain". —Helmholtz

2.1 Introduction

All most all kind of Life on our Earth, as our knowledge, gained from numerous sources, especially from nature, leads us up till today that it exists as well as flourishes, largely because of photosynthesis, the process through which light energy is converted into chemical energy, especially, by plants, algae (Blakenship and photosynthesis bacteria etc.[1–11]). Because of its undoubted importance for all life on earth, molecular mechanisms of energy transport in photosynthetic light-harvesting complexes have been a subject of study for decades. We define this process, adopting a somewhat narrower definition of photosynthesis: "Photosynthesis is a process in which light energy is captured and stored by an organism, and the stored energy is used to drive cellular processes." Thus higher plants, algae, and phototropic bacteria use solar energy to synthesize high-energy molecular species that power life. More than 10 quadrillion photons of light strike a leaf each second. Incredibly, almost every visible photon (those with wavelengths between 400 and 700 nanometers [nm]) is captured by pigments and initiates the steps of plant growth. It is now quite an well known phenomenon that every possible kind of existence, for example, animal as well as human life, are heavily dependent on plant life, either directly or indirectly. This, in turn, follow and depend on certain kind of precisely developed mechanisms, for example, the process of photosynthesis which, being one of the basic criteria for the survival as well as flourishing of of biochemical machines, also plays a responsible role in carrying tasks, essential for plant kingdom, s survival and flourishing in our planet.

For this purpose, all of these bacteria, algae, and plants employ a molecular machinery, especially developed and suitable for their own special, specific kind of functions, for example, in capturing, as well as, transforming solar energy into a proton gradient across the cellular membrane. This membrane, being responsible for a multitude of cellular processes, plays also a critical role by participating in driving the process. The amount of knowledge regarding photosynthetic systems, accumulated up till now, likely evolved into presently existent enormous diversity, out from a humble common ancestor, because, in many respects, all of these systems employed the same physical principles. Not only that, these principles have been found to be universal in many respects.

Started by Baptista van Helmont[7] who partially discovered photosynthesis, this process made a long journey ranging several centuries. Later on, Joseph Priestley carried out an experiment showing indirectly and almost unknowingly that showed that plants produce oxygen, thus giving the birth of an idea called photosynthesis. However, it was Jan Ingenhousz (1730–1799),[12] a Dutch, best known scientist for showing that light is essential for photosynthesis which made him recognized and remembered as the person who discovered photosynthesis. By now, the process of light harvesting involved in the photosynthesis, has been decided as being the primary source of energy in the biosphere, responsible for the existence of life on Earth. As we are quite aware, all the diverse manifestation in Biological system, already been noted by now, is incredibly complex. This ever existent complexity hardly diminishes even when we focus on processes occurring in the cell level, for example, the process of respiration or the action of enzymes, dealing the problems as the averaging process, had been much more encouraged taking statistical one, rather than considering the problem from the point of quantum mechanical (or coherent) phenomena.

Basically, the objective of photosynthesis is to produce biological energy, which is generated in direct ratio to photo-induced charge-separation reactions occurring in reaction-centre complexes. This process is initiated, starting with the absorption of a single photon by a pigment molecule which culminates with the production of adenosine-5′-triphosphate (ATP). The process of Hydrolysis of ATP, producing adenosine-5′-diphosphate (ADP), releases the stored chemical energy, initially created by the proton motive force that drives ATP synthase. This hydrolysis reaction is used throughout the cell to drive energy requiring metabolic processes. Photosynthesis can be represented by the general equation:

$$2H_2A + C0_2 \overset{light}{\rightarrow} CH_20 + H_20 + 2A$$

where H_2A is the source of hydrogen ions and electrons used to reduce carbon dioxide to carbohydrate (CH_20), and A is the oxidation product. In plants, algae and cyanobacteria, water is the reductant and oxygen is the oxidation product (oxygenic photosynthesis) in the process of photosynthesis. Thus this process is responsible for the conversion of light energy in a successive manner into a more stable forms of energy storage. There, the electronic excitation executes a random walk among tens to hundreds of molecules in the antenna complexes until it is either trapped by a reaction center (RC) or decays to the ground state. Or, more specifically, at this juncture, we could elaborate the process like this: the light harvesting starts with an absorption of photons by chlorophyll or carotenoid pigments, which is embedded within an ensemble of proteins; the systems of these pigments and proteins being called light harvesting complexes (LHC). The most crucial step in light harvesting is the conversion of the electronic excitation into a charge-separated state across the cellular membrane where this process takes place in a pigment-protein complex, known as the reaction center (RC). Therefore, the efficiency of light harvesting depends on the rate of each energy transfer step (hop), compared with the lifetime of the excited state, which, for chlorophylls, is typically a few nanoseconds. Tangibility in light harvesting is less compared to many other biological processes, but, classically, it can still be viewed as a hopping of electronic excitation from one molecule to another in the spirit of a theory reported by Förster in 1948.

Starting with the process of the light absorption, terminating ultimately, into a nanosecond-lived electronic excitation, in turn, the above mentioned process leads to a charge-separated state, with the ultimate goal of reaching to a membrane proton gradient. Or, at this juncture, more specifically, we could elaborate this process as: the light harvesting starts with an absorption of photons by chlorophyll or carotenoid pigments, embedded within an ensemble of proteins; the systems of these pigments and proteins being called light harvesting complexes. The most crucial step in light harvesting is the conversion of the electronic excitation into a charge-separated state across the cellular membrane which takes place in a pigment-protein complex, known as the reaction center (RC). In 1932, Emerson and Arnold being the first group demonstrated categorically that there exists a specific ways of cooperation among hundreds of pigment molecules, in light-harvesting. This fact has fascinated, especially, the biophysicists since then. Though Photosynthesis is based on the interaction between living matter and the sun's radiation field, mainly visible light, their interaction involving the electrons of biological macro-molecules, especially, all kind of plants together with bacteria, algaquantume etc. These phenomena lead us, accordingly, to the light absorption, basically governed by quantum physics. It appears, during the course of biological evolution, photosynthetic lifeforms, in their own indigenous way, learned to exploit quantum physics, particularly, under the circumstances of physiological temperature. But, in describing these phenomena caused by the influence, related to the various kind of strong thermal effects, arising under these circumstances place a challenging task before us, all of which factors, since then, makes the quantum biology of photosynthesis an active as well as a fascinating research area.

In 1929, Bohr,[13] while delivering a lecture to the Scandinavian Meeting of Natural Scientists, entitled "The atomic theory and the fundamental principles underlying the description of nature": focusing on the successes of quantum mechanics in describing the nature of the atomic and subatomic world, he moved on to consider whether it might have something to say in biology: "Before I conclude, it would be natural at such a joint meeting of natural scientists to touch upon the question as to what light can be thrown upon the problems regarding living organisms by the latest developments of our knowledge of atomic phenomena which I have here described." So, this way, the remarkable efficiency of photosynthetic light harvesting has been a mystery since the 1930s. The above statement had influenced Pascuel Jordon[14] who wrote then, arguably, the first scientific paper on quantum biology in 1932, in the German journal "Die Naturwissenschaften" with the title "Die Quantenmechanik und die Grundprobleme der Biologie und Psychologie" ("Quantum mechanics and the fundamental problems of biology and psychology")

Schrödinger pioneered this idea in his (1944) book "What is Life?[15] It should be remembered that at that time when Förster's work was published, Schrödinger's seminal book[15] stating that physical theory be applied to describe biological processes was still fairly new. As delivered by McFadden (2018),[16] summarized in a series of beautiful lucid lectures, Schrödinger hypothesized this pioneering idea in his book (1944) stating that "*quantum mechanical effects in living systems may be evident, and indeed active, if the number of particles in the system of interest is very small, hidden in the average response of a very complex system*". In fact, Emerson and Arnold[25] was the first who reported some unique characteristics, found in their path breaking experiment during measurement of the maximum oxygen evolution from the suspensions of a green algae, Chlorella pyrenoidosa, used to be considered as a model organism that time. In that experiment, after light flashes being applied repeatedly, light flashes saturated briefly, separated by an optimum dark period, This they observed, in contrary to the expectations of the results, i.e., it was expected that time that about 2,400 chlorophyll (Chl) molecules were required to produce one oxygen molecule only. This pioneering experiment triggered an serious inquiry and hence starting an revolutionary period, related to the quantum biological aspects of photosynthesis. In summery, this experiment and corresponding observations pointed to a very important acts of cooperation in energy harvesting of pigment molecules present in photosynthesis. Not only that, these facts have also been noticed for the production of one oxygen molecule, acting among many chlorophyll (Chl) molecules of green algae.

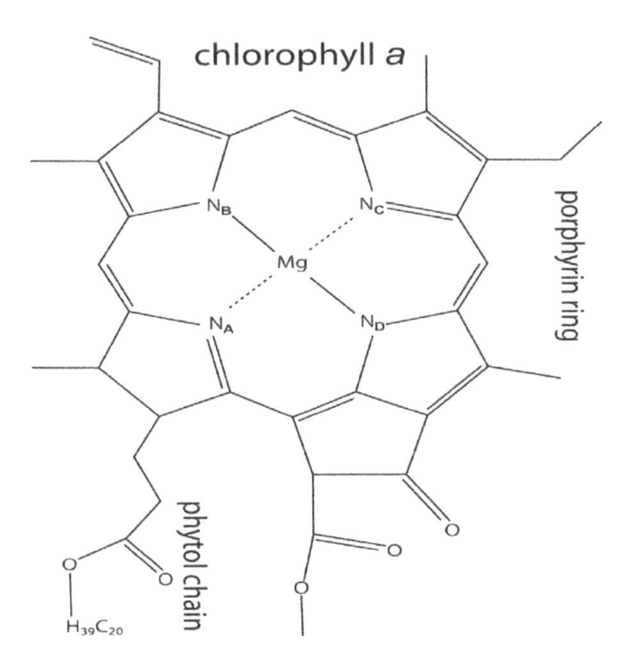

Figure 2.1. Structure of chlorophyll a molecule. The transition dipole moment of the lowest (Qy) state lies approximately in the plane of the porphyrin ring, along the vector connecting the NB and ND atoms.

To be precise, together with such an understanding, we state Photosynthesis, in general, as encompassing various processes in living cells following which life forms utilize sunlight to drive chemical synthesis. For example, the primary processes of light-harvesting, i.e., transformation of electronic excitation energy into a membrane potential includes the splitting of water into oxygen, abstracting electrons, added to molecules of nicotinamide adenine dinucleotide together with the phosphate (NADPH+) at a high redox potential. The role of membrane potential, in such cases, is to drive the synthesis of adenosine triphosphate (ATP), responsible for fueling many of the processes in living cells. For example, in plant photosynthesis, NADPH+ and ATP are needed for the synthesis of sugar and starch, the most widely known products of photosynthesis. Because of its fundamental importance in cellular energetics, photosynthesis has been the subject of great evolutionary pressure, because of the fact, i.e., amidst a deep overall similarity, many variants being developed in the competition connected to the habitats, depending on their efficiency, i.e., not for only survival but also for growth. This way, absolutely vital and the basic process involved, is to convert the energy of photon into a more stable energy forms, i.e., in the form of pigments with electronic excitations. The next step follows the process of the charge separation which exist across the cell membrane. As has already been observed, Photosynthetic organisms have also been found in having features of supra-molecular assemblies. These, by turn, contain hundreds of cooperating protein sub-units that, starting from the process of absorption of a photon, convert the short-lived electronic excitation. This important characteristics not only transforms into more stable forms of energy in a gradual increasing manner but also into the form of ultimate stable chemical bonds [Van Amerongen *et al.*,[17] Blankenship,[9] Cogdell RJ, *et al.*,[19] Şener, MK. & Schulten, K.[20]].

Again, under changing environmental conditions and that also keeping high reliability in nano-scale machinery, some of the basic aspects, related to the vital characteristics of biological cells, are expressed through unique combination of high efficiency and adaptability. After obtaining substantial progress in atomic resolution structures, proper opportunities have been achieved in explaining these biological functions of cellular machinery, including those of the underlying physical mechanisms. Some of quite astonishing facts have been observed in such findings, for example, i.e., even at room temperature, fluctuations of site energies have little effect on the calculated excitation lifetime and associated quantum yield. Also, the efficiency of the system has been found to be robust against 'pruning' of individual chlorophyll. It should be pointed out that considering an optimal condition, and comparing the quantum yield as well, the arrangement of chlorophylls has been observed, to be connected to an ensemble of randomly oriented chlorophyll. However, as mentioned by the group,(Melih Sener & his group), they observed a narrow channel of interval within which the quantum yield change, in such an ensemble. This fact made the researchers quite surprised, i.e., it appears, the demand for highest possible efficiency of light harvesting has been shaped by the evolution of photosynthetic species in an amazing way, starting from bacteria to plants. But, added to that, some other extraordinary and interesting facts appear to us as quite surprising as well as extraordinary too, i.e., despite the presence of such a great variation in architecture as well as in the environment associated with photosynthetic phenomena, it displays a substantial amount of common structural themes, determined by the physics of energy transfer only.

During such a junction of many queries and consequent seminal contributions, the physical chemist Theodor Förster[21,22] opined, out of his pioneering research and observations that observed electronic excitation are transferable by means of a physical process, i.e., incoherent, radiation-less scattering process. Now, under the prevailing necessary condition which is possible to be operated with the range extending from a donor to an acceptor molecule, the resonance could exist in between excited states of donor and acceptor. He applied the idea of quantum physics, in explaining the related biological processes. In this juncture, it should be mentioned that, in 1941, Oppenheimer,[23] though largely unnoticed, related the physical basis of the energy transfer process meant for a donor-acceptor pair, but, not that much in explicit way, as pointed out in FRET formula of Förster. The basics of this model have been presented by him in a short abstract of a talk at a Physical Society meeting, dealing with nuclear physics. It was Knox[24] who presented an account of Arnold and Oppenheimer's contributions in detail where he stated: "*The resonant transfer would take place, driven by the strong near zone field of the initially excited molecule's transition moment. As an irreversible process, such a mechanism would have a rate proportional to R^{-6}, where R is the intermolecular separation*". Importantly, n 1956, through the observations of depolarization of chlorophyll a fluorescence, Arnold together with E. S. Meek[25] provided the first clear evidence for the existence of excitation energy migration among chlorophyll a molecules.

Under such an retrospective condition, when Förster published his path breaking works, the seminal book of Schrödinger's was still fairly new which stated about the possibilities, to the application of above physical theory in the related biological processes. The theoretical framework developed by Förster, followed by others, explained in details about the process of related electronic excitation migration in the photosynthetic apparatus of plants, algae, and bacteria which starting from the light absorbing pigments follows up to the so-called reaction centers (RC). There, light energy, is utilized for the eventual conversion into chemical energy. It was Förster who was pioneering person in unifying the theoretical models of strong to very weak coupling. The title of his manuscript was "*Delocalized Excitation and Excitation Transfer*" where he pointed out that a very important, but common characteristics, may exist in many system, i.e., the ground state properties could be found as additive in additive in nature, but the behavior of the corresponding excited state properties can be of quite different nature, i.e., from what could be obtained out of the simple summation of the individual component properties, "*it is among a few crucial characteristics which allow an illuminating demonstration of the fact that properties of living beings ultimately rely on and are determined by the laws of physics*".

At such a juncture, in light harvesting, from bacteria to plants, *this very demand for availing highest possible efficiency*, establishes the fact, i.e., in having shaped the evolution of photosynthetic life forms together with its applicability in the development of this framework. Förster and coworkers worked hard in order to establish this fact, i.e., giving stress on the basic principles of quantum physics which showed the successful applicability of these principles, in different forms of life, besides others, reigning all possible photosynthetic lifeforms on Earth. As a result, following Förster's theory, we could not only explain the beautiful colors in plants, algae, and photosynthetic bacteria but also, have been able to point towards a few telling signs about the success in applications of Förster's theory. To do that, we must consider the crucial roles, played by quantum theory in order to fuel and sustain the life on Earth, taking the sunlight as the sole source of energy. Also, this theory governs exclusively the detailed ways about the migration of electronic excitation occurring in the process of photosynthesis. Especially, the process which are being executed and followed by the photosynthetic apparatus of plants, algae and bacteria, together, throws the light in explaining the facts connected to the execution of the process. This starts from the active part taken by the pigments, followed by the crucial role of performance played by the reaction centers (RC), after which, finally, the light energy becomes capable of taking active part in the process of eventual conversion into chemical energy. Though, in photosynthesis, a great variation in architecture, found on the basis of quantum physics of energy transfer, it was the first theoretical formulation of pair transfer which successfully identified the inverse sixth power distance dependence, i.e., the process itself, now being almost universally associated with, was made by Förster in 1946.

Förster first realized that the role of pigments which has a crucial role to play in the process of photosynthesis. To mention a few, many important bases are found for the transference of excitation transfer, which, are simultaneously responsible as well as absolutely essential in photosynthetic light harvesting. Within the present arena of observations, it has already been revealed that, pigments, being scaffolded by intricate protein structures, are able to carry out their role in an extremely efficient way and that again in an amazing hierarchical assemblies. All of these facts encouraged, Förster, might be, to make a famous statement "*quantum effects rule how nature harvests sunlight*". After a long, strenuous journey of research, present conclusions have also reached to the same understanding, might be, with a little addition, i.e., we could comment that, following Sener, indeed, *he was definitely right, but, real role of quantum physics is even more beautiful than what he could envisage*, even though, theoretically, we are quite aware about the underlying fundamental processes in nature, namely, harvesting of sunlight by photosynthetic life forms, together with its applicability in the development of this framework. Subsequent experimental verification of Förster's[26] idea, has been performed, first, by Latt and his group.[27] All of these efforts, when recast in quantum mechanical terms, have produced a theory of 'radiationless' energy transfer. Not only that, it has found a substantial amount of success and enthusiasm for taking the efforts for furthering the queries, producing more and more successful applications for well over half a century. This theory has still been continuing to be widely valid, although subjected to certain conditions, not originally understood but, gradually unfolding before us as a result of continuing theoretical as well as experimental improvisations together with sophistication of many kind of modern experimental techniques, of course, based on equally diversified ideas, applied, in doing so.

In the following section, Förster theory of excitation transfer has been summarized including recent extension. In doing so, substantial help (including different pictures for the necessary demonstrations) have been taken related to the theory, together with beautiful pictures of illustration and necessary equations developed in a detailed and exhaustive way, from the Melih Sener & Schulten's group.[27–29, 39, 120] They established the relevance of their approach by demonstrating the applications of the theory in photosynthetic systems, proposed by them. In doing so, they took the proper factors, necessary in order to explain almost all possible intricate steps in great detail, especially found in purple bacteria, developed different approaches following different school of thoughts related to particular specimen involved, taking the other potosynthetic cynobacteria, and plants etc. In order to address this challenge, they investigated theoretically, in utmost detail, especially, those problems, related to the functioning of protein-pigment complex in photosystem I (PSI) of the cyanobacterium Synechococcus elongatus. All the related mechanisms coming within the limitation and range, applicable to Förster's theory, have been taken care of, out of which results, they noticed some of the following interesting characteristics, connected to these biological cells, for example:

(1) These cells contain nano-scale machinery, always exhibiting a unique combination of high efficiency, quite unique to the working of the system,
(2) Possessing high adaptability, following the change of environmental conditions,
(3) Not only that, such grade of their fantastic performance posses high reliability which characteristics make them perfectly effective and suitable for the process,
(4) Follows a few among the various important quantum mechanical aspects.

We are already aware by now that for the chlorophyll aggregate, an effective Hamiltonian is essentially needed in order to describe the excitation transfer dynamics associated with the spectral properties of PSI. This crucial and vital problem has been studied in utmost detail, by many groups, especially as stated earlier, by Melih Sener, together his group and also, by several other groups, especially, that of Olaya-castro' *et al.*,[31] Mohseni,[32] Izhiaki and Flemming[33] to name a few of them. Each of the above mentioned group have considered many of crucial kind of problems related to photosynthesis but taking different samples, under different physical conditions but in an utmost exhaustive way. Not only they explored applicability of different mathematical tools in these photosynthetic processes but importantly, applied different quantum mechanical phenomena, successfully, in explaining the detailed intricacies of the photosynthetic processes, basically responsible for the extraction of the most needed energy from the SUN, absolutely necessary for the survival of every possible kind of the living organisms.[35–37,136,141] Sener's group[39] took the advantage of the results obtained from the contemporary experimental ventures, some of the valuable information regarding high resolution structure of the purple bacteria. The structure, thus obtained, exhibit an aggregate of 96 chlorophylls, surprisingly, almost in a perfectly complete manner. These bacteria have been found to posses, not only an extra but also a very important characteristics. They are not only electronically coupled but in addition to that, simultaneously, they also function as well, playing the role of an antenna complex, by taking part also in light-harvesting process. For example, two trans-membrane protein-pigment complexes for light-harvesting, named photo-systems I and II have been observed being employed in the Oxygenic photosynthetic species, playing quite a crucial role in the photosynthesis. Interestingly, it has been found that in such cases, typical role of Photosystem I (PSI) is to play as a ubiquitous protein-pigment complex, which absorb sunlight, found, for example, in green plants, algae and cyanobacteria, located in the bacterial membrane.

To begin with, we discuss about the process of light absorption by chlorophylls which poses two conceptual challenges to quantum biologists. Firstly, due to coupling between vibrational and electronic degrees of freedom, the chlorophylls develop their broad absorption characteristics but, only when close to physiological temperature. A key factor for having a broad absorption, it must be capable of utilizing the continuous solar spectrum which is one of the key factors to biological function again. Added to this, the role of temperature in shaping chlorophyll spectrum is of great relevance from the biological perspective. The spectra of chlorophylls in bacterial photosynthesis are also shaped by chlorophyll–chlorophyll interactions. For example, the light-harvesting proteins of the purple bacteria contain rings of closely (about 10^0 Å centre–centre distance) spaced chlorophylls forming excitons after light absorption. The excitonic interaction affects the light absorption characteristics of chlorophyll rings present in the light harvesting proteins. Next point to be mentioned is PSI's quite important role containing the antenna complex, reaction center (RC) as well as the electron transfer chain. Importantly, all of these are contained in the same protein which plays a crucial as well as vital role in the process of photosynthesis. Added to this characteristics, this complex, has been observed to be responsible for the use of the solar energy, needed for the transference of electrons across the cell membrane.

Many groups, especially to be mentioned of Ishizaki and Fleming, exclusively studied this problem considering different kind of structural motifs while explaining this process. In their path breaking works Mino Yang & Flemming (2003),[34] modeled the energy transfer and trapping kinetics in PSI by developing rather a new approach, instead of applying Förster theory in a quite simplistic form. They applied self-consistent complex transfer process consisting of heterogeneous coupling, employed experimentally determined spectral densities in calculating the energy transfer rates. As a result, they have been enabled in reproducing the absorption spectrum as well as fluorescence decay time components of the complex at room temperature in a reasonable grade. Importantly, they also discussed the roles of the special kind of chlorophylls (for example, red, linker etc.) molecules. With the help of quite strenuous calculations, they availed the exact expression for the trapping. In doing so, they calculated the time for both the trappings and that of detrapping, i.e., the role of the intrinsic trapping time as well as mean first passage time all together. Importantly, their group, also successfully determined the slowest steps of the arrival, for the case of primary electron donor containing only two dominant steps,

(1) Transfer-to-reaction-center, and
(2) Transfer-to-trap-from-reaction-center;

i.e., the estimated intrinsic charge transfer time, approximately, of the order of 1.7 ps.

Their contribution also includes understanding as well as presenting the underlying different working processes in photosynthesis. Finally, they reported an unusual, surprising scenario, i.e., tracing of the presence of different sophisticated and complicated quantum mechanical processes or, simply to say, presence of all possible shorts of basic aspects of quantum characteristics in photosynthetic system. Added to these surprising observations, all of the associated functions have been observed to be amazingly active, profoundly diverse as well as traceable abundantly in nature. Also, many of such features associated with different quantum mechanical aspects, observed in having their own ramifications too; for example, these characteristics can be followed as well as explained in a specific manner, starting even from their point of origin. However, it is interesting to mention how Förster successfully adopted all of these aspects in developing the equation for phenomena of FRET between two pigments which can be expressed in a quite simple form as

$$\nabla \times k_{ij} = \frac{1}{\tau_0}\left(\frac{R_0}{r_{ij}}\right)^6 \tag{2.1}$$

which, later on, we will try to explore and explain in a more elaborate way.

Starting initially with the concepts, depicted by Förster, the above equation has been developed with the aim of expressing, importantly, overlapping (resonance) between two factors, in particular, namely,

(1) the inter-molecular distance
(2) donor and acceptor states

Quite a interesting point should be pointed out at this stage of development, important and surprising as well, i.e., these two processes must go together, following the rule with respect to the relative orientation of the so called transition dipole moments of the molecules.[21,22] Started first by Förster, these being both fascinating as well as quite challenging subject to be explored, has been observed and followed by the gradual increase of the theoretical interests and developments related to this. FRET have been quite successful, in explaining the applications as well as many important aspects of photosynthetic phenomena, for example, taking the case of purple bacteria, all the basics of the bacterial energy harvesting and others. Their successful research of observations and applications have been found to be successful in explaining the intricacies related to photosynthetic phenomena, thus producing vast reaching effects and enthusiasm as well among the researchers, associated with this field of research. As has already been known now, the main thrust behind all of these studies, done that time, was to explain, properly, some of the quite sensitive, as well as path breaking observed phenomena.

But, till then, even being path breaking attempts had to face some of serious questions, for example, the viability of explanations related to those observed phenomena. However, most important results obtained from all of such earnest and serious studies, did culminate, ultimately into, not only to an interesting but also a convincing and a very exciting field of research as well, opening up many of novel and unique opportunities, in following further explorations, developing serious interests among the researchers. Not only that, related to these exclusive interesting processes,[136,141] a new era of the possibilities and probabilities have also been started in order to apply many of quantum mechanical aspects. Importantly, many of modern, the then famous laboratories of life science have successfully been persuaded in continuing to develop these methods and techniques with the aim of exploring further on; might be, possible mysteries which could yet been kept in store by nature.[35] The physics of the excitation migration process, as followed in the applications and explanations by Förster's, is fairly well understood, describable within a reasonable degree of approximation by constructing an effective Hamiltonian. Considering all of these characteristics, i.e., all the successes and various applications of FRET, following Melih Sener,[39] we can repeat one of his many of famous comments, i.e., during the evolution of photosynthetic organisms, Förster's methodology of treating his formula for bacterial energy harvesting, played the role of a "design constraint".[18]

2.1.1 *Photosynthetic Unit (PSU)*

A multiplicity of different antenna structures, antenna complexes as well as different kind of pigments and polypeptides can be found in the various photosynthetic organisms. This happens due to presence of, mainly, diverse as well as selective environmental conditions, particularly light, and metabolic restrictions which dominates the evolution of antenna structures. Metabolic and structural constraints have also been observed as well, determining and coordinating at the same time, the evolution of pigments and antenna polypeptides. As the antennae are the part and linked as well to highly differentiated membranes, their diverse structures as well as that of the antenna polypeptides are found to be adjusted structurally to the various types of lipid membranes. Out of already investigated results, finally, it can safely be stated that a photosynthetic unit (PSU) is shared as a common feature with most of the photosynthetic species, containing peripheral pigment-protein (antenna) complexes, surrounding RCs. Or, in other words, being surrounded by chlorophylls and carotenoids, this can be termed as the biological equivalent of a solar cell, working with high efficiency and high adaptability. Thus, after excitons being created by initial absorption of sunlight in peripheral pigments, electronic excitations, being spread over several neighboring pigments, exhibit strong quantum mechanical coherence as has been reported over long ago.[37–40] Created by pigments, unless and until harvested at a RC, energy being stored inside the excitons, remains untapped. This could be emphasized as being the crucial reason for, not only an effective but also an efficient harvesting of light energy, i.e., it becomes a 'must' in demands for the transference of excitons to RCs. Associated with this process, there exists another crucial factor, i.e., absolute limit of the decay time of excitons, typically, must be limited within the range of tens of picoseconds. At the RC, the excitation energy transform into a more stable form ultimately, through electron transfer.[40,41] Importantly, as governed by Förster's formalism, this precondition, also have been found to be needed very much, for the rapid migration of energy among pigments, thus imposing constraints, not only on inter-pigment separation but also on the pigment energy levels. The presence of such thermal disorder appears as if the nature have adapted itself to be quite active in utilizing this kind of disorder, already existing at physiological temperature. For example, this disorder, acting as a facilitator, facilitates efficiently the transference of excitation between pigments which becomes a vital factor to be maintained in turn, in order to maintains a broad spectral resonance. This phenomenon is not only crucial but also a key factor in formulating the whole process.[42–44]

In the PSU, the time needed for the conversion of the energy coming from sunlight into a membrane potential with corresponding 95% quantum yield, is typically 100ps, thus providing the necessary energy for furthering metabolic processes. Due to the multi-component architecture of PSU, many options can be found avaiable in controlling the flow of the energy into PSU. This typical characteristics, i.e., the control mechanisms employed, need to be elucidated thoroughly which, together with various other reasons, motivated researchers in following extensive, rigorous research related to the physical mechanism, i.e., how the energy, captured from the sun, can be transferred between two pigment molecules. Not only that together with this, the constituent proteins, present in PSU, have also been determined for several species. For example, sometimes pigment-protein complexes have been observed being augmented by additional satellite complexes, expressed depending on physiological conditions (Sener *et al.*[46] In both the cases of oxygenic and anoxygenic species, reaction centre (RC) is the place where a charge separation[30,47] is initiated. However, in our present endeavor, we discuss only those cases, found in light harvesting systems, which are not only ubiquitous in nature but also has already been studied exhaustively. Considering such cases (Fig. 2.2), supra molecular architecture of PSUs has been studied systematically together with the associated details related to their constituent proteins. The supramolecular organization of the PSU in purple bacteria has recently been considered as a subject of exhaustive studies. Thus, following Föster, based on AFM[48–50] and EM[51] data, studies related to the geometrical details needed for the proper description of energy transfer across an entire photosynthetic membrane, has been obtained.

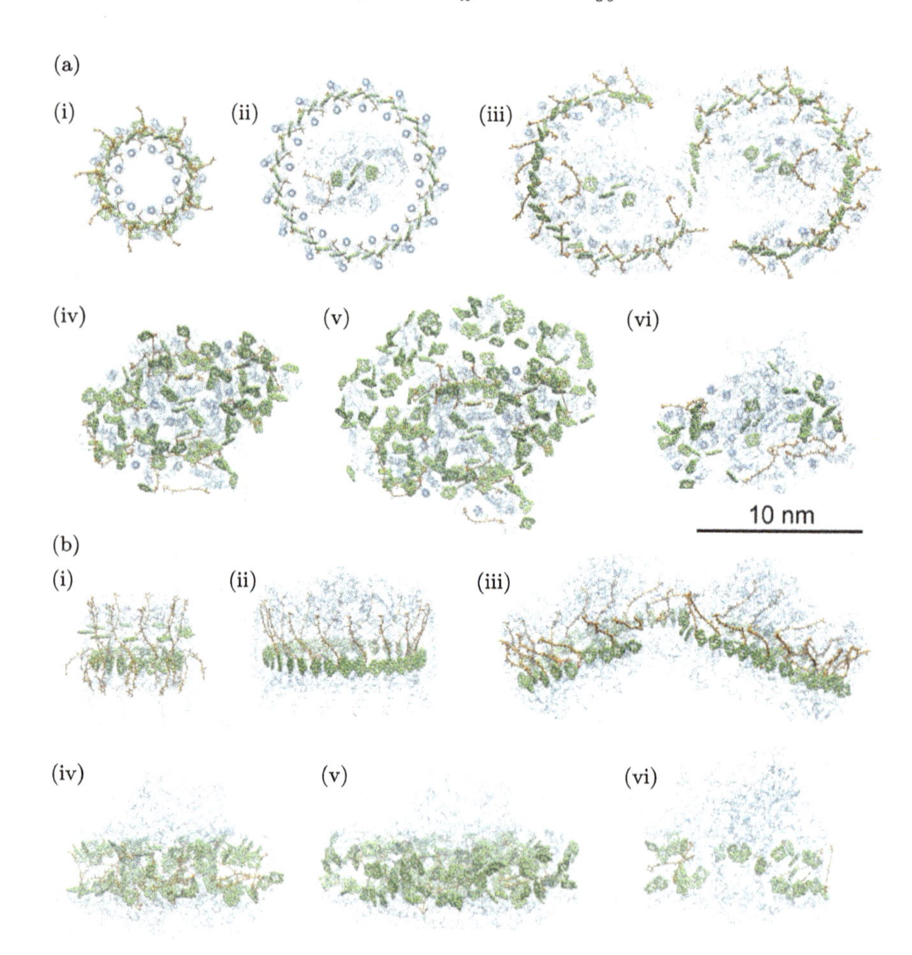

Figure 2.2. Pigment organization across different photosynthetic systems. (A) Top view (perpendicular to the membrane plane) and (B) side view (along the membrane plane) of pigment-protein complexes (i to vi) LH2,[40] RC-LH1 monomer,[41] RC-LH1 dimer model,[68] cyanobacterial PS1,[136] plant PS1 with Lhca subunits,[35] and PS2,[36] respectively. Protein components are shown as transparent blue traces to highlight the Chls and BChls (green; shown only as porphyrin rings) and the carotenoids (orange).

This contains thousands of bacteriochlorophylls (BChls).[52] Some of these unique characteristics have also been discovered for several species which can be categorized (see Fig. 2.2 from MEIH-SENER), for possessing the typical characteristics of their own. Förster's theory, quite satisfactorily, connects the findings of the aforementioned basic structural motifs in a collective way. For example, in the case of cyanobacteria and plants, the well known cases of oxygenic photosynthetic species, containing two pigment-protein complexes, i.e., photosystem I (PS1)[48–50] and photosystem II (PS2).[52,53] These have been found to cooperate when using light reactions and water. On the contrary, in the case of anoxygenic purple bacteria, the PSU have been found to contain the peripheral light harvesting complex II (LH2)[52–54] only, which in turn are coupled excitonically, to a RC-light harvesting core complex I (RCLH1).[53–55]

Interestingly, sometimes additional satellite complexes can also be found with the purpose of augmenting these pigment-protein complexes. However, their expression have been noticed to be quite dependent on physiological conditions,[55] coming into effect only when in need. Eventually, the production of the charge gradient have been observed for the utilization of ATP production, in turn, at the ATP synthase.[56,57] As a next step, the production of charge gradient occurs at RC where the charge separation is finally initiated.[58,59] From the recent discovery of supramolecular organization of the PSU in purple bacteria, based on AFM[48–50] and EM[52] data, these findings have been established in a confident way. Here, the geometrical details obtained, give the crucial information about thousands of bacteriochlorophylls (BChls), present across the entire photosynthetic membrane. Following Förster and also Sener *et al.*; (2010), all of these information have been found to be quite essential and fruitful, in describing the details of energy transfer process in photosynthesis.

2.1.2 *Excitation Migration Process & Förster's Theory*

As already been stated, the Photosynthetic organisms, being not only the fundamental but also an essential fuel for their survival, uses successful harvesting of sunlight for utilizing it's energy, i.e., for metabolism as well as nourishment and development. However, key features of the basics and essential cause for the development of an efficient apparatus for harvesting sunlight had conceptually been established following quite a long, as well as strenuous journey of research works. Thus, in order to do justice to Förster's numerous elegant theoretical analysis of great importance, we should mention following his publications (for details, Clegg *et al.*[60] and references there in) which can be cited, especially, those of M.Kasha, R.S. Knox,[61] Pearlstein[62] and others.

Among many of crucial and vital points, found in Förster's article, established the fact that the basic physical laws behind the functioning of the different energy transfer mechanisms, extending from strong coupling all the way through to quite-weak coupling (sometimes called intermediate), are not only related but also derivable from the same fundamental principles. He described this kind of variation related to the modes and dynamics of excitation transfer have been described in detail by him considering the relationship between the energies of interaction as well as the spectral widths of the electronic and vibrational spectral bands. In his pioneering contributions, he distinguished three types of coupling by studying in detail, i.e.: strong, weak and very weak. The theory, connected with these processes have been developed and applied successfully by Förster, followed by other dedicated researchers who elaborated clearly how the migration process of electronic excitation occurs in the photosynthetic apparatus of plants, algae, and bacteria as well, starting from light absorbing pigments. Their observations included, description of a quite interesting, typical characteristics of these processes, starting from the harvesting of sunlight, continuing its journey towards the final destination, until the so-called reaction centers (RC) is reached, where they cooperate for the proper utilization and absorption of light energy, eventually converting into chemical energy. However, in explaining this excitation migration process, the essential starting point is to describe as well as develop a proper definition of excitation transfer dynamics, followed by the application of an effective Hamiltonian for the chlorophyll aggregation, essentially needed for the description of the spectral properties of PSI.

At this point, some observed facts, crucial and astonishing as well, needs to be mentioned. Firstly, fluctuations of site energies have been observed to affect very little on the room temperature of the calculated excitation lifetime and hence associated quantum yield. Interestingly, certain facts, observed in the system, relates to the robustness of the efficiency of the system, for example, against 'pruning' of individual chlorophylls. At this point, it appears quite surprising in finding the quantum yield, related to it, taking into account of above findings as well and eventually, comparing with corresponding ensemble of randomly oriented chlorophyll. This, points to the fact that an optimal condition is found to be considered, in availing an efficient arrangement of chlorophylls. It should be mentioned that, in such an ensemble, as noticed by different groups, (especially, Melih Sener's group), only a narrow channel of interval has been observed for allowing the quantum yield change. Starting from bacteria to plants, i.e., as far as it is related to the evolution of photosynthetic species, amazingly, it should be noted that light harvesting appears to be influenced and shaped itself, just depending on the demand for highest possible efficiency.

As pointed out first, by Förster, another quite typical and extraoqdinary characteristics associated with this process is: even if there is an huge presence of variation in architecture, photosynthetic phenomena, displaying surprisingly common structural themes, are basically determined by the physics of energy transfer. Let us mention some of Förster's basic ideas in verbatim: especially, two of his sentences sum up the topic of his paper: *"The reason for this differences in the spectral and photochemical properties from the separate components is that in excited states the electronic excitation is not completely localized within one or the other of these components. The excitation may be completely delocalized and spread out over the whole system or, in a less drastic way, it may be localized only temporarily, but transferred from one component of the system to another."*

In the following section, taking together some glimpse of the recent developments, we summarize the Förster's theory of excitation transfer. In doing so, we have taken substantial help (especially, including different pictures for the necessary demonstrations), applying the same form of the necessary equations, developed in a detailed and exhaustive way by the above group.[26,28] In the following sections, the Förster theory will be presented related to their relevant applications in photosynthetic systems.

2.2 Different Stages in Photosynthetic Processes and Förster's Theory

It has already been stated as well as explained that the preliminary process in light harvesting starts, just after the light absorption, by transferring the accumulated light to the excitation, then to a reaction center (RC) as final destination. In this respect, so far as our present state of knowledge goes, in case of purple bacteria, most of all components of the PSU are studied in detail, at least at the scale of atomic resolution but we have been able in determining the structure of the RC through X-ray crystallography only. A very important finding resulted out of these detailed studies, stating that "a RC contains only 4 bacteriochlorophyll (BChl) molecules". This, being unable in availing a sufficient number with an optimal rate, further BChls are needed for the necessary amount of the sun light. In fact, in so-called light-harvesting complexes, further BChls have already been found to be present in RC in an organized manner. In essence, all of these act only as a basic starting point of transference in facilitating the separation of charge across the cell membrane. But, the vital and essential part in constructing the representative pathways for excitation transfer, includes another important factor, i.e., adding an expansion for excitation lifetime. However, this must be done by considering repeated trapping and subsequent detrapping events.Another much crucially needed requirement is related to the achievement of robustness and optimality of excitation transfer which is based on mean first passage times measurement. However, ultimately the quantum yield plays the role of deciding factor in assessing the above mentioned criteria, crucially needed for the fulfillment of all the essential and vital activities.

In the present case, following, especially, Melih Sener group's works on Photosystem I (PS1), the case of purple bacteria has been considered because of the simplicity in their system, so far as their construction mechanism involved. However, several examples of light harvesting networks, having varied nature and vast range are available, starting from the case of a single protein being placed between the pigments. Their applications have been quite successful in finding almost hundreds of proteins and thousands of pigments. Many of them have been detected even among extensive light harvesting networks. Taking the advantage of such kind of efficient apparatus, materials which are more complex, comprising various kind of light absorbing atomic or molecular components (chromophores), have been identified. Not only that, interestingly, optically well characterized absorption and fluorescence bands have also been observed which contain electromagnetic radiation of light. Added to that quite interestingly, these chromophores have been found commonly followed by a spatial translation of the absorbed electromagnetic radiation, lying in between different, though usually closely separated, chromophores. This is an important factor, among various possible roles played by the different mechanisms through which the transfer of energy from the antenna structures to the reaction centers (RC) occurs. Since long, it has been known that the chromophores posses very high number density in the photosynthetic antennae. More recently, various results obtained from many kind of structural studies, have discovered that, in many aspects, this constituent possess an exceptionally high degree of organization in which, might be, being an evolutionary gift from the nature, one can find exquisite and precise placement of Chl molecules relative to each other in many photosynthetic antenna systems. Many Chl molecules are also found quite close together, for example, only approximately (8–20)Å between neighboring Chl molecules, found to be held specifically at relative orientations which are present in large macromolecular complexes. These are hence quite expected in having a range of interaction energies, starting from strong to very weak.

2.2.1 *RET: Resonance Energy Transfer*

At the present stage of research developments, it has already been well-established fact that the initial light-harvesting process leads to the process of collecting energy at reaction centres (RC) which act is performed with almost perfect efficiency. In a photosynthetic system, this mechanism, fundamentally underlying the energy migration, is a quantum-mechanical one, being known as excitation energy transfer (EET) or resonance energy transfer (RET). A quite interesting factor, worth to be mentioned here, is that this process takes place, maintaining such a speed so that the underlying process is completed well before the occurrence of any effective thermal degradation in such materials. Another crucially important point to be mentioned here is that the process of photo-excitation is immediately followed by immediate relocation of the acquired electronic energy, accomplished by a mechanism known as Resonance Energy Transfer (RET). Resonance Energy transfer has been established, long before understanding of modern quantum mechanical molecular systems which, by then, had been observed and noted as sensitized luminescence. In 1982, Agranovich and Galanin[63] reported in their pioneering work that, normally, when a molecule becomes electronically excited due to the absorption of a photon, it is the luminescence only which is observed, emitted by another photon, happening within about a nanosecond. This phenomenon corresponds to fluorescence or much later for phosphorescence. However, one of their crucial observations stated that the possibility of swapping of its excitation with the molecule concerned, can happen only in the presence of another molecule, having similar excitation energy. Not only that this occurs within a distance of "tens of nanometres" which mechanism can be expressed as:

$$D^{\star} + A \rightarrow D + A^{\star}$$

$D^{\star}$, here, represents the excited state donor of the energy whereas, (D) being the ground state. Similarly, $(A^{\star})$ being the excited state acceptor, A being the ground state acceptor. This way, the excitation of D sensitizes that of A. Hence, this fact, i.e., reaching to a clear and sensible understanding of the long range interaction, happens in the RET process, but not within the reach of classical mechanics, involving matter–radiation interaction. In 1948, Förster[22] developed the first quantum-mechanical theory for RET and derived the following celebrated rate with the expression

$$k_F = \frac{9000(ln10)k^2}{128\pi^5 N_A \tau_D n_r^4 R^6}\left(\int d\tilde{\nu}\frac{f_D(\tilde{\nu})\epsilon_A(\tilde{\nu})}{\tilde{\nu}^4}\right)$$

Here, $f_D(\mu)$ depicts the donor, with normalized emission spectrum, $\epsilon_A(\mu)$ being the coefficient of molar extinction of the acceptor, R represents the distance in between the donor and the acceptor, n_r, the refractive index of the medium. In the above expression, N_A is meant for the Avogadro's number, τ_D being the spontaneous decay life time of the excited D. Here, k^2, is considered as the oriental factor, usually taken as $2/3$ in Förster's Rate expression. The particular form of the above form of equation has influenced in profound ways. Not only that the above expression has found an huge applications producing successful consequences, involving, especially the contributions to the fields of chemistry, physics together with biology; especially related to luminescence properties.[61–64] It is quite important to be noticed that Förster considered the relationship especially between the energies of interaction, in order to describe the variation, present in the modes and dynamics of excitation transfer, including the spectral widths present in the electronic as well as, also in vibrational spectral bands.

2.2.2 *Light Harvesting System*

Recent progress, especially, obtained from the above mentioned results, exploit the knowledge of atomic resolution structures in providing an atomic level explanation related to the biological functions associated with the cellular machinery inside a bacteria which includes that of underlying inside physical mechanisms. Quite important outcome of this knowledge brings an important opportunity, i.e. bringing the opportunity of studying the entire light harvesting system, consisting of a complex nanometric aggregate of transmembrane protein. Out of such kind of important experimental observations, the results obtained, a recognized fact has been achieved, i.e., these spectrum, being tuned through the local as well as excitonic interactions, may include the disorders, present in the light harvesting system, displaying a hierarchy of integral functional units as well. For example, purple bacteria has been found employing the easiest as well as a simple structure, compared to others in harvesting sun-light. Recent studies related to x-ray crystallography and electron microscopy has given the details through a combination of modeling, giving the detailed knowledge of the atomic structure for a main protein constituent, present in purple bacteria. Surprisingly, the presence of such a highly symmetrical architecture, found in these experimentally established structures, has been observed, clearly displaying a very close interplay of biological functionality with that of quantum physical processes. However, this fact is the result of highly symmetrical architecture and quite close interplay of biological functionality with that of associated quantum physical processes, however, demonstrating an illuminating fact i.e., basically and ultimately as well, all living beings rely "ON" and "ARE" determined by the laws of physics. As a quite simple but vital example, we mention the continued investigations regarding the light harvesting systems, present in purple bacteria.

However, the outcome of these clear results enthused researchers a very strenuous but effective and fruitful voyage, especially related to the biological research. But as an very consequence, these results, have also been enthusiastically followed, searching for the presence of any possible kind of functionality, not only related to the architectural elements but yet to be recognized. Finally, when with the help of such kind of serious, thorough, refined observations, followed by the theoretical analysis of related elements, these attempts ultimately achieved the goal, related to physical mechanisms as well been successful in achieving the goal by exploiting the knowledge of corresponding organisms, together with the cellular functioning. However, other idea connected to semiclassical models, have also been investigated thoroughly invoking the idea of 'hopping' of the excited-state populations occurring along discrete energy levels.[18,19] The excited state dynamics, initiated by light absorption, plays the central role in the primary reactions of photosynthesis. A crucial point, necessary to be remembered at this point is that excitation energy, if not successfully transferred away from the excited chromophores, within the excitation lifetime, relaxes back to the electronic ground state. This is caused by the underlying process which is either via emission of a photon (radiative decay) or happens through various nonradioactive processes. In this way, the photosynthetic machinery is quite capable for the successful control of the non-radiative relaxation rate. These characteristics point to the fact that it is quite capable of increasing itself under stress conditions (i.e., high light). However, this very important part is played through the adjustment of electronic properties of chromophores together with their interactions. But, on the contrary, under optimal conditions, it can achieve energy with >90% efficiency transfer. That is why it becomes necessary to understand some background related mechanistic aspects in photosynthesis, together with the initial events following photoexcitations of photosynthetic complexes.

2.2.3 *Reaction Center (RC)*

In harvesting, conversion of short-lived excitation energy is a crucial as well as a primary step in the photosynthetic processes as this can result in forming charge gradient, to be used more leisurely by a living cell. Reaction center refers to the sites in the pigment–protein complexes from which electron transfer reactions of the process of photoexcitation drives. In a protein complex, the reaction centers (RC) are found to be among some vital steps, taking place, as primarily, in the presence of sun rays, solar energy is converted into energy with the help of chemical bonds only. Again, existing as aggregates of protein in the intracellular membranes of these bacteria, the photosynthetic units (PSU) not only absorb sun light but utilize its energy for the synthesis of ATP. This light harvesting component involves ring-shape proteins, surrounding directly in the form of satellite rings i.e., the so-called reaction center (RC). The structure of these proteins have been established, followed by the combination of experimental and sophisticated computational methods, establishing the fact that proteins provide a scaffold for a hierarchical aggregate of chlorophylls and carotenoids which funnel electronic excitation towards these reaction centers (RC). This contains, mainly, complexes, beautifully arranged, peripheral pigment-protein (antenna), surrounding RCs. As already stated, the crucial as well as basic function of the RC is to receive excitation energy which can take place either by absorbing a solar photon directly, or by excitation transfer from the pigment molecules of nearby light harvesting complexes, the next step of which is to convert that energy into a charge-separated state. The RC contains four bacteriochlorophyll (BChl) pigments and two bacteriopheophytins (akin to a BChl without its magnesium) that absorb light.

The light energy, after being absorbed by all of six pigments, is eventually delivered as an excitation to the central pair of the four BChls. the termed being called as a special pair (SP = Chl1 + Chl2). The recent experimental findings has established the fact that the light absorbed by pigments, possessing higher electronic excitation energy (Chl3, Chl4 and Ph1, Ph2) lead to coherent (excitonic) oscillations between excited states by involving these pigments only. But, interestingly, this happens before settling into those particular pigments which have lower

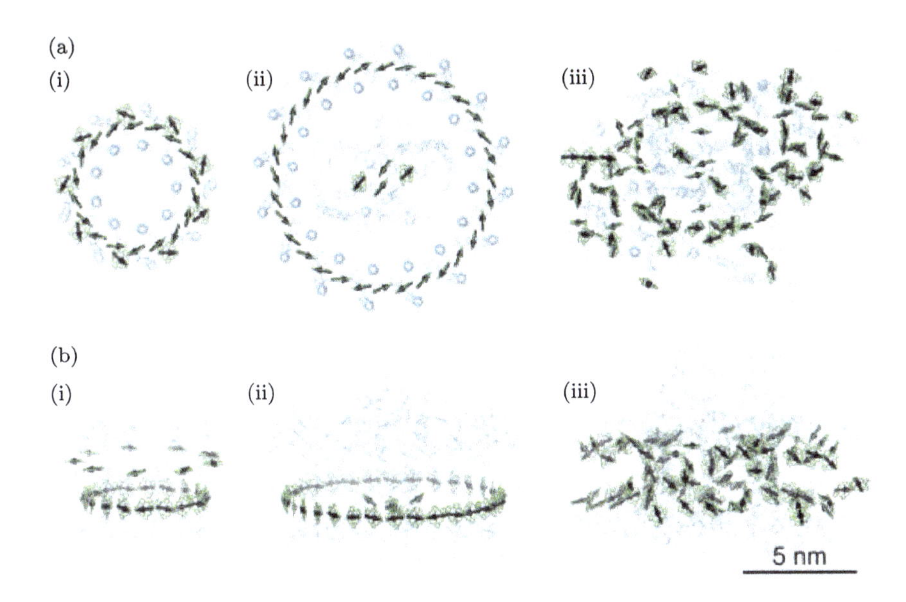

Figure 2.3. Orientations of the transition dipole moments of the constituent BChl and Chl molecules. (a) Top view (perpendicular to the plane of the membrane) and (b) side view (along the plane of the membrane); (i) LH2, (ii) monomeric RC-LH1, and (iii) cyanobacterial PS1.

electronic excitation energy only, namely Chl1 and Chl2 of the SP. Then the next step followed is when the SP initializes charge separation by transferring an excited electron through Chl3 towards a nearby pigment, i.e., a bacteriopheophytin (Ph1), followed by subsequent transference of electron to a permanently bound molecule of quinone, Q1. This, as a final step, lead to a second exchangeable quinone, Q2. Interestingly, this process, i.e., the transference of the series of electron, is established within about a hundred microseconds of a charge separated state $Q2{+}SP{+}$.

But, in the next stage, this energy, generally absorbed at pigment level, are not utilized there but, on the contrary, often transferred over hundreds of Ångströms, to reach to a reaction center (RC), with the purpose of initiating the transference of electron in order to form a membrane potential. Following Förster's path breaking works, researchers connected to this venture, offered some of the pioneering results, explaining especially, how, in the photosynthetic apparatus of plants, the electronic excitation takes the journey of migration, algae, and bacteria from light absorbing pigments to so-called reaction center (RC). The much efficient and diverse roles, played by the excitation can be termed as an sophisticated engineering tool, possessing all of the necessary ingredients and characteristics, demanding an outcome of light harvesting, to be with highest possible efficiency. For this very reason, scientists think this part, playing a responsible role for the evolution of photosynthetic species where the concerned process starts from bacteria, extending up to plants, even though, possessing a great variety in its architecture. All of these ventures display a few important points to some of the common structural themes whose basics might have guided Förster in describing the underlying physics of energy transfer by applying quantum mechanical principles.

2.2.4 *Photosynthetic Pigments*

The pigments, present in chloroplasts or photosynthetic bacteria, play a vital role in functioning of the process of photosynthesis. These act as the primary part of the components in a system of light harvesting usually known as base of the photosynthetic process which is responsible for the conversion of the energy of a photon absorbed into an electronic excitation. Many kind of pigments can be found depending on the services played by them; say, accessory pigment; chloroplast pigment; antenna pigment etc. The vast majority of chlorophylls, in fact, instead of being found in the reaction centers, are found to be used solely for light harvesting which means that they absorb sunlight, then transfer to the RC the corresponding energy through a network of intervening chlorophylls with an efficiency approaching 90% (van Grondelle *et al.*[36]; van Amerongen *et al.*[17]; Van Amerongen H, Valkunas L. & Van Grondelle R.[17] The result of such absorption cross section of the reaction centers is effectively increased by a factor of 200–300. Added to this fact, these light-harvesting chlorophylls allow a higher flux of excitation for their arrival at each RC in comparison to the isolated RCs. Thus, similar to the low to average sunlight, each chlorophyll also receives about 0.1–1 excitation per second, enhancing the effective absorption cross section of RCs which is quite relevant.

Due to the turnover time of the electron transfer chain being around 10 milliseconds, the above two facts nicely match, i.e., the light-harvesting antenna have been observed in allowing the reaction centers of photosystem I and photosystem II to operate at their optimal capacity, in which, electron transfer chain, however, have been observed to become saturated at higher light intensities. This fact leads to unwanted recombination reactions and the formation of triplet states in the RCs and light-harvesting chlorophyll molecules appears to be a problem, particularly in the reaction center of photosystem II. Basically, however, this is avoided, by switching off the light-harvesting antenna in a process called non-photochemical quenching (Non-photochemical quenching). Importantly, Chlorophylls are found to be essential pigments for all phototrophic organisms. Specifically, for example, in case of photosynthetic purple bacteria, these are anchored within protein complexes. Even though not each of these pigments carry out photochemistry but the vast majority of them function as antennas, collecting light, delivering energy to the RCs as well where the photochemical reactions take place. Or in other words, after collecting energy, it concentrates in a receiver by converting the signal into a different form, i.e., being purely a physical process, the most interesting part of this system is not meant for participating in any chemical reaction. Instead of that it depends on a weak energetic coupling of the antenna pigments only. This coupling works by an energy transfer process, involving the migration of electronic excited states from one molecule to another.

Thus the antenna system can conceptually be described as quite similar to a satellite dish. Specially, the light-harvesting complexes I and II (LH1 and LH2) play vital role which should be mentioned, i.e., the LH1 complex surrounds the reaction center (RC), forming the RC-LH1 core complex. Thus, pigments, here, are not only responsible for the energy conversion of an absorbed photon into an electronic excitation but also used both for the transportation of an electron across the cell membrane as well as for the production of the voltage gradient. Again, another quite important role, played by the pigment is that not all of the pigments get involved directly in the process of charge separation, but, instead, their main function is to transfer their excitation energy to reaction center (RC) in order to facilitate the transference of charge.

At this stage, we describe the light harvesting system, typically, as comprising of a reaction center (RC) which is surrounded by an array of peripheral antenna pigments. However their location depends on what will be their typical roles, needed to be played for the success of that role. Clayton once wrote[8] *"The pigment aggregate acts as an antenna, harvesting the energy of light quanta and delivering this energy to the reaction center as well"*. For example, their location, might be, in separate antenna complexes, excitonically coupled to the RC core complex or these might constitute parts of a fused photosystem which posseses both the network of peripheral pigments and a RC. Location of these pigments might be in separate antenna complexes, being excitonically coupled to the reaction center core, or they may constitute parts of a fused photosystem containing both the network of peripheral pigments as well as RC. The essential activity of these light-excited pigments is to pass its excitation energy onto a nearby pigments quite rapidly which, after arrival, deposit their energy into the RC within tens of picoseconds, to be utilized for charge separation, heading back again to the starting point, for later steps of photosynthesis to be followed.

Anyway, photochemistry is not carried out by each of the pigment. Instead, the vast majority of these pigments function as antennas, collecting light in order to deliver energy to the reaction center where the photochemical reactions take place. Conceptually, the antenna system can be described as similar to a satellite dish, in fact, collecting and then concentrating it in a receiver where the signal is converted into a different form. Being purely a physical process, this system is not meant for participating in any chemical reaction, instead depends on a weak energetic coupling of the antenna pigment which works by an energy transfer process involving the migration of electronic excited states from one molecule to another. Though various pigment- protein complexes are found to serve as antennas in photosynthesis organisms, they must meet many critical requirements, for example, absorb visible or near infrared light strongly and the excited states generated by this absorption must be sufficiently long-lived. Not only that they must relatively be stable molecules, added to that, in order to migrate to RCs, they can also be packed together in a particular way, thus providing paths for executions. Importantly, they also should have ways for the deactivation of potentially destructive side products, for example, triplet states and singlet O_2. Also, some antennas undergo structural modifications by optimizing their operation, as a result of response to large variations which occurs in the intensity or wavelength of the incident light. Especially, Chlorophylls, bacteriochlorophylls and also phycobilins lend themselves well to being antennas.

Theoretically, in a single statement, it can be stated that in peripheral pigments, some kind of electronic excitation is being created at the initial stage of absorption of sunlight for substantial period. Spreading over several neighboring pigments this exhibits strong quantum mechanical coherence. Different research groups have reported this problem, spanning more than a decade.[35,37,39] It is surprising to observe how efficiently these organisms fuel their mechanism with light energy where an efficient apparatus, possessing such kind of key features, has been developed just for the purpose of harvesting sunlight since for a long time of evolution. For example, thousands of pigment molecules (bacterio chlorophylls and carotenoids) are contained in Bacterial photosynthetic membranes, non-covalently bound to proteins, thus forming well organized pigment-protein complexes.

Thus, in short, the photosynthetic process aided by pigment molecules can be described as follows : first, photon is absorbed by a photosynthetic pigment molecule from sunlight, Hu, Thorsten & Ritz, *et al.*,[40] following which, as next, the process takes the immediate steps by transferring as well as trapping them too, i.e., the excitation energy, followed by the separation at the RC. Then, added to this responsibility, the vital next step, followed by excitation energy, is to generate the process by which voltage gradient is followed after this energy being used for feeding the transport of an electron across the cell membrane. Here, most important characteristic point to be noticed is the role played by pigments;- instead of directly getting themselves involved in the process of charge separation, when the excitation energy reaches to a reaction center (RC), thus eventually facilitating the charge transfer.

Importantly, it is to be noted that in a system of inter-pigment with couplings, comparatively weak than the pigment-environment coupling, Förster theory holds; but instead, Redfield theory has been established to be more effective whenever pigment-environment coupling is weak compared to that of inter-pigment. Again, when the case of a typical pigment-protein complex considered, i.e., whenever there is presence of coherence in between nearby pigments, limited applicability of both the Förster and Redfield approach of approximations have been found. In such case, the application of both the non-perturbative and non-Markovian formalism have been observed to be more effective. However, in recent years, the development of a fully non-Markovian description of quantum dynamics in a noisy environment, have been successfully applied, in order to characterize the system-environment coupling to an arbitrary order. This has been done by utilizing a hierarchy of auxiliary density matrices by which, the associated coherence dynamics has been studied, successfully, in the case of Fenna-Matthews-Olsen pigment-protein complex of green sulfur bacteria, together with DNA and pigment-protein complexes of purple bacteria.[66–68] The results show, that in a cluster, for example, underlying the cluster-cluster rate, light absorbing Chls spread over the excitation coherently.

Not only that, the concerned cluster relax within 1–2 ps, thus following the Boltzmann distribution in the case of populated exciton states.

Importantly, as reported in,[19] using such kind of cluster-cluster rate implies immediate relaxation, i.e., the short calculated relaxation time. It is to be noticed that this kind of excitation process in migration occurs in PSU of the purple bacteria rather well, i.e., location of these pigments could either be in separate antenna complexes. Not only that these, being excitonically coupled to the core of reaction center (RC), may constitute parts of a fused photo system, which, interestingly named so, pointing out to the reason that this system may contain both a network of peripheral pigments as well as that of a reaction center (RC). However, as a result of extensive experimental studies, obtained from the atomic resolution geometries of pigments in the PSU, a much more complicated picture has been found than that considered by Förster, where, he assumed that the PSU contains "identical" copies of individual pigment molecules only.

The PSU, in fact, has been found to have three types of pigments. Among them, individual BChls are found only a small part of the pigments, the larger portion are found to be organized in rings of closely coupled BChls, together with a third type of pigments, i.e., the carotenoids. These members, have not only maintain close contact with both the individual BChl's and BChl rings, but also play a vital part so far as the performance of the photosynthetic process is concerned. But, importantly, among all of such pigments, only a very few bacteriochlorophylls (BChls), take part directly in photochemical reactions in the primary rection site. Thus in a nutshell, a light harvesting system can be defined as comprising, typically, of an array of peripheral antenna pigments surrounding a (RC), having the possibility of having different location, depending on their purpose of works. For example,

(1) It is possible for BChls to be located in separate antenna complexes, coupled excitonically to the reaction center core.
(2) Or, they may be constituted with the parts containing both the network of peripheral pigments together with that of a reaction center (RC).

However, due to the imposition of some constraints among pigments, the rapid migration of energy follow some constraints on the separation between pigment as well as on the pigment energy levels which are governed by Förster's formalism. Surprisingly, at this point, adaption by the nature appears to act the most active way, i.e., as far as the utilization of the thermal disorder, existing at physiological temperature. This disorder facilitates efficiently the excitation transfer in between pigments. This is basically

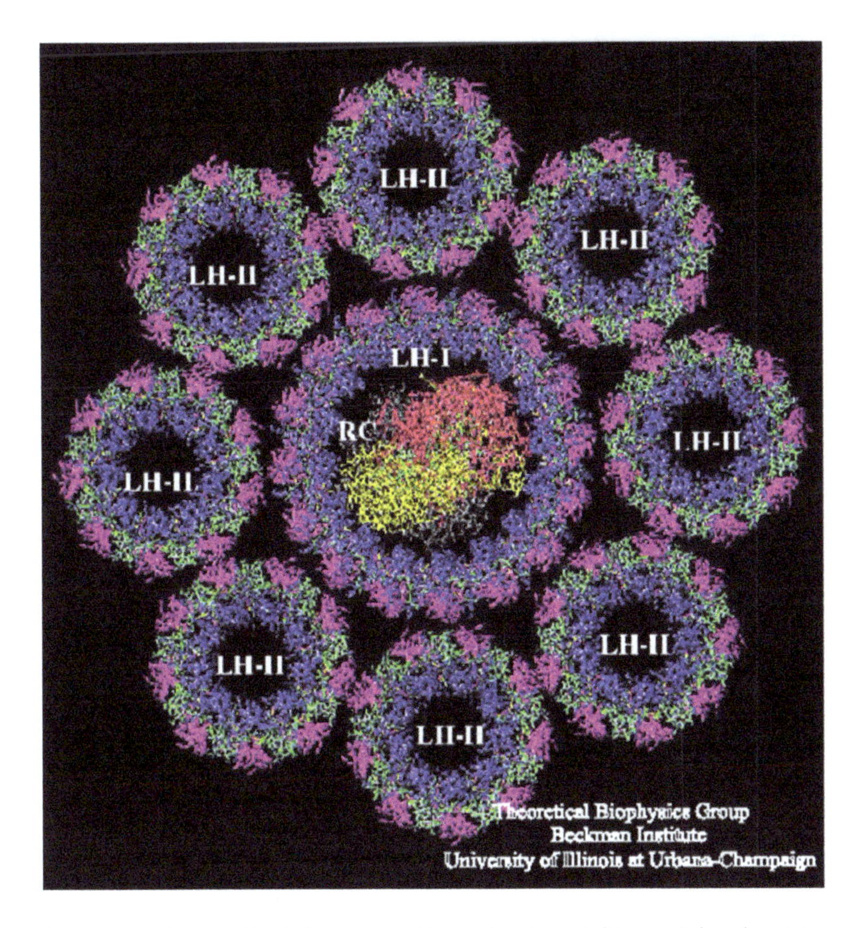

Figure 2.4. Photosynthetic Unit https://www.ks.uiuc.edu/Research/psu/psu.html.

maintained by a broad spectral resonance which is a key factor in Förster's formulation.[41–43] Again, quite an interesting fact should be mentioned here, found, related to the circular arrangement of pigments in purple bacteria. This appears as a consequence, rather related to the requirements of simple assembly and aggregation, rather than having any kind of specific functional role. Because, we are quite aware that the effects of thermal disorder erase any kind of symmetry artifacts that would be left over, especially, if the system were to function at cryogenic temperatures. This raises, rather quite an important question, i.e., whether the more closely packed and seemingly random chlorophyll network, found in cyanobacteria and plants, serves a particular functional role or whether it can safely be regarded as a 'random bag of chlorophylls'. Are there overarching organizational principles for the geometry of chlorophyll networks?

2.2.5 *Chromophores*

In recent times, exceptionally high degree of organizations has been discovered in many photosynthetic antenna systems,[69,71,81] having typical structural studies and precise placement of Chl molecules relative to each other. Among them, one kind of antenna systems, i.e., light harvesting antenna consists of one or more chromophores, consisting of quite small molecules, being capable of absorbing light, the protein or proteins to which these chromophores are bound. The dominant energy-transfer pathway occurs on the subpicosecond timescale across the largest energy gap in each of the proteins, from central to peripheral chromophores. Photosynthetic energy conversion is initiated by photoexcitation of chromophores that are bound at high concentration in light-harvesting complexes. This excitation energy is conveyed efficiently among the chromophores to a membrane bound reaction center where charge separation occurs. It is to be noted that photosynthetic organisms use three classes of chromophore only, i.e., the chlorophylls and bacteriochlorophills, the phycobilions and the carotenoids. These chromophores are bound to specific proteins, among most of which are members of seven major protein families: the core complex family, i.e., the protobacterial antenna complexes (LH1, LH2 and LH3), chlorosomes, the FMO protein, phycobilisomes, the LHC superfamily, together with preidimin-Chl-a protein. Photosynthetic energy conversion is initiated by photoexcitation of chromophores that are bound at high concentration in light-harvesting complexes. This excitation energy is conveyed efficiently among the chromophores to a membrane.

The chromophores, present at quite a high number density in the photosynthetic antennae, are composed of vast arrays of chlorophyll, together with other highly colored molecules.[36,72,73] Among them, many of Chl molecules between neighboring Chl ones, only of approximately 8–20 Å, are found to be very close together which indicates that these molecules have been found to posses specific relative orientations in large macromolecular complexes. But, even if various pigment- protein complexes are found, serving as antennas in photosynthesis organisms, the system are required to fulfill meet many other critical requirements, for example, required to absorb visible or near infrared light strongly. Added to that, the excited states which has been generated by this absorption, rather, must be sufficiently long-lived, as they must relatively be stable molecules so that these can be packed together in a particular way, as, it is a necessary requirement for providing paths in order to execute migration process to the reaction centers (RC). Also, they should have ways for the deactivation of potentially destructive side products, for example, triplet states and singlet O_2. Besides this, importantly, it has also been observed that some antennas also undergo structural modifications, for the requirement of the optimization of their operation needed in response to large variations in the intensity or wavelength of the incident light. Hence, starting from strong to very weak, a range of interaction energies, is expected. These processes also exhibit absorption of incident light within certain wavelength bands, only to be absorbed by them by exploiting their electronic wave function.

As we are aware by now, the quantum mechanical aspect of light harvesting can be traced to the assumption, considered in in Förster's theory which states that electronic coupling between the molecules is extremely weak (compared to line broadening). However, in light-harvesting complexes, the chromophores (e.g., chlorophyll) are packed at a high density. As a consequence, the average center-to-center separation of neighboring molecules has been observed typically, only about 10 angstroms; that is why, the electron couplings are moderately large. As a consequence, a good chance is there of finding the excitation, simultaneously, on more than one chromophore. But, the excitation could be spread having a defined amplitude across those molecules. This particular factor plays a very crucial role by carrying information on the relative sign of the excitation wave on each molecule. One important ramification of this is that several chromophores can act cooperatively in the absorption and transfer of electronic excitation.[17,74–76] When chromophores do so, it not only means that there is a good chance of finding the excitation, simultaneously, on more than one chromophore but also the excitation is spread across those molecules, having a defined amplitude. This factor, i.e., amplitude carries the vital information on the relative sign of the excitation wave on each molecule. Interestingly, quantum jump has also been noted occurring in this process, ranging from a ground state to an excited state which, after being produced, causes the action of excitation to be relaxed rapidly, to the lowest-energy state of the system.

This way, the relaxation of excitation energy happens by going back to the electronic ground state again. But, within the excitation lifetime, this relaxation does not occur unless and otherwise, successfully transferred

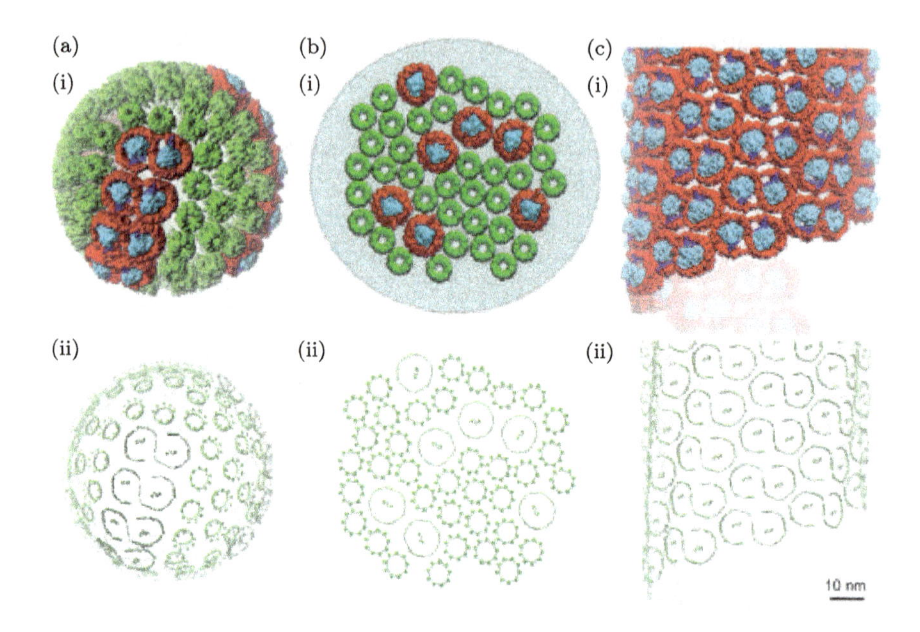

Figure 2.5. Structural models for (a) spherical, (b) lamellar, and (c) tubular photosynthetic chromatophores from purple bacteria. Shown above are the constituent proteins: LH2 complexes (green) and RC-LH1 complexes (LH1 in red; RC in blue). Shown below are the associated BChl networks. The lamellar patch (b) is shown embedded in a lipid membrane, containing a total of nearly 23 million atoms including water (not shown).

away from the excited chromophore and that also again, within the excitation lifetime. Not only that, this is dependent on the two processes:

(i) either this happens through a process, i.e., the emission process of a photon (radiative decay);
(ii) or through the possibility of the presence of various non-radiative processes.

At this point, a very vital point should be mentioned here, i.e., the photosynthetic machinery is capable of controlling the non-radiative relaxation rate, i.e., by adjusting electronic properties of chromophore. This could happen either by increasing itself under conditions of stress, caused by high light, including the interaction associated with that, or decreasing itself under optimal conditions which could reach >90% efficiency of energy transfer, to be stored in the form of an exciton. But, this remains untapped until harvested at a RC. Again, in order to have an efficient harvesting of light energy, the 'transfer' of excitons to RCs is absolutely essential, needed well within the decay time of excitons. The typical range lies within tens of picoseconds (around a nanosecond). This is followed by the essentially important step at RC. There, the excitation energy is finally transformed into a more stable form through electron transfer.[43,44] But, the process of *inter-pigment separation* imposes constraints. Governed by Förster's formalism in order to have rapid migration of energy among pigments, as well as on their energy levels itself, this is essentially needed.

2.2.6 *Light Harvesting (LH) Antenna in Photosynthesis*

The light-harvesting antenna is a particular structural organization, present, whereas the whole antenna contains multiple LHs, possessing various kind of spectral characteristics. For example, to start with, one or more chromophores, may be called small molecules, absorb light and then the protein or proteins to which the chromophores are bound. In almost all the cases, the pigments are bound to proteins in highly specific associations. In addition to chlorophylls, common antenna pigments, also include carotenoids and phycobilin pigments, found in phycobilisome antenna complexes. However, photosynthetic organisms use only three classes of chromophore : (1) chlorophylls; bacteriochlorophills (2) phycobilions and (3) carotenoids. Most of the PSUs contain two types of LHs only, for example, in purple bacteria, these are usually known as B875 (LH-I) and B800–850 (LH-II) complexes, found in vivo research works of absorption maxima at the near-infrared[77–79] zone. More specifically,[69,71,80] the arrangement in which the reaction center is surrounded by a ring of LH1 antenna protein subunits was originally proposed on the basis of the appearance of native crystalline array of *Rh.viridis*. But LH-II has not been found in direct contact with the RC, but their vital role remains in the energy transference to it, which can be traced via LH-I.[36,81–83] For some specific bacteria, LH-III, a third type of light-harvesting complex has also been found to exist, for example, Rhodopseudomonas (Rps.) acidophila and Rhodospirillum (Rs.) molischianum strain DSM 120.[84] The number of LH-IIs and LH-IIIs in the PSU varies according to growth conditions, specifically, light intensity and temperature.[85]

In these antennae, in order to capture sunlight by RC, enlargement of the cross-section has been noticed which possess light-absorbing chlorophylls itself, but, quite importantly, photons absorbed by the RC chlorophylls are found insufficient for the saturation of its maximum turnover rate. Whenever when exposed to direct sunlight, chlorophylls are found to absorb at most at the rate of 10 Hz, but in dim light, this becomes around 0–1 Hz[95]). However, it has been observed that the associated chemical reaction of the RC has been found to be able to 'turn over' at 1000 Hz, so that, LHs is capable of fueling excitation energy to the RC, thus keeping the RC running at an optimal rate. In the light harvesting system, the role of Photosynthetic bacteria plays a vital and quite important role which have evolved to be a quite pronounced energetic hierarchy. Again, just to mention one of many roles played by bacteria, quite an interesting example could be mentioned, i.e.,: in Purple bacteria, absorption of light occurs in a spectral region complementary to that of plants and algae, mainly at wavelengths of approximately 500 nm through carotenoids and above 800 nm through BChls. It should be noticed at this point that most of BChls have been observed in serving as light-harvesting antennae. These capture the sunlight, the next step of which followed by the process of funneling electronic excitation towards the RC.[65,67,68,87,88,108] Importantly, by convention, the photosynthetic RC together with associated LHs, are named collectively, by convention, as the photosynthetic unit (PSU),[90] i.e., the organization of pigments into PSUs in which multiple light-harvesting antennae serve the RC. This has been adopted by all photosynthetic organisms.[67,68,90–94] Their activity lies in collecting light from a broader spectral range and using that energy much more efficiently. In this activity, the light-harvesting antennae contributes by enlarging the cross-section in order to capture sufficient sunlight by the RC. It is important to note at this point that even though the latter possesses light-absorbing chlorophylls itself, photons absorbed by the RC chlorophylls have been observed, not quite sufficient in saturating its maximum turnover rate. For example, when exposed to direct sunlight, chlorophylls absorb at a rate of at most 10 Hz whereas in dim light, at a rate of 0–1 Hz.[95] However, the chemical reaction of the RC can 'turn over' at 1000 Hz whereas, LHs fuel excitation energy to the RC, keeping it capable of running at an optimal rate.

Among all kind of such outstanding research works, for the key electronic excitations in the PSU, ultimately we have been able to achieve the knowledge about many energy levels. It is important to note that in pigments, the outer LHs absorb at a higher energy than do the inner ones. Just for an example, the LH-II, which surrounds LH-I, absorbs maximally at 800 and 850 nm whereas LH-I, surrounds the RC in turn, absorbs at a lower energy (875 nm),[36,77,83,96] With the presence of such an excellent and extremely efficient arrangements, gifted by nature, there is production of energy cascade, serving in funneling electronic excitations from the LH-IIs

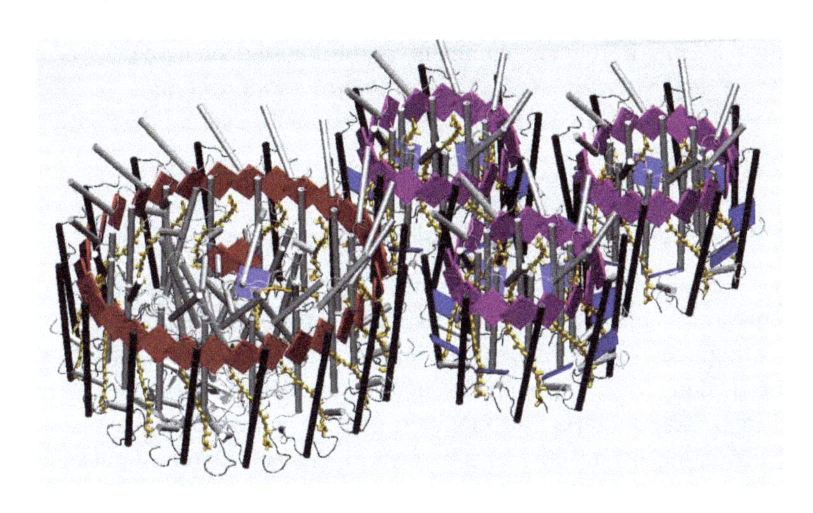

Figure 2.6. The photosynthetic unit (PSU) is the biological equivalent of a solar cell. The PSU consists of the photosynthetic reaction center (red) and of circular light harvesting complexes surrounding the RC. A protein scaffold (black and white cylinders) fixes the pigments in the light harvesting complexes: carotenoids are shown in yellow, bacteriochlorophylls are colored according to their absorption maxima in blue, purple, and red.

through LH-I, leading to the RC. Experimental observations through time-resolved picosecond and femtosecond spectroscopy, revealed that excitation energy transfer, occurring within the PSU, proceeds with an ultrafast timescale, nearing unit (95%) efficiency.[98,99] Importantly, for the energy of the excited LHs, less than 100 ps is spent, in reaching the RC. That is why this range of timescales for elementary energy transfer steps, range from femtoseconds to picoseconds.

At this point, it is quite important to note that, although an $8.5AI$ resolution projection map observed by electron microscopy was reported for LH-I of Rs . rubrum (Karrasch *et al.*1995,[100] the structure of LH-I is not yet known to atomic resolution. However, the structures mentioned as well, found from the above kind of investigation, provide detailed knowledge of the organization of chromophores in the photosynthetic membrane. Even if quite active field of research already been continuing, a new wave, focused on more and more rigorous theoretical and experimental investigations have been found in demand for bacterial photosynthesis, where the different kind of experimental findings recorded, have been under the continuous scientific scrutiny, with the help of latest theoretical developments.[9,41,101–109] For example, we should mention Clayton's opinion in this regard where he made a very important comment regarding the photosynthetic unit, i.e., "*Like van Niels' concept of photochemical oxidoreduction, the idea of the photosynthetic unit has become a cornerstone of current descriptions of photosynthesis.*"[110]

2.2.7 *Carotenoids and Its Role in Photosynthesis*

Besides having other functionaries, the LH2 complex contains light harvesting carotenoids. Carotenoids, also called tetraterpenoid, are essential organic pigments being distributed among photosynthetic bacteria, some produced along with chlorophylls, mainly by plants and algae, normally found in several bacteria and fungi. In general, carotenoids, having wavelengths ranging from 400–550 nanometers (violet to green light). These pigments are essential in photosynthetic organs which, along chlorophylls, causes the sample to be of varied colors, specially, deep orange, yellow or sometimes even red. Importantly, in plants, after being absorbed by carotenoislgae, these serve two key roles in photosynthetic organism: by absorbing light energy, for the use in photosynthesis, thus acting as a protector for the chlorophyll from photodamage and also as a precursors of plant hormones in non-photosynthetic organs of plants. Not only that Carotenoids have also been found essential in oxygenic photosynthesis, i.e., they stabilize the pigment–protein complexes,, being active in harvesting sunlight and in photoprotection. They are present as carotenes in plants and their oxygenated derivatives, xanthophylls.

Not only that, being the most widely distributed pigments in nature, they also act as accessory light-harvesting pigments, effectively extending the range of light absorbed by the photosynthetic apparatus. Secondly and importantly, they also act as a contributor by performing an essential photoprotective role, i.e., by quenching triplet state chlorophyll molecules, scavenging singlet oxygen and other toxic oxygen species, formed within the chloroplast which. This can be found in photosynthetic bacteria, some species of archaea and fungi, algae, plants, including animals.

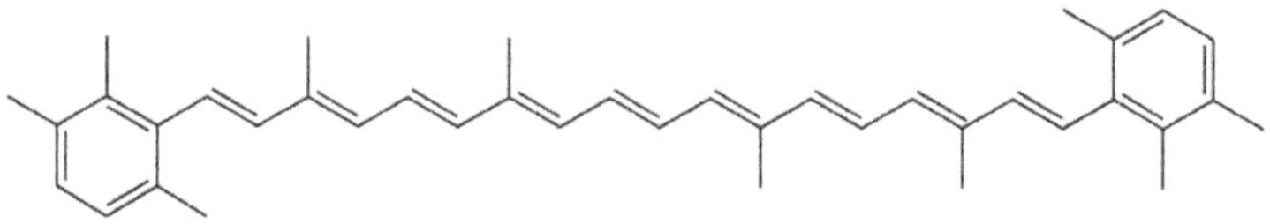

Figure 2.7. General structure of Carotenoids.

Most carotenoids consist of eight isoprene units with a 40-carbon skeleton. Their general structures commonly consist of a polyene chain, having nine conjugated double bonds with an end group, at both ends of the polyene chain. The function of the carotenoids and BChls of the antenna, basically, is to feed the RC with excited state energy. Carotenoids harvest light energy and transfer this energy to chlorophylls through singlet–singlet excitation transfer. This kind of transfer is a lower energy state transfer being used during photosynthesis. It also absorbs excessive energy from chlorophylls through triplet–triplet transfer and release excessive energy by polyene vibration. Importantly, this kind of transference, i.e., triplet–triplet transfer is a higher energy state essential in photo-protection. Not only that, carotenoids act as light-absorbers in the blue-green region of the spectrum in light-harvesting complexes where, absorption of a photon is followed by rapid singlet excitation energy transfer to bacteriochlorophyll (BChl). Thus, in addition to their light-harvesting role, by acting as a photoprotecting antenna complexes, they also prevent the formation of photo-oxidizing singlet oxygen. This act is performed by quenching BChl triplet states through triplet excitation transfer. Light-harvesting and photoprotection by carotenoids has been studied in detail by AJ Young[111] as a result of which it has been observed that carotenoids have two important roles in photosynthetic organisms, i.e., not only they participate in different types of cell signaling, also, they are able to signal the production of abscisic acid, which regulate plant growth, seed dormancy, embryo maturation and germination, cell division and elongation, floral growth, and stress responses. Carotenoids have also been found to play the role of precursors in case of some aroma compounds.

2.3 Förster's Theory for the Fluorescent Resonant Energy Transfer (FRET) Process in Pigment-Protein Complexes

Following section, is organized, primarily and particularly, in order to describe as well as explain briefly the essential and important applications of Förster's theory in photosynthesis, taking the case of purple bacteria. To avail this, we have followed, primarily, many of path breaking works of Sener and Schulten's group, almost in each and every possible way, i.e., basically followed their works and connected mathematical details together with many beautiful illustrations. We tried to organize the following chapter, primarily giving thrust to the applications of Förster's theory, connected to the present particular problem, i.e., theory of excitation transfer. The organization of this chapter is as follows:

To start with, we will try to explain the Förster's theory of excitation transfer, summarily. Together with its applicability to biological pigments, as a next, we have discussed other common structural motifs, existing among light harvesting proteins which can be related to the physics of energy harvesting processes, existing among light harvesting proteins.

Theory of resonant energy transfer (RET), as invented and developed by Förster, is basically related to the fundamental process in nature which, in essence, underlines the process by which harvesting of sunlight is materialized in photosynthetic life forms. Here, Photosynthetic organisms, possessing supra-molecular assemblies, are essentially featured as containing hundreds of cooperating protein around hundred subunits, converting the short-lived electronic excitation. This excitation, even after being the results from the absorption of a photon, transform increasingly into more stable forms of energy which, ultimately, takes the form of stable chemical bonds. Förster and others developed a theoretical framework which elaborates how the electronic excitation process starts its journey by migrating in the photosynthetic apparatus, for example, bacteria, algae and plant together, from light absorbing pigments to their ultimate destination, i.e., reaction centers (RC) where, light energy is to be utilized for its eventual conversion into chemical energy. Light harvesting is less tangible than many other biological processes, but, classically, still can be viewed as a hopping of electronic excitation from one molecule to another– in the spirit of a theory reported by Förster in 1948.

It is now an well established fact that the pigments can be stated as the primary components of a light harvesting system, responsible for converting the energy of an absorbed photon into an electronic excitation. Together this, also at the same time, these perform the transportation of an electron across the cell membrane, resulting in a voltage gradient. But, generally, most of these pigments do not get themselves directly involved in the process of charge separation.

Instead, eventually, the energy, corresponding to their excitation, is transported to a reaction center (RC), facilitating the charge transfer. This way, a light harvesting system could be described as comprising, typically an array of peripheral antenna pigments which surrounds a reaction center (RC). However, their locations might be traced in separate antenna complexes, either, excitonically coupled to the RC core, or constituting parts of a fused photosystem which is done, generally, by containing both a network of peripheral pigments and a reaction center (RC). As stated earlier, many of famous groups performed research works in detail, both experimental and theoretical; for example, Sener and his team carried out detailed theoretical studies, related to the functioning of the protein-pigment complex photosystem I (PSI) of the cyanobacterium Synechococcus elongatus together with the proper applications of these results, related to the observed studies. Studying the structure of this complex with high resolution, they showed that totally, 96 chlorophylls, electronically coupled, function as a light-harvesting antenna complex. It should be pointed out in this regard that in the Oxygenic photosynthetic species, two trans-membrane protein-pigment complexes are employed for light-harvesting, named photosystems I and II. Specifically, the Photosystem I (PSI) is a typical ubiquitous protein-pigment complex, found in green plants, algae and cyanobacteria, located in the bacterial membrane. These are responsible for absorbing sun light, using its energy for the transference of electrons across the cell membrane. This way, PSI possess this typical characteristics, quite importantly and effectively, also contain the antenna complex, reaction center as well as the electron transfer chain, all within the same protein.

Starting with this background, let us start summarizing preliminary explanation of the Förster's theory of excitation transfer, which has found successful applications to biological pigments, as also in the energy transfer characteristics of the theory, with the aim of examining and asses the role of structures, in case of photosynthetic pigment-protein complexes. As the associated physics underlying the energy harvesting process is intimately related to the common structural motifs among light harvesting proteins, the connected issues have also been studied and discussed.

2.3.1 *Principles of Excitation Transfer (FRET) in Pigment-Protein Complexes*

We try to depict how the tuning of pigment energies and their coupling result in the formation of delocalized excitons, whose spatial structure is used to direct energy transfer in the FMO complex. This excitonic level-to-level transfer has recently been fully mapped out with the help of two-dimensional electronic spectroscopy (2DES). More specifically, electrostatic interactions with the protein and solvent environment tune local pigment excitation energies (termed site energies), and interaction between these energetically varying local states results in a ladder of excitonic states, where the higher energy states are localized toward the peripheral antenna complexes, the lower energy excitons being close to the photosynthetic reaction center (RC). The protein and solvent environment not only act to tune the energy of the collective excitations but also play an essential function as the thermal bath into which excess energy can be dissipated. This efficient dissipation of excess energy, enabled by coupling between the excitons and vibrations, is crucial for fast and efficient energy transfer among the excitonic states. We must emphasize here that the warm, wet, and disordered environment of pigments in biological systems is far from the situation, found in strongly coupled, highly ordered, solid-state systems where excitations can be delocalized over the whole crystal. The inter-pigment coupling strength is often of the same order of magnitude as the interaction with the environment (bath), which, in combination with static disorder, results in a tendency to localize the excitation over a small number of pigments, even incase of strongly coupled antenna complexes (e.g., order three to five pigments for LH1 and LH2).

As a starting point, as discussed in detail by Şener and his group in their works,[118] for the present case, we consider Förster theory, primarily, based on excitation. In order to describe that transfer, an effective Hamiltonian formulation has been constructed. In such cases, i.e., when the normal light conditions is such that the flux of photons (~10 photons/chlorophyll/s) results to be quite low, the excitation process can satisfactorily be modeled by considering a specific case of migration in a pigment network, i.e., considering that a single chlorophyll excitation can model the process satisfactorily. Hence, for example, let us consider a case in terms of exciton states but with only a single pigment, having lowest excited electronic state, but, instead, all the other pigments in their ground state. Thus, a basic set can be constructed for an effective Hamiltonian which is follows:

$$|i\rangle = |\phi_1, \phi_2 \cdots \phi_i^* \cdots \phi_N\rangle, \quad i = 1, 2, \ldots, N. \tag{2.2}$$

In the above expression, N is meant for the number of pigments whereas ϕ_i and ϕ_i^* denote the ground and first excited states of the i-th pigment respectively. In the present case, we follow Scheer[103] by naming a molecule of chlorophyll with lowest excited state a to be so-called Q_y state. For such case, an effective Hamiltonian can then be expressed as

$$H = \begin{bmatrix} \epsilon_1 & H_{12} & \ldots & H_{1N} \\ H_{21} & \epsilon_2 & \ldots & H_{2N} \\ \vdots & \vdots & \ddots & \vdots \\ H_{N1} & H_{N2} & \ldots & \epsilon_N \end{bmatrix}$$

where ϵ_i depicts the excitation energy for pigment i and H_{ij} for the electronic coupling i between pigments i and j.

By now, it has already been an established fact that only the laws of Quantum physics governs the process of light absorption, the same being followed in case of the subsequent transfer of energy in a biological system. At the same time, however, this very fact also relates to the Coulomb interaction in between electrons which belong to "spatially separated" pigments. Förster, the pioneer in this approach, first estimated an approximate interaction between the pigments, sufficiently separated. This can be expressed in terms of a process, called transition dipole-transition dipole which leads to the multiple order of the relevant Coulomb interaction.[21,22] As a result of this, transition of an electronically excited pigment P_i^* (the donor) occurs back to its ground state P_i. Here, $P_i^* \rightarrow P_i$ can be characterized through a transition dipole moment d_i and an excitation energy ϵ_i. Following the same, in case of a ground state pigment P_j (the acceptor) which undergoes the transition to its electronically excited state $P_j \rightarrow P_j^*$, which is expressed as transition dipole moment d_j and excitation energy ϵ_j respectively, the joint transition, induced by the interaction energy, can be expressed as,

$$P_i^* P_j \rightarrow P_i P_j^* .$$

Two quite important facts should be noted here. Among them, first one is the coupling between two pigments contributing two terms, namely,

(i) a term, introduced by Förster,[22] corresponding to a direct Coulomb term,
(ii) the next, an electron exchange term.[112]

Generally, this exchange is considered as negligible, the vital cause playing behind this fact being the distance between a pair of chlorophylls which is large enough for a typical network of chlorophylls. These results cause the coupling to be dominated by a Coulomb term[113] which, in the lowest order approximation can be expressed as,

$$V_{ij} = \frac{\vec{d_i} \cdot \vec{d_j}}{r_{ij}^3} - 3\frac{(\vec{r_{ij}} \cdot \vec{d_i})(\vec{r_{ij}} \cdot \vec{d_j})}{r_{ij}^5} \tag{2.3}$$

Here, $\vec{r_{ij}}$ denotes the vector, connecting the center of P_i to that of P_j. These have been expressed in the valuable works by Melih Sener and his group, details of which available in one of their various famous works.[114]

But in the case, when the measured distance of the Mg-Mg coupling in the pigments are found to be typically too close (<10 Å), for example, in case of the measured distance of the Mg-Mg coupling, the underlying multipole expansion (Eq. (2.3)) breaks down. Because, at this stage, electron exchange becomes quite relevant. This [112] results into the condition requiring complementary description of the FRET coupling V_{ij}.[113] Under such condition, the excitation transfer rate, in between donor pigment i and an acceptor pigment j, is expressed as

$$k_{ij} = \frac{2\pi}{h}|V_{ij}|^2 J_{ij}, \quad J_{ij} = \int S_i^D(E) S_j^A(E) dE \tag{2.4}$$

In the above, $S_i^D(E)$ and $S_J^A(E)$ are the spectra for donor emission and acceptor absorption respectively. These are determined by calculating the coupling of the pigment to the states, caused by the vibrational motion, due to the presence of thermal motion of the bath. The integral J_{ij}, hereby, accounts for the spectral overlapping. At this stage, the normalization condition have been applied as $\int S(E)dE = 1$, in order to normalize the spectra.

Now, importantly, when identical pigments are considered, emission and absorption spectra are generally related to one another by the so-called Stokes Shift.[103] In such case, by averaging over all possible orientations of d_i and d_j in Eq. (1.4), the transfer rate, can be expressed as

$$k_{ij} = \frac{4\pi}{3\bar{h}} \frac{d_i^2 d_j^2}{r_{ij}^6} J_{ij} \tag{2.5}$$

This transfer rate had been expressed by Förster in his original formulation,[22] as:

$$k_{ij} = \frac{1}{\tau_0}\left(\frac{R_0}{r_{ij}}\right)^6 \tag{2.6}$$

where τ_0 has been meant for the relevant fluorescence lifetime of participating pigments, having a typical value i.e., of the order of a nanosecond. The corresponding R_0, can be expressed as

$$R_0 = \left(\frac{4\pi}{3\bar{h}} d_i^2 d_j^2 \tau_0 J_{ij}\right)^{\frac{1}{6}} \tag{2.7}$$

which is caused through loss due to fluorescence or internal conversion. here, the term R_0, defined as the Förster radius, depicts the inter-pigment separation, approximately 90 Å for Chls, meaning that the energy transfer is 50%. In fact, this is the amount of energy, expected to succeed without suffering any loss due to fluorescence or internal conversion.

2.3.2 *Migration of Energy through Pigments and Extension of Förster's Theory*

The next crucial query surfaced at this stage of pioneering theoretical contribution of Förster, raising the important question in the research front, is: scientifically, how to describe, particularly those systems which contain particularly, clusters of strongly interacting pigments. A great many subsequent serious works have been pursued, describing this order scientifically, i.e., those clusters of strongly interacting pigments. As a result, the original theory of Förster needed to be extended substantially, in order to take into consideration of the proper roles, played by those very pigments which interact strongly among themselves, as found out by a large competant research group[111] and also many other important ones.[61] An effective Hamiltonian matrix has been constructed then, taking into consideration those clusters pigments which poses elements, responsible for the coupling energy between electronic excitations. It should be noted that the light absorption and subsequent transfer of excitation energy are the first two steps of photosynthetic processes to be carried out by protein-bound pigments, i.e., the Bacteriochlorophils (BChls); the photosynthetic bacteria can be mentioned among one of them. These BChls are anchored in light harvesting (LH) complexes. For example, the case of light harvesting core complex 1 (LH1) could be mentioned as it is directly associated with reaction center (RC), forming RC-LH1 core complex. Out of the results obtained from the extensive research works, by now, it is already an well known fact now that the process of photosynthesis is initiated only when protein-bound pigments absorb light energy.[9,17,19]

Typical example, can be mentioned, for example, as stated already, in photosynthetic purple bacteria, pigments are found to be placed, specifically, within protein complexes, i.e., for instance, the light-harvesting complexes I and II, i.e., LH1 and LH2. As a next step, this excited energy, possessed by the corresponding excited pigments, passes its excitation energy onto nearby pigments quite rapidly, i.e., very next arrival of energy happens within tens of picoseconds at the RC. By turn, this action occurs as a starting point for the later steps of photosynthesis, as an immediate next step, for the utilization in the charge separation. The LH1 complex surrounds the reaction center (RC), forming the RC-LH1 core complex. These pigments are found typically held together in a PSU, within a single light harvesting complex and also with many other light harvesting complexes. So, in order to consider these clusters, containing many of such light harvesting complexes, we need to construct an effective Hamiltonian matrix, by considering those elements, accounting for the coupling energy between electronic excitation in cluster pigments. These, importantly, possess quite a specific criteria to be fulfilled i.e., typically, in a PSU, all of such pigments must held together by a single light harvesting complex only.

Next, taking into consideration, all of such facts, an effective Hamiltonian H, is constructed for two pigments, specifically, where the basis states are expressed by building a 2×2 matrix for two pigment clusters which acts on the basis states $|1\rangle$ and $|2\rangle$, expressed by

$$H = \begin{bmatrix} H_{11} & H_{12} \\ H_{21} & H_{22} \end{bmatrix}$$

and

$$|1\rangle = \begin{bmatrix} 1 \\ 0 \end{bmatrix},$$

$$|2\rangle = \begin{bmatrix} 0 \\ 1 \end{bmatrix}$$

This specific case needs an important point to be mentioned. For example, taking a case containing N number of Chls or BChls, the Hamiltonian H can be expressed as an $N \times N$ matrix acting on states $|1\rangle, |2\rangle, \ldots, |N\rangle$. But, considering the present case, the state $|i\rangle$ describes the cluster containing all the pigments in their ground state only. This is an important exceptional case because the state for pigment i is in the lowest electronic excited state.

As is shown in Figure 2.1, in the case of Chl and BChl molecules, the lowest excited state is the so-called Q_y[101] state havng a transition dipole moment, approximately along the vector connecting NB and ND atoms of the Chl porphyrin ring. The effective Hamiltonian H can then be expressed as[20,39,40,114]

$$H = \sum_{i=1}^{N} \epsilon_i |i\rangle\langle i| + \sum_{i \neq j}^{N} V_{ij} |i\rangle\langle j| \tag{2.8}$$

Here, ϵ_i denotes the energy for the Q_y-excited state of Chl_i whereas V_{ij} depicts the coupling energies given in equation 3.

In such circumstances, an important fact to be mentioned in the presence of the coupling terms V_{ij},the above expression, i.e., excitation of a single Chl leads rapidly to a sharing an excitation which is coherent among all of N Chls. These coherent states of exciton, defined as the eigenstates $|\tilde{n}\rangle$ of the Hamiltonian matrix H, can be expressed as,

$$H|\tilde{n}\rangle = E_n|\tilde{n}\rangle$$

E_n being the eigen values. The time evolution of exciton states can be determined from the Hamiltonian, expressed by

$$|\psi_i(t)\rangle = \sum_{n=1}^{N} \alpha_n \exp\left(-\frac{i}{h} E_n t\right) |\tilde{n}\rangle$$

Here, α_n are the coefficients, dependent on the initial state of values due to excitation i.e., the state of initial excitation of NChls. An important point to be noted here, i.e., the above description neglects not only the role of couplings of the electronic excitation of the pigments but also the vital role of their environment. Under the prevalent physiological conditions, caused by the presence of system-bath coupling, usually spread over N Chls, the population of eigen states $|\tilde{n}$ affects the exciton states $|\psi(t)\rangle$ in order to be localized over a few pigments with the purpose of approaching the thermal motion. So, at this stage, approximately follows Boltzmann distribution is followed, having the probable occupancy which can be expressed as

$$\rho_n = \frac{\exp\left(\frac{-E_n}{k_B T}\right)}{\sum_m \exp\left(\frac{-E_m}{k_B T}\right)}$$

But, the case becomes quite different whenever "weak coupling" limit is considered. In such cases, the corresponding energy migration occurs following stochastic sequence of energy transfer which is to be considered. As a result, in such cases, the exciton becomes localized to individual Chls in a pigment network and so, following the Förster theory, the process like stochastic sequence becomes applicable for their description. This process, termed as hopping phenomena, can be formulated as $Chl - Chl$ hopping having the probability $p_i(t)$, where Chl_i, electronically excited at time t, leads to a master equation.[13,56]

$$\frac{d}{dt}p_i(t) = \sum_j K_{ij}p_j(t), \tag{2.9}$$

and,

$$K_{ij} = k_{ji} - \delta_{ij}\left(\sum_k k_{ik} + k_{diss} + \delta_{i,RC}k_{CS}\right) \tag{2.10}$$

Here,

(1) k_{ij} are for the Förster equation;
(2) $diss$, is the excitation dissipation rate, due to fluorescence or internal conversion
(3) K_{CS},the charge separation rate respectively at RC,
(4) $\delta_{i,RC} = 1$, if Chl, i belongs to a RC or otherwise, 0 in all other case.

In overall, the description of the above equation, relates the process of stochastic migration which in general occurs, in the process of charge separation, reaching from higher energy pigments to lower ones, the ultimate destination, i.e., RC, responsible for causing an event.

Accordingly, migration of the initial excitation proceeds stochastically, generally, from higher energy pigments to lower energy ones, until this reaches a RC causing a charge separation event. Notably, some photosynthetic proteins, i.e., PS1 contain Chls which absorbs light at energies lower than the RC Chls. These participate in charge separation, being remained excitonically connected to the RC at physiological temperatures.

But, sometimes, at energies lower than RC Chls, i.e., even at physiological temperatures, it becomes possible for the absorption of light energy. But, even in such cases also, a few number of photosynthetic proteins (for example, $PS1$) containing Chls, participate in charge separation, remaining excitonically connected to RC. In such particular cases, the average excitation life time τ together with quantum efficiency q, can be computed from the above formulas, meant for the probability of the excitation, leading to separation, given by:

$$\tau = -\frac{1}{N}\langle I|K^{-1}|O\rangle \tag{2.11}$$

and,

$$q = -\frac{1}{N}k_{cs}\langle RC|K^{-1}|O\rangle \tag{2.12}$$

Here, $|O\rangle$ represents the initial state whereas, $|I\rangle \equiv \sum_i |i\rangle$ and $|RC\rangle \equiv \sum_i \delta_{i,RC}|i\rangle$.

2.3.3 *Failure of Hopping Model*

In the case of the large inter-pigment couplings, i.e., (>100 cm^{-1}), this Chl-Chl hopping model finds difficulties in explaining the excitation migration process properly and successfully, i.e., in the cases of the sufficiently large enough inter pigment couplings (i.e., >100 cm^-1; here, energies being in terms of reciprocal wavelengths). So, when the case of strongly coupled Chl clusters are to be considered, for example, in case of the purple bacterial LH2, as explained in detail by Sener & his group,[29] it becomes important and vital to consider the case of delocalization of exciton over multiple *Chl* clusters. This kind of situation may arise due to the typical fact, i.e., even in the presence of strong intra-cluster pigment coupling, the inter-cluster coupling has been found to be quite weak. This kind of typical situation causes the "in between transfer" of energy. For this reason only, before the transfer occurs to the acceptor cluster, the Boltzmann equilibrium of donor cluster states is assumed. This way, the modified advanced form of Förster's theory can successfully be applied in describing the transfer energy in between the clusters. Accordingly, the same method has been followed in calculating the transfer rate between two individual pigments k_{ij}. Here, the energy transfer rate k_{DA} in between donor pigment cluster D and an acceptor pigment cluster A, is given by

$$k_{DA} = \frac{2\pi}{\bar{h}} \sum_{m \in D} \sum_{n \in A} \frac{e^{\frac{-E_m^D}{k_B T}}}{\sum_{l \in D} e^{\frac{-E_l^D}{K_B T}}} |V_{mn}^{DA}|^2 \int dE S_m^D(E) S_n^A(E) \tag{2.13}$$

In the above equation, V_{mn}^{DA} represents the electronic coupling energy between donor exciton state m and that of acceptor state, i.e., n. The associated energy levels of the donor and acceptor pigment clusters are

E_m^D and E_l^D respectively, being also the corresponding expressions for the associated emission and absorption spectra. Under the above considerations, in a network, consisting of M pigments, excitation life time for the network can be expressed as:

$$\tau = -\frac{1}{M}\langle I|K_{cl}^{-1}|O\rangle \tag{2.14}$$

Here, $M \times M$ matrix, i.e., K_{cl} is defined in a similar fashion, in terms of inter-cluster transfer rates i.e. k_{DA}, as given in above equation. It is important to note here that, in application of the quantum coherent sharing in the advanced Förster theory, thus developed, the analogous consideration i.e., the idea of individual weakly coupled pigments in a cluster, followed by their corresponding energy migration has been described analogously. Added to this, the theory has also been applied in treating the corresponding coherent quantum mechanical sharing of excitation, among pigments. This approach has taken vital prominence, being applied in case of many future developments. It is quite interesting to note that though the physical role, played by quantum coherence states had been pointed out long before, i.e., in 1997, during the determination of structures of the light harvesting complexes, this idea of quantum mechanical coherent sharing of excitation has taken the prominence again, after this extension only.[116,117] Interestingly, in such extension, energy migration is described among pigment clusters, taking account coherent sharing of excitation which is quantum mechanical in nature.

2.3.4 *Repeated Detrapping of excitations in Excitation Migration among Reaction Centers*

Generally, the termination of the process of excitation migration, neither necessarily happens with the arrival of the excitation to a charge separation site, nor, after arrival at RC, every excitation, results in a charge separation event. Thus the nature of the phenomena demands a finite probability from the charge separation site which in turn, dependent on the ratios of the charge separation rates only, including the total detrapping rate. In such a situation, energy transfer away from the RC, competes against to that of the charge separation process at RC, both having similar time constants, lying within the range of a few picoseconds. It is an well established fact that a loss of light harvesting efficiency in such events, generally occurs. This probability causes a few choices: for example, either some amount of this excitation may escape back to the chlorophylls in the periphery, or, some dissipation might have already been caused or otherwise, migrating back again to a reaction center (RC). So, by nature, this kind of migration process with excited states, is able to expand its role, i.e., can play the role in migration, by trapping, followed by events of detrapping and retrapping. From the experimental studies, related to purple bacterial RC-LH1 complexes, the probability for this kind of exciton detrapping events has been observed as 20–30%[118] for a particular event, i.e., from the RC back into the pigment array, causing generally, a loss of light harvesting efficiency. A quite interesting fact should be mentioned here, i.e., at the site of initial receiving, the sharing of detrapped excitations by other RCs has been observed experimentally, thus, producing an advantageous situation whenever the RC remains nonfunctional.[56] On the contrary, another interesting fact is that, in PSU, where all Chls are shared by the excitation, the reduction of radiation damage has also been noticed, because of the consequent effect, i.e., due to the spreading of its effect over many pigments. In such a situation, i.e., in a system of multiple RCs, when describing the energy migration, the process must be augmented by the process of repeated detrapping and retrapping events. Thus, the pigment network, having multiple RCs, can have the probability Q_{ij}, depicting the excitons, detrapped in, say, at RC_j to be retrapped at RC_i again, will be[30,45]

$$Q_i j^X = -k_{CS}\langle RC_i|K_{cl}^{-1}|T_j\rangle \tag{2.15}$$

Here, $|T_j\rangle$ represents the state of the event of detrapping at RC_j. The probabilities for the detrapping are responsible in estimating the sharing of excitation among dozens of RCs which extends an entire photo-synthetic vesicle.[56] These events of socalled retrapping and detrapping is summed up ultimately in the so-called sojourn expansion.[30,45]

2.4 Applications of Förster Theory

2.4.1 *Transference of Energy in the Presence of Thermal Motion*

Almost full dependency on physiological temperature is an vital factor for the very existence of Photosynthetic life forms. Hence, any kind of disorder, especially, caused by the ever presence of thermal motion in the surrounding, can affect this process. In all probability, photosynthetic organisms, have been influenced by the evolutionary developments, especially related to the atmospheric aspect in such an important way, so that, it has not only already developed but also continuing up till now to follow that, perhaps being persuaded, for their nourishment together with their urge for survival. The result of such endevour has developed means for coping with thermal disorder as well, at the same time, also for the successful utilization, because, it is absolutely necessary for the successful maintenance together with a high degree of light harvesting efficiency. Following an interesting observation done by Melih Sener and his group, an example could be put here: after detail observations, they pointed out an interesting phenomena i.e., the quantum efficiency, as shown in the PS1 complex (Fig. 2.2(a).iv, (b).iv), is over 0.9, found by, however, at physiological temperature, can drop to 0.5 or less.[30, 45, 48, 119] Not only that, this turns out to be, surprisingly, wavelength dependent at cryogenic temperatures.[30, 45, 48, 119] In fact, this drop in efficiency is the resultant effect, caused by the presence of a resonance, but turned weak J_{ij}, this being a key factor in the Förster's original formula. In such cases, spectral line-shapes have been observed to be narrower at lower temperatures, because, pigments of long wavelength, becoming excitonic traps, thus cause the significant reduction in the transfer rates of excitations to neighboring pigments.

It has already been an established fact that spectral properties of photosynthetic systems have been observed quite sensitive to thermal effects. So, considering this criticality, dependent on the temperature associated, an effective Hamiltonian H (Eqn. (2.8)) has been developed at a finite temperature, for the absorption spectrum of a pigment-protein complex. Among many models, following molecular dynamics trajectories,[42, 114] these phenomena can be described by applying the so-called dynamic disorder model of having fluctuations in the site energy, in terms of ensemble average, describes the thermal effects over many realizations of a system. It should be mentioned that due to the presence of static disorder, another model, i.e., a static model, can be introduced in explaining these thermal effects, by considering the ensemble average over many realizations of a certain system.[119] At this juncture, in order to construct an expression for the Hamiltonian H, for the description of the effects of thermal motion on energy transfer, an account for the coupling of pigments with environment, i.e., a thermal bath of harmonic oscillators is to be added to the system Hamiltonian H (Eqn. (2.8)),[42, 121–124] as a consequence of which the effective Hamiltonian (H_B) can be expressed as

$$H_B = \sum_{\xi=0}^{\infty} \left(\frac{p_\xi^2}{2m_\xi} + \frac{m_\xi \omega_\xi^2 \chi_\xi^2}{2} \right) \tag{2.16}$$

At this stage, a system-bath coupling can be written, i.e.,

$$H_{SB} = \sum_{j=1}^{k} W_j \sum_{\xi=0}^{\infty} c_{j\xi}\chi_\xi \tag{2.17}$$

Here, W_j represents the coupling operators, describing k as independent forms of coupling, existing in between system and environment. As a next case, the role of vibrational states x_ψ is to be considered for the purpose of constructing the model containing the thermal motions of the environment, around the pigment. Importantly, these states being responsible, simultaneously, both for the donor emission as well as that for the absorption spectra, thus influence also the overlapping. Next, as it is important to consider the contribution from the bath also, we do the average of the possible bath degrees of freedom,[56] calculating the time evolution of the exciton states. This way, expressing it in terms of reduced density matrix $\rho(t)$, we get

$$\rho(t) = tr_B\left[\exp\left(-\frac{i}{h}\int_0^t dsL_T(s)\right)\frac{e^{-\beta H_B}}{tr_B(e^{-\beta H_B})}\right]\rho(0) \tag{2.18}$$

In the above expression, $L_T = [H_T, .]$ and $H_T = H + H_B + H_{SB}$ and $\beta = \frac{1}{k_B T}$.

Next, assuming Boltzmann distribution of the bath vibrational states (i.e., the states caused by the vibrations, present in the environment due to the presence of thermal noise), the time evolution of the reduced density matrix, $\rho(t)$ can be calculated,. In order to do that, we need to take trace over the bath degrees of freedom, in case of the total system propagator, i.e.,[59(b)]

$$exp\left(i\int L_T(s)ds/\bar{h}\right)$$

assuming a Boltzmann distribution of vibrational states. Under these present developments, an important opinion could be arrived stating that in all such cases, Förster theory holds only in the case of a weak inter-pigment couplings, when compared with the coupling surrounding pigment environment. But, in a particular case, i.e., in the case of weak pigment-environment coupling in comparison to that existing in the inter-pigment arena, the application of Redford theory, then, has been proven to be quite successful.[107,111,124,125] However, another typical situation may arise when there is coherence between nearby pigments.

At this juncture, we could arrive at an conclusion stating that in photosynthetic complexes, transfer of electronic energy can usually be described in one of two perturbation limits. As a first case, when the electronic coupling is small, compared to that of electron–nuclear coupling, the original localized electronic state can appropriately be presented, considering the inter-pigment electronic coupling and by treating the case perturbatively, yielding the Förster theory (Förster, 1948[22]).

In the opposite limit, i.e., in the case of small electron–nuclear coupling, the case should be treated taking perturbation into consideration, in order to obtain a quantum master equation. In achieving this limit, the most commonly used approach is that of the Redfield equation, in the literature of photosynthetic EET (Redfield, 1957, 1965[125]). Even after the great success of the general conventional Förster theory, especially, predicting the energy transfer from phycocyanins to chlorophyll, in cyanobacteria and elsewhere, it can provide a complete description of energy transfer, only when a few cases of photosynthetic light-harvesting complexes are considered. The reason behind such a failure, i.e., offering proper explanation in order to explain the complete picture of energy transfer, is due to the presence of the collective character of the excitation in photosynthetic complexes. This ineffectiveness, responsible for not being able to offer the proper as well as adequate pictures of the process of energy transfer, is the cause of the failure of Förster theory in this respect. The generalized Förster theory (Novoderezhkin and Razjivin, 1994, 1996[129(a)&(b)]; Sumi, 1999[126]; Scholes and Fleming, 2000[127]; Jang *et al.*, 2004[130]) considers energy transfer between clusters with an arbitrary degree of delocalization. But, it is applicable only for a narrow window, restricted only to weak inter-cluster interactions.

2.4.2 *The Generalized Förster Theory*

The classical Förster theory has been further generalized with the purpose of the inclusion of more realistic models of excitonic couplings and coherent effects which is observed within either donor or acceptor subunits. The purpose of such application is meant for the systems of weakly coupled domains of delocalized states, found to be made of strongly coupled pigments. The modern multichromophoric Förster resonance energy transfer theory considers both the donor and acceptor, each as a small group of molecules, coherently excited. Also, dynamics of incoherent hopping between these groups seems to provide a satisfactory description for the dynamics of energy transfer in energetically well-separated components, for example, the LH2 complex of purple bacteria. In such cases, either the Redfield theory or its modified version may satisfactorily provide the necessary explanation for the relaxation within these domains.[125]

In general, the Förster theory, essentially is applied in explaining properly the transfer of the excitation energy, starting from the excitonic state of one domain to a neighboring one and extended further, taking into account the pairwise interaction of pigments from each domain which includes also the contribution of each pair. Not only that this has been weighted by the coefficients of the contribution of each pigment, related to the state of the corresponding exciton domain. In the Förster theory, these kind of pairwise contributions, then summed up or when the two domains are far enough apart compared to their spatial extensions, can successfully be used by taking the effective average transition dipoles of the domains. Again, in case of closer aggregates (domains), each pair of pigments must be taken into account. However, whenever the absorbency and fluorescence line shape functions are needed for the inter-domain Förster calculation, difficulty arises in taking into account both the excitonic and excitonic-vibrational couplings in these domains.

2.4.3 *Redfield Theory and Photosynthesis*

Redfield theory, as stated earlier, has been developed with the aim of dealing with the problems related to strong excitonic coupling together with weak exciton-vibrational coupling. In this case, i.e., exciton relaxation between excitonic delocalized states, the rate constant are determined in terms of the coefficients. These are calculated by taking the summation of the molecular product of wave functions which make up the exciton wave functions.In such cases, N different energy levels are considered for N participating molecules which makes up the excitons. The rate constant for this transition has been found to be proportional to a sum, taking over all pairs of molecules of negative exponential. Not only that, this very rate constant is the function of the distance in between the different chromophores, (however, this is an approximation and the corresponding electronic overlap factor might be there and hence, should be taken into account). As a next, in each exciton state, each member of the sum is to be weighted by the product of the coefficients, related to the molecular wave functions. Together with it, an important factor, involved in this rate constant, is to be considered. This, specifically, represent the efficiency by which energy can be exchanged in between the proteins and the chromophores. Now, the re-organizational energy can be related to the coupling of the local protein vibrations which is an integral over the spectral density. Being weighted by the frequency, the spectral density is nothing but the weighted density of states of the protein vibrations, coupled to the local transitions of the pigments. Under these circumstances, Förster's theory, i.e.,FRET can successfully be applicable only if the re-organizational energy turns out to be large enough, compared to the intermolecular coupling.

2.4.3.1 *The Modified Form of Redfield Theory*

The modified form of Redfield theory has been developed with the aim of studying strong exciton-vibrational coupling which includes strong excitonic coupling itself in order to take both the couplings into account but not invoking perturbation. With this aim, the modified Redfield theory considered the nuclear reorganizational effects when estimating the interaction between delocalized states. As observed, the electric density has been found to be different for the different excitonic states. Hence, the nuclei relax toward the positions which are meant for new equilibrium. Not only that, for strong excitonic-vibrational coupling, it becomes necessary to take into account the exciton-vibrational coupling. In order to achieve this, the diagonal elements of the excitonic-vibrational coupling has been considered in describing the reorganizational effects. These factors have been neglected by the normal Redfield theory, i.e., the standard Redfield theory results reappears if all these diagonal strong exciton-vibrational coupling is neglected. A rate constant for the modified Redfield theory was first given by Mukamel and coworkers,[128] Jang, Newton & Silbey,[130] Novoderezhkin *et al.*,[131] etc. It is to be mentioned that the name as 'modified Redfield theory' was introduced by Yang and Fleming[132] only.

2.4.4 *Reconciliation of Dissipative Quantum Mechanical Methods with Förster Theory*

The approximations, inherent in Förster and (modified) Redfield theories, prevent each of them from being universally applicable to all light-harvesting systems. In an effort of conciliation, recently, by using a hierarchy method,[129,130] there have been substantial rise in the development, for the use of the dissipative quantum mechanical methods, with the aim of directly computing the time evolution of the density matrix. This approach has been applied in explaining exciton systems in the FMO complex, in green sulfur bacteria[30] and also in the LH-II complex of purple bacteria. However, it should be noticed that there has been exponential rise in computational cost of the aforementioned hierarchy method with the number of pigments involved. That is why this method has to be restrained, and not been found not to be feasible to study systems containing more than a few dozen pigments. In such situations, generalized Förster theory has been found to successfully deliver similar results when the hierarchy method has been applied for calculating energy transfer in between a pair LH-II complexes. In such cases, thermal disorder present a constant challenge for the study of proteins in light-harvesting, especially, associated with quantum biology. Anyway, considering random matrix theory, these kind of effects can be viewed, in terms of a dynamic disorder model which involves trajectory averages or a static disorder by taking the ensemble averages.

Interestingly, in several cases of Photosynthetic organisms, related to efficient light-harvesting function, there seems to be an evolution, for not only coping up with, but also for exploiting quite successfully the effects, caused by thermal disorder. In such a situation, i.e., in the standard Redfield theory,[125] explicitly, all of exciton couplings are taken into consideration, allowing a description of all possible types of exciton relaxation/migration. These, normally, are found within the strongly coupled antenna complexes. This arrangement also includes coupled dynamics of the populations and coherence in between the exciton states. Another important theoretical approach, considered here, is to describe the related dynamics on the pure exciton basis and by including the exciton–phonon coupling, the corresponding relaxation between exciton states is accounted for an off-diagonal perturbation. This way, by including strong coupling of excitations to a few vibrational modes, the standard Redfield approach can be explained in a generalized mode, where, in such a system, the description of relaxation is based on the idea of electron-vibrational eigenstates. The approach, adopted by Novoderezhkin *et al.*,[131] described in detail the related electron transfer. He applied it to a coupled coherent nuclear motion in the bacterial RC, considering long-lived vibrational coherences and coupled exciton-vibrational relaxation in LH1.

At this point, some of the typical characteristical aspects are needed to be mentioned, i.e., light harvesting antenna structures contain chromophores of very high concentrations, i.e., levels reaching up to $0.6M$ in some pigment-protein assemblies. Not only that, in light-harvesting system, inter chromophoric distance between neighboring chlorophyll molecules can vary between 5 and 20 Å, giving the results with variations in the strength of intermolecular coupling. This fact directly influences the quantum mechanical nature of the energy-transfer mechanism which has been the basis for the advancement towards extended Förster's original theory. Thus, in order to address the predictions, related to energy transfer in light-harvesting complexes, four principle modifications for the theories have been found to be essentially needed.

As the intermolecular separation is close, as a first approximation, the associated electronic coupling must be calculated without invoking the dipole approximation. Secondly, while dealing with a breakdown in the dipole-dipole approximation, solvent screening of the electronic coupling needs a reconsideration hand-in-hand. As a third point, the presence and role of molecular exciton states acting as a excitation donors and acceptors needs to be considered. As stated earlier, the generalized Föster theory (GFT) or the modified Redfield theory has been typically suitable for doing this. Also, importantly, in such kind of typical conditions, both the Förster and Redfield approximations have been found applicable but, only within a certain limitations, i.e., a formalism which is simultaneously non-perturbative and non-Markovian in nature, is needed to be applied[126,129–132] in such kind of cases.

Again, in explaining photosynthetic phenomena in a noisy environment, quantum dynamic description having the nature of fully non-Markovian quantum dynamics, is necessary for such particular situation. In such cases, by utilizing a hierarchy of auxiliary density matrices, the necessity arises for being characterized to an arbitrary order, especially, in a system-environment coupling. Surprisingly, even when DNA+[90] included, this method has been successfully employed in studying the coherence dynamics, for example, both in the case of the Fenna-Matthews-Olsen pigment-protein complex of green sulphur bacteria, together with those in pigment-protein complexes of purple bacteria.[56,137,138] For example, the related calculations show that, in a cluster, light absorbing Chls coherently spread the excitation over the cluster within the corresponding relaxation time of 1–2 ps. But quite interestingly, this activity is obeyed by the Boltzmann distribution, that also in the case of populated exciton states underlying the cluster-cluster rate.[59b]

In fact, as reported in the works of Strumpfer,[139] this kind of cluster-cluster rate, indicates the process of immediate relaxation, thus the calculated relaxation time can be acieved. It is important to mention that after detailed studies, Schulten, together with his group,[56] finally concluded that the energy transfer dynamics is not affected due to the changes in the BChl organization, and the accurate transfer rate is achievable through the application of generalized Förster theory.

2.4.5 *Robustness & Optimality of a Light Harvesting System (LHS)*

From the extended studies, followed up till now, all possible form of the biological system, present in our earth, are destined to face two major challenges, i.e., environmental change followed by competition. Interestingly, the photosynthetic process, has been found to solve successfully this problem through adaption i.e., by changing external conditions, or robustness. In fact, all biological systems must cope up with the environmental change and competition which, being two major challenges, are intrinsically and mutually related. Among them, typically manifested are all biological systems which must cope up with adaptability towards changing external conditions or robustness of a system. This occurs in terms of a parameter insensitivity, related to its dynamics together with graceful degradation of its components, whereas, the fact that competition itself drives a system towards optimality so that the efficiency of the system, will find itself at an evolutionary disadvantage.

Thus, in general terms, it becomes difficult in quantifying both the robustness and optimality, especially, in case of arbitrary biological system. Because, it appears extreme difficult in finding an fitness landscape, based on which one could judge the adaptability. This typical characteristics of the system manifests itself by being insensitive towards the concerned parameters which are related to its dynamics and the graceful degradation of its components follows, but, driving the competition of the system, on the contrary, towards an optimality resulting into a less efficient system which eventually find itself at an stateof evolutionary disadvantageous. However, in case of the quantum yield of the excitation migration process, a light harvesting system provides a natural measure of its efficiency, in terms of the quantum yield of the excitation migration process, however, in a crude way, certain measure of its efficiency can be achieved. Thus, so far as the quantum yield concerned, it becomes a difficult task in modeling the process which possesses the diverse effects of various perturbations, for example, thermal disorder or loss of individual components. Again, as mentioned earlier, different questions may arise regarding the optimality of the geometry, for example, the chlorophyll network employed which can only be investigated and analyzed, only by generating ensembles of alternative network configurations.

In a light harvesting system, the first point of argument can not conclude quantum yield of excitation migration to be a best measure of robustness and optimality but, on the contrary, termed merely as the simplest one. Ideally, we need to consider in detail, all of such possible aspects, for example, the aspects of regulation, synthesis, assembly, together with the corresponding possibility of repair within the light harvesting apparatus, together with the inclusion of other possible processes. In this respect, one should consider the process of charge transfer before judging the adaptability of a light harvesting system together with role of photo-protection. In fact, the excitation migration can not act as a rate - limiting step in a so called typical way as the dissipation rates are much lower than excitation transfer rates as far as the context of the overall photosynthetic function is concerned. In such cases, the quantum yields, investigated up till now, have been noted to be high enough, for typical chlorophyll networks. However, these results, quite a small changes on the quantum yield, are caused by the disturbances on the network. However, even with such substantial amount of shortcomings, an investigation of observed excitation migration process, caused due to the influence of external perturbations, provides substantial insights to the design principles of a light harvesting complex. Following Sener and his groups's work,[29, 46, 47] presented here, an glimpse on their results related to the robustness and optimality of the chlorophyll network of cyanobacterial PSI can be achieved. The similar results have also been arrived at by Yang *et al.* a.[132]

The facts, as reported by several researchers, shows very little amount of quantum yield in the network of PSI changes which could be caused through fluctuations of chlorophyll site energies or through selective loss of individual chlorophylls. Due to the case of parameter insensitivity, the aspect of graceful degradation showing two major manifestations of robustness has been noticed. While considering the case of insensitivity, the site energy fluctuations (cf. Fig. 2.7(a) of the above file), together with consistently high quantum yields are found to be as a consequence of the broad line shapes of pigments. This is the reason for the maintenance of a significant overlapping in the transference of resonant energy, even facing the situation when chlorophyll site energies are displaced randomly. In such cases, the tolerance, observed against loss of individual chlorophylls (cf. Fig. 2.7(b)) is basically the consequence of the Förster radius which has been found to be sufficiently large compared to the typical inter-chlorophyll distances. Thus, even with the pruning of individual components, the network maintains efficient excitation transfer. Interestingly, it has been observed that the same results can be obtained, also for the case of simultaneous pruning of a large number of chlorophylls, even though, the loss of the corresponding cross section, the relative quantum yield of the pruned system remains high, the reason behind which is the slowness of dissipative processes. However, as expected, the pruning of the chlorophylls, especially, closest to reaction center (RC), produces the highest impact on the quantum yield results.

At this stage of the development, a few other questions also arise: for example; "For efficient excitation transfer, is there any optimization in the geometry of the chlorophyll network in PSI?"

In other words, the important query arises before us, are about the particular positions and orientations of individual chlorophyll which might pose as a critical factor, affecting the light harvesting function. In fact, these kind of queries arise naturally where, the existence of contrast, seemingly, lie in between random arrangement of chlorophylls in PSI with the symmetrical arrangement of chlorophylls, found in LH2.

Here, following the works of Sener and Schulten, in Figure 2.8(a) & (b), the distribution of quantum yields across an ensemble of alternative network geometrieshas been presented which is generated by random reorientation of chlorophylls. But, Sener and Schulten's works have showed that taking an ensemble of such kind as an example, the quantum yields, has been found to vary only within a narrow interval. On the contrary, the individual chlorophyll orientations have already been observed to be not that much critical for maintaining a reasonable light harvesting efficiency. At this juncture, related to the apparent optimality depicted in Fig. 2.8, an interesting question has been raised by Sener & Schulten, i.e., "is this optimality a genuine result of competitive evolution or only a computational artifact?" Considering sufficiently large number of generations, minimal amount of reproductive advantages can possibly be multiplied in becoming discriminating, as a resultant effect. However, substantial amount of examples has been reported in connection with competitive advantage, but, without displaying the phenotypical differences in the growth competition experiments (Ouyang *et al.*, 1998[142]). Based on such findings and arguments, Schulten and his group opined that even after 1 billion years of divergent evolution, the astonishing and remarkable conservation of the geometry, observed in the chlorophyll network, in PSI of

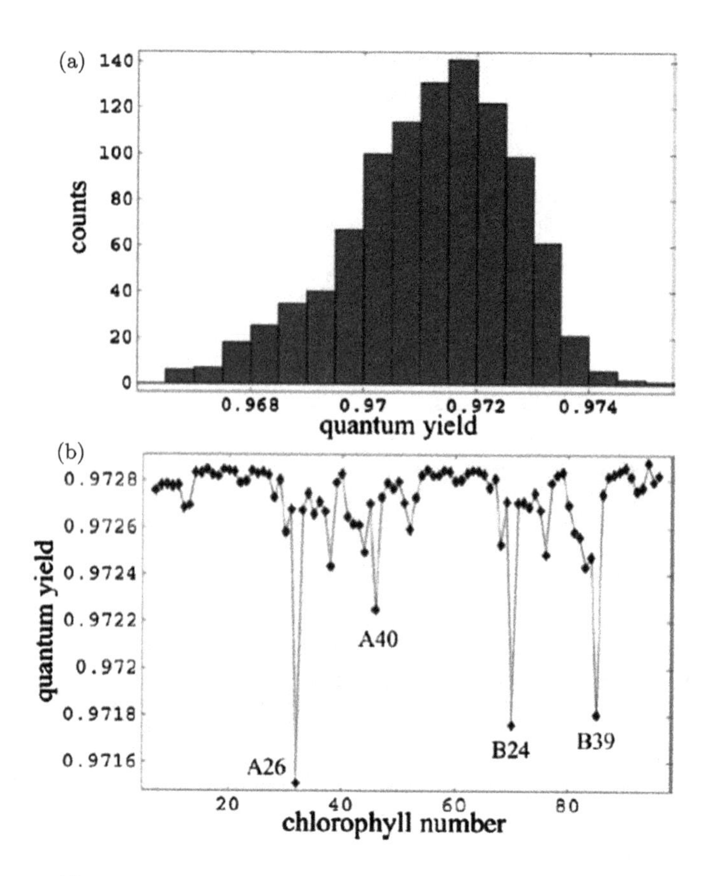

Figure 2.8. (a) Robustness against fluctuations of site energies. The histogram shows the distribution of the quantum yield over an ensemble of 1000 chlorophyll configurations generated by randomly displacing the chlorophyll site energy within a width of 180 cm-1. (b) Robustness against the pruning of individual chlorophylls. The average quantum yield of the remaining chlorophyll network is shown as a function of the pruned chlorophyll for all chlorophylls except for six central chlorophylls. The chlorophylls whose deletion has the highest impact on quantum yield are indicated.

cyanobacteria and plants, points to a very important fact, i.e., a degree of optimality has been reached prior to the divergence of the two organisms.

2.4.6 *Influence of Förster Energy Transfer Theory (FRET): Architecture of Light Harvesting Systems*

Up till now, we have followed Sener and his group's exhaustive works following which, we tried to explain the related problem i.e., how far

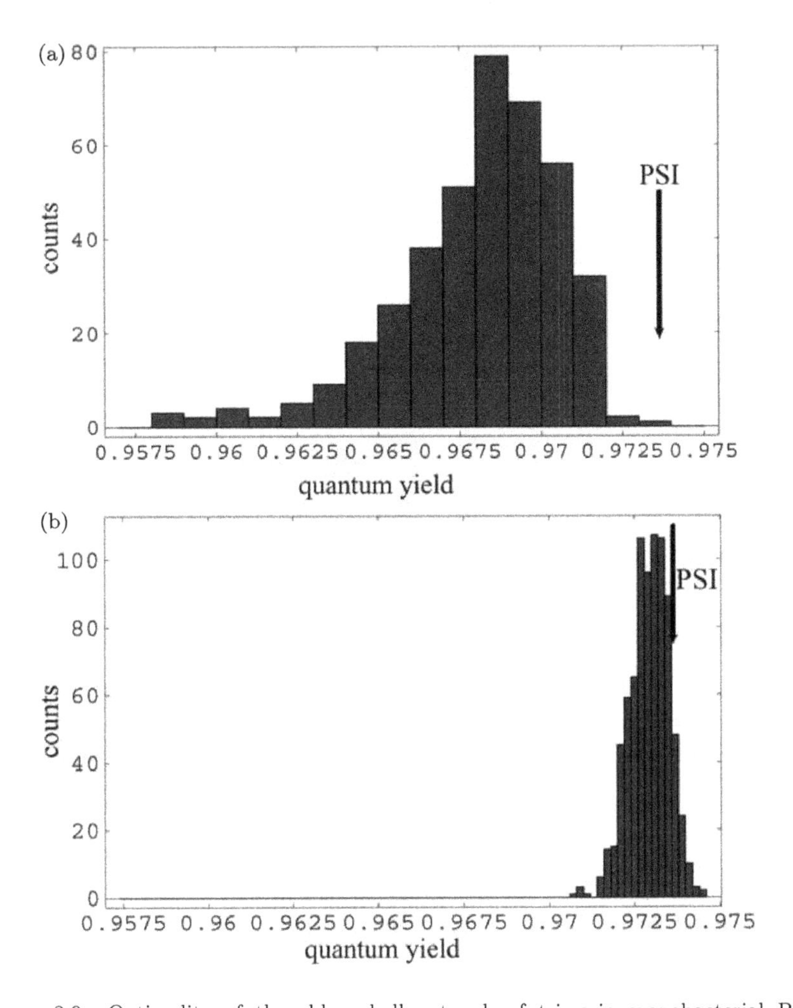

Figure 2.9. Optimality of the chlorophyll network of trimeric cyanobacterial PSI. Histograms show the distribution of quantum yield over an ensemble generated by randomly rotated chlorophylls. The quantum yield of the original configuration is indicated by an arrow. (a) All chlorophylls, including the reaction center chlorophylls are reoriented (400 configurations) (b) All chlorophylls other than the six central chlorophylls are reoriented (800 configurations).

effective the influence of Förster Energy Transfer theory has, on the explanation of Photosynthetic phenomena. As has been already noticed by now, a remarkable variety can be observed in the display of Photosynthetic species, found, in their physiology as well as in all kind of the structure and supramolecular organization, found in their constituent light harvesting proteins. But, interestingly, the reflection of common structural motifs can very well be noticed in Förster's theory of energy transfer where, the key characteristics are geometrical properties, i.e., K, R_0, and rapid decay, i.e., r^{-6}, together with the optical ones, namely the spectral overlap J_{ij}. As we know now, the geometrical term K accounts for the effect of pigment orientation on transfer rate whereas the Förster radius (R) (approximately 90 Å for Chls), along with the r^{-6} decay of transfer efficiency, place bounds on pigment separation and associated pigment packing density. The resonance term J_{ij}, as calculated following Förster's idea, points to the fact that it is necessary for the pigments to be spectrally matched for this kind of efficient transfer. To be specific, it could be mentioned that due to the detailed, exhaustive research works, done especially by Melih Sener and his group, here we are now, with some of the extremely valuable information, about purple bacterial light harvesting system.

2.4.6.1 *Role of Fluorescent Energy Transfer as a Physical Constraint: Effects on Shaping of the Organization of Photosynthetic Systems*

Many of the of critical features, dealt by Förster's theory of enegy transfer (FRET), can be connected to Photosynthetic phenomena, for example, pigment orientations, inter-pigment separation and spectral resonance, to name a few. This theory deals with all of those processes, applying a very critical and most exhaustive methods, starting from the ways and outs in them. Among them, each can be mentioned having proper prominence in the organization of photosynthetic systems which have influence on the evolution of photosynthetic species. In the following, some of them with common structural motifs, have been mentioned, explained, followed through the light of Förster's theory.

To start with, it could be observed that the first findings from Förster's theory proposes a high energy transfer efficiency according to Eq. (2.7), demands a small inter-pigment separation compared to R_0, i.e., a large pigment density. Next, out of detailed studies, especially by Sener and Schulten's group, it has been noticed that plant and cyanobacterial PS1 feature a greater pigment density than its purple bacterial counterpart. On the contrary, for the scaffolding of comparable pigment networks, less protein content is found to be used in the oxygenic systems. In fact, out of such valuable works, many pioneering as well as surprising results came out, for example,

(1) 1 Chl per 27 amino acids has been observed in the features of cyanobacterial PS1.
(2) An assembly of a RC-LH1 complex in purple bacteria, have been found which possessing two accompanying LH2s. It features 1 BChl per 40 amino acids which are being distributed covering a large volume.[20]

The next important finding shows us quite a bunch of surprising results which has been about the optimality with respect to Ch1. It has been found out that even after showing apparent disorder, the geometry of pigment networks in cyanobacterial PS1, quite surprisingly, represents an optimality with respect to Chl orientations and site energy fluctuations. Not only that, whenever overall excitation transfer efficiency is used as a fitness criteria, a robustness against damaged pigments has also been noticed.[137,147,148] Secondly, another vital as well as an important point to be noticed is that the interval over which the fluctuation of quantum efficiency occurs, caused by the perturbation of the PS1-Ch1 network physiological temperature, is rather narrow (0.95–0.97).[20] However, related to the geometry, with which the optimality is concerned,[20] this gives rise to a narrow interval of efficiency, most prevalent in the pigments surrounding the RC.[144] Not only that, surprisingly, it is found to deliver nearly the highest possible value for the efficiency.

2.4.6.2 *Role of Thermal Disorder*

The role of thermal disorder, found in photosynthetic system, is another important point to be mentioned as this kind of disorder plays a active role in effective broadening of pigment absorption profiles, needed for the high energy transfer efficiency. This has strong resonance values J_{ij}. The crucial part played by the concerned photosystems is the employment of a variety of pigments which have different absorption peaks with the purpose of harvesting light energy over a broad range of wavelengths which causes a funnel-like energy flow. Another interesting point to be noted here is that all of the pigments absorb light, but virtually and that also at the edge of the water absorption spectrum which largely overlapps with the visible spectrum. This happens, particularly, only for those species which evolve in water. Not only that in such cases, wavelengths of light, absorbed by water are inaccessible.

2.4.6.3 *Role of Reaction Center and their associated Electron Transfer Chains (ETCs)*

In light harvesting systems, RCs and their associated electron transfer chains (ETCs) are found to be surrounded which could be termed as a cordon sanitaire, [term used by Sener & Schulten group]. Basically this means an absence of non-ETC pigments in the immediate vicinity of the ETC.[28,146,147] This appears to be important and necessary also, because of theirconnection with charge separation at RC. This appears to be important and necessary for the maintainer of an high electron transfer efficiency along the ETC. If any kind of the 'leaking' of electrons occurs out of the ETC, it would require an overlapping of electronic wave functions, indicating its exponential decay with pigment distance.[112] But, interestingly, it has been found that the energy transfer into the RC has an $\frac{1}{r^6}$ dependence (c.f. Eq. (2.7)) on pigment distance,[22] thus, winning over the greater distances.

As a result, the efficient energy transfer into the RC is to be maintained while, at the same time, loss of electrons can be prevented.[20,28] The pigment gap is readily discerned in Fig. 2, ii, iv, v, and vi.

It is quite interesting to note some of design principles which are further physics-based, appearing to be quite prevalent in the evolution of photosynthetic systems. Such kind of substantial amount of interesting examples can be traced, for example,— the protection from photodamage demands few crucial activities, for example;- proximity between Chls and carotenoids, associated repair modularity of components, assembly of supramolecular aggregates, and graceful degradation with unavoidable radiation damage[28] which are necessary for the energy transfer. All of these characteristics can successfully be explained by applying Förster theory which, not only, is the fundamental process, essential for physiological function but also, nonetheless, appears to have profound influence on the architecture of the photo-system.

2.5 Conclusion

In recent times, due to the availability of an sufficient and increasing number of atomic resolution structures, together with those of the data meant for ultrafast spectroscopy, obtained from the different photosynthetic systems, it has been possible to compare the details of various light harvesting mechanisms. However, again, different kind of modeling in later time, have been provided by multi-subunit light harvesting systems. Because, both these kind of theoretical and experimental facts, i.e., multiple protein-pigment complexes, interact systematically, resulting to an intense as well as detailed studies which have taken the pieces together regarding the the evolutionary history of photosynthesis.

At this present stage of scientific developments, taking into account of different light harvesting systems, this topic of research has been transformed to be a fascinating challenge. Out of various kind of comparative studies, quite an interesting aspect can be be noticed to be played by nature, in making the process of photosynthesis a successful tool for the survival as well as nourishment of plant life, i.e., as if, following the physics of the FRET process, it is the evolution of photosynthetic life forms through which the nature herself have been influenced herself, as has been described and developed by Förster.

In the present era, the scientists, the idea that living matter should be studied with the same laws of physics as inanimate matter, is obvious. But, in retrospect, at the time when Förster's work was published, it was still in a rather formative stage of application, i.e., when Schrödinger's seminal book[15] stated that physical theory should be applied to describe biological processes as well. The pioneering works of Förster, together with his co-workers aforesaid, followed this revolutionary idea, placed at that time and has started this challenge, explored all kind of probable and possible ideas existing during that period of research.

Förster, in his pioneering work proposed as well as applied quantum physics for the description of fundamental biological process. In retrospect, it could also be mentioned that Förster has been the first in establishing that it is quantum physics which governs all photosynthetic lifeforms on Earth. Together with his group, they explored and explained in great detail, all the possible paths, i.e., how the migration of electronic excitations follow these paths during migration. Not only that, also they explained the intricacies of the processes, following up to the proper execution. Thus, together with other facts, what and when we see today, for example, the beautiful colors and other astonishing characteristics in plants, algae etc., makes us capable of studying the the energy of sunlight. This points nothing but to a telling sign of success of Förster's theory in applying quantum physics at work only. This journey has subsequently been followed in exploring the photosynthetic apparatus of plants, algae, and bacteria as well. He realized the importance of the extremely vital roles played by pigments– how beautifully and cleverly as well, these pigments are being scaffolded, carry out their role of hierarchical assemblies in a amazing way, following intricate protein structure. Not only that, important bases, take crucial parts in excitation transfer - absolutely an essential factor in photosynthetic light harvesting.

At that time, Förster realized that excitation transfer among pigments, what we see today, is essential in photosynthetic light harvesting, that the pigments, scaffolded by intricate protein structures, play the crucially important role in amazingly designed hierarchical assemblies. It was Förster who first specifically applied the concept of Quantum mechanics in his theory stating that, "*basically, quantum effects rule how nature harvests sunlight*". Quoting "Sener", today we know that "*he was definitely right, but the real role of quantum physics in photosynthesis is even more beautiful than he could envisage*".

Next important point, worth to be mentioned is that Förster's approach emphasized on the difference in FRET and exciton transfer. For this case, he introduced the possible contributory role of vibration which states — "*if the coupling energy is large compared to the width of the vibrational levels, then the transfer would proceed extensively, follow the exciton model before the occurrence of next collision*". Not only that, in FRET, in general, in the electronic absorption as well as in emission, the spectra are quite broad, and consequently the probability that the donor and acceptor molecules will simultaneously have the same transition energies, is quite small (expressed in the overlap integral). This fact decreases the rate of energy transfer dramatically, the direct consequence of which fact causes the coupling energy to be quite small compared to the width of the vibrational energy levels. The rate of transfer has been observed to be proportional to the square of quantum mechanical amplitudes, as a result of which there are no quantum interference effects. The reason behind this fact is that in such cases, the excitons having broad spectral widths cause the probability to be quite small, needed for being the donor and acceptor molecules as well and also simultaneously at identical locations within their spectral bandwidths. But this condition is required to allow energy transfer to happen (electronic transitions of the donor and acceptor must be of equal energy to obey energy conservation). At the same time, interestingly, in the limit of large coupling energy, the transfer would be completely excitonic. Interestingly, in FRET, the excitation energy is said to "hop" from one molecule to the other in a stochastic manner (i.e., like in a diffusion-like process). However, in the limit, when the width of the vibrational level becomes large enough compared to the coupling energy, this causes the progression between collisions to be very small, causing the transfer to end up proceeding according to the FRET mechanism. Because in the FRET limit, the vibrational relaxation becomes complete. Not only that the energy levels of the excited state, also being very broad, causes the FRET transfer process to be moderately slow i.e., around nanoseconds, compared to excitonic transfer $(10 - 10^3)$ femtoseconds.

An important characteristic of normal intermolecular FRET is the excitation energy being completely localized on either the donor or acceptor molecule, thus the total excitation wave function is involved at any time, only to the excited state of either the donor or the acceptor molecules, and the other partner being in the ground state. Added to this, this theory also explained execution of the process as well, starting from light absorbing pigments to so-called reaction centers (RC) after which stage, light energy, finally is capable of acting for the eventual conversion into chemical energy. It should be remembered that at the time when Förster's work was published, Schrödinger's seminal book;[15] stating that physical theory be applied to describe basic biological processes was still fairly new. Förster tried to explain the biological processes by applying quantum physics. He has shown, not surprisingly, that it is the basic principle of quantum physics whose different applications in life form, besides others, reigns also in all photosynthetic life of Earth. What is perhaps surprising is how well Förster's theory manages to capture the essential physics of the energy migration process in photosynthetic light harvesting. Duysens[35] and others could apply Förster's theory to describe energy transfer in photosynthetic species even before detailed structures of the constituent proteins were available. Today, Förster theory is used to study energy transfer not only in massive networks of photosynthetic proteins found in nature and resolved in atomic level structure, but also in artificial light harvesting systems.

This also points towards a few telling signs about the success in applications of Förster's theory together with quantum theory, the fueling of life on Earth with the energy of sunlight.The beautiful colors due to plants, algae, and photosynthetic bacteria are a telling sign of Förster's theory and quantum physics working together, fueling life on Earth with the energy of sunlight. While Förster realized that excitation transfer among pigments is essential in photosynthetic light harvesting, we see today the signature of those very characteristics i.e, the pigments, scaffolded by intricate protein structures, carry out the role in amazing hierarchical assemblies. It was him who specifically stated how nature harvests sunlight following the rule of quantum effects which pointed us not only knowing about his revolutionary works but also about its immense consequences today. The physics of the excitation migration process is fairly well understood and is described to a reasonable approximation by Förster theory within the context of an effective Hamiltonian formalism.[39]

Notably, nature herself appears to be influenced in the evolution of photosynthetic life forms by the physics of the FRET process as described by Förster. For today's scientists, the idea that living matter should be studied with the same laws of physics as inanimate matter is obvious. In retrospect, at the time when Förster's work was published, Schrödinger's seminal book[15] stating that physical theory be applied to describe biological processes was still fairly new; The beautiful colors due to plants, algae, and photosynthetic bacteria are a telling sign of Förster theory and quantum physics at work, fueling life on Earth with the energy of sunlight. While Förster realized that excitation transfer among pigments is essential in photosynthetic light harvesting, we see today that pigments, scaffolded by intricate protein structures, carry out the role in amazing hierarchical assemblies.

In that regard, the works of Förster can be termed as unique, pioneering, and, if not, revolutionary in this field of research. The theoretical framework developed by Förster and others describes how electronic excitation migrates in the photosynthetic apparatus of plants, algae, and bacteria from light absorbing pigments to so-called reaction centers where light energy is utilized for the eventual conversion into chemical energy. However, it is very important to report that, in 1941, Oppenheimer,[23] though largely unnoticed, related the physical basis of the related energy transfer process, but not that much in explicit way for the FRET formula of Förster, meant for a donor-acceptor pair. Let us conclude with a famous statement of Förster which says: "Effects rule how nature harvests sunlight". Ultimately, after a long strenuous journey of research, today we have reached to that point of understanding stating that, indeed, he was definitely right, but taking the words from Schulten, we conclude "the real role of quantum physics is even more beautiful than what he could envisage".

References

[1] Joseph Priestley; (1733–1804); Experiments and Observations on different kinds of Air. (1774–1777).

[2] Barnes CR; (1893); On the food of green plants; Bot Gaz; **18**; 403–411.

[3] Barnes CR; (1896); Photosyntax vs photosynthesis; Botanical papers at Buffalo; Bot Gaz 22: 248.
[4] Wurmser R.; (1929); The Energetic Efficiency of Photosynthesis; Nature; vol 124; 912–913.
[5] Hill R.; (1937); "Oxygen Evolved by Isolated Chloroplasts". Nature. 139 (3525): 881. doi:10.1038/139881a0.
[6] Van Niel; (1941); (a) The bacterial photosyntheses and their importance for the general problem of photosynthesis; Adv. Enzymol.; 1; 263–328. (b) Van Niel, e.B.; (1962); The present status of the comparative study of photosynthesis. Ann. Rev. Plant Physiol. 13, 1–26.
[7] Baptista Van Helmont; (1580–1644); Partington, J. R. (1936); "Joan Baptista Van Helmont". Annals of Science. 1 (4): 359–84 (359).
[8] Clayton R.K. & Sistrom W.R.; (1978); The Photosynthetic Bacteria; Plenum; New York.
[9] Blankenship, R., M.T. Madigan & C. Bauer (eds.); (1995); Anoxygenic Photosynthetic Bacteria; Kluwer Academic, Netherlands.
[10] Ort Dr. & Yocum CF; (1996); Volume 4 of the Advances in Photosynthesis and Respiration Series: Ort DR and Yocum CF (Eds.); Oxygenic Photosynthesis: The Light Reactions, Itista van Helmont; alchemist, physician and philosopher. London: William Rider & Son.; 46.
[11] Emerson Robert & W. Arnold; (1932); "A Separation of the Reactions in Photosynthesis by Means of Intermittent Light,"; Journal of General Physiology; vol 15, no. 4 (1932), 391–420; written with W. Arnold; "The Photochemical Reaction in Photosynthesis," ibid., 16, no. 2 (1932), 191–205, with W. Arnold; "Photosynthesis," in Annual Review of Biochemistry, no. 6 (1937), 535.
[12] Inzenhousz; (1730–1799); (a) Beale and Beale, Echoes of Ingen Housz, (2011); (b) Jan Ingenhousz; Experiments upon Vegetables, Discovering Their great Power of purifying the Common Air in the Sun-shine, and of Injuring it in the Shade and at Night- To Which is Joined: A new Method of examining the accurate Degree of Salubrity of the Atmosphere, London, 1779; From Henry Marshall Leicester and Herbert S. Klickstein; A Source Book in Chemistry 1400–1900; New York, NY: McGraw Hill, 1952. Excerpts. Retrieved 24 June 2008.
[13] Niels Bohr; (1929); Collected Works Volume 6; 1985; Pg; 219–221, 223–253.
[14] Pascuel Jordon; (1932); "Die Quantenmechanik und die Grundprobleme der Biologie und Psychologie": Quantum mechanics and the fundamental problems of biology and psychology"; "Die Naturwissenschaften".

[15] Schrödinger, E; (1944); What Is Life?; Cambridge University Press; New York.
[16] Johnjoe McFadden, Jim Al-Khalili: The origins of quantum biology; Proc Math Phys Eng Sci. 2018 Dec; 474(2220): 20180674. doi: 10.1098/rspa.2018.0674. Epub 2018 Dec 12.
[17] van Amerongen, H.; Valkunas, L.; van Grondelle;2000; R. World Scientific; Singapore.; van Amerongen, H.; Valkunas, L.; 2000; Molecular Mechanisms of Photosynthesis. Blackwell Science; Malden, MA.
[18] Blankenship, RE.; (2002); Molecular Mechanisms of Photosynthesis. Blackwell Science; Malden, MA:
[19] Cogdell RJ, Gall A, Kv öhler J.; (2006); Quart Rev Biophys. 2006; 39:227–324.
[20] Sener, MK.; Schulten, K.; (2008); The Purple Phototrophic Bacteria, vol. (28) of Advances in Photosynthesis and Respiration; Hunter, CN.; Daldal, Fevzi; Thurnauer, Marion C.; Thomas Beatty, J., editors. Springer; p. 275–294.
[21] Förster T. (1946); Naturwissenschaften.; 6:166–175.
[22] Förster T. (1948); Zwischenmolekulare Energiewanderung und Fluoreszenz; Ann Phys (Leipzig). 1948; 2:55–75.
[23] Oppenheimer, JR; (1941); Internal conversion in photosynthesis. Phys Rev; Proceedings of the American Physical Society; p. 158.
[24] Knox, R. S.: (1996); "Electronic excitation transfer in the photosynthetic unit: Reflections on work of William Arnold", Photosynth. Res., 48, 35–39.
[25] Arnold & Meek; (1956); Arnold, W., and Meek, E. S.; 1956, "The polarization of fluorescence and energy transfer in grana", Arch. Biochem.Biophys, 60, 82–90 (1956).
[26] Förster T.; (1965); Delocalized excitation and excitation transfer, In: [Modern Quantum Chemistry; Part III; Action of Light and Organic Molecules; O. Sunanoglu (eds), Academic, New York, 93–137.
[27] Latt, S.A., Cheung, H.T. & Blout E.R.; (1965); Energy transfer: a system with relatively fixed donor-acceptor separation; J. Amer. Chem. Soc.; 995–1003.-27.
[28] Schulten, K; (1999); From simplicity to Complexity and back; Function, Architecture, and Mechanism of Light-harvesting Systems in Photosynthetic Bacteria; Frauenfelder, H.; Deisenhofer, J.; Wolynes, PG., editors. Berlin: Dahlem University Press; 1999. p. 227–253.
[29] Şener, M.; Schulten, K.; (2005); Energy Harvesting Materials. Andrews, David L., editor. World Scientific; Singapore: p. 1–26.
[30] Şener, M., & Schulten, K., 2002; "A general random matrix approach to account for the effect of static disorder on the spectral properties of light harvesting systems.", Phys. Rev. E, 65, 031916–031940.

[31] Alexandra Olaya-Castro, Chiu Fan Lee, Francesca Fassioli Olsen, Neil F Johnson; (2008); Efficiency of energy transfer in a light-harvesting system under quantum coherence; Physical Review B; vol 78; (8).
[32] Masoud Mohseni, Patrick Rebentrost, Seth Lloyd, Alán Aspuru-Guzik; (2008); Environment-Assisted Quantum Walks in Photosynthetic Energy Transfer; Journal of Chemical Physics 129, 174106.
[33] Ishizaki A, Fleming GR.; 2009; Theoretical examination of quantum coherence in a photosynthetic system at physiological temperature; Proceedings of the National Academy of Sciences; 106: 17255–17260.
[34] Mino Yang, Ana Damjanović, Harsha M. Vaswani, and Graham R. Fleming; (2003); Energy Transfer in Photosystem I of Cyanobacteria Synechococcus elongatus: Model Study with Structure-Based Semi-Empirical Hamiltonian and Experimental Spectral Density; (2003); Biophys J.; 85(1): 140–158. doi: 10.1016/S0006-3495(03)74461-0.
[35] Duysens LNM.; (1951); Nature.; 168:548–550. [PubMed: 14875152].
[36] Van Grondelle, R, Dekker JP, Gillbro T, Sundström V.; (1994); Biochim Biophys Acta.; 1187:1–65.
[37] Truong K, Ikura M. Curr Opin Struct Biol.; (2002); 11(5):573–578. [PubMed: 11785758].
[38] Rasnik I, Myong S, Cheng W, Lohman TM, Ha T.; (2004); J Mol Biol.; 336:395–498. [PubMed: 14757053].
[39] Melih Şener, Johan Strümpfer, Jen Hsin, Danielle Chandler, Simon Scheuring, C. Neil Hunter, and Klaus Schulten; 2011; Chemphyschem; 12(3): 518–531.
[40] Ritz T, Hu X, Damjanović A, Schulten K.; (1998); Excitons and excitation transfer in the photosynthetic unit of purple bacteria; J Luminesc.; 76–77:310–321.
[41] Hu X, Ritz T, Damjanović A, Schulten K.; (1997); J Phys Chem B. 1997; 101:3854–3871.
[42] Hu X, Damanovitch A, Ritz, T. & Schulten K.; (1998); Architechture and function of the light harvesting apparatus of purple bacteria; Proceedings of the National Academy of Sciences of the United States of America; Published - May 26 1998; vol 95, issue 11; 5935–5941.
[43] Damanovitch A, Kosztin I, Kleinekathoefer U, Schulten K.; (2002); Phys Rev E.; 65:031919.
[44] (a) Xu, D.; Schulten, K.; 1992; The Photosynthetic Bacterial Reaction Center: II Structure, Spectroscopy and Dynamics. In: Breton, J.; Vermeglio, A., eds. NATO Sci Ser A. Plenum Press; New York: p. 301–312.
[45] Xu D, Schulten K.; 1994; Chem Phys. 1994; 182:91–117.

[46] Şener MK, Lu D, Ritz T, Park S, Fromme P, Schulten K.; 2002; J Phys Chem B.; 106:7948–7960.
[47] Şener MK, Park S, Lu D, Damjanović A, Ritz T, Fromme P, Schulten K.; 2004; J Chem Phys.; 120:11183–11195.
[48] Melkozernov AN, Lin S, Blankenship RE, Valkunas L.; 2001; Biophys J.1; 81:1144–1154.
[49] Fromme P, Witt HT, Jordan, Klukas O, Saenger W, Krauß Norbert.; 2001; Nature; 411:909–917.[PubMed: 11418848].
[50] Byrdin M, Jordan P, Krauß N, Fromme P, Stehlik D, Schlodder E.; 2002; Biophys J.; 83:433 –457.[PubMed: 12080132].
[51] Ben-Shem A, Frolow F, Nelson N. Nature.; 2003; 426:630–635. [PubMed: 14668855].
[52] Marcus RA. J Chem Phys. 1956; 24:966–978; Marcus RA. J Chem Phys. 1956; 24:979–989.
[53] Bahatyrova S, Frese RN, Alistair Siebert C, Olsen JD, van der Werf KO, van Grondelle R, Niederman RA, Bullough PA, Otto C, Hunter CN.; 2004; Nature.; 430:1058–1062. [PubMed:15329728].
[54] Scheuring S, Sturgis JN, Prima V, Bernadac A, Levy D, Rigaud J-L.; (2004) Proc Natl Acad Sci USA.; 91:11293–11297.
[55] Olsen JD, Tucker JD, Timney JA, Qian P, Vassilev C, Hunter CN.; (2008); J Biol Chem.; 283:30772–30779. [PubMed: 18723509].
[56] Qian P, Bullough PA, Hunter CN.; 2008; J Biol Chem.; 283:14002–14011. [PubMed: 18326046].
[57] Sener MK, Olsen JD, Hunter CN, Schulten K.; 2007; Proc Natl Acad Sci USA.; 104:15723–15728.
[58] Strümpfer J, Schulten K.; 2009 J Chem Phys.; 131:225101. (9 pages). [PubMed: 20001083].
[59] Sener M, Strumpfer J, Timney JA, Freiberg Arvi, Hunter CN, Schulten K.; 2010 Biophys J.; 99:67–75. [PubMed: 20655834].
[60] Clegg, Robert M *et al.*; (2010); Proceedings of SPIE — The International Society for Optical Engineering 7561.
[61] Kasha, M.; 1963; "Energy transfer mechanisms and the molecular exciton model for molecular aggregates", Rad. Res., 20(1), 55–70.
[62] Pearlstein, R. M.; 1982; "Exciton migration and trapping in photosynthesis", Photochem. Photobiol., 35, 835–844.
[63] Agranovitch V; Galanin M *et al.*; 1982; Electronic excitation energy transfer in condensed matter; 9e9656a1034511aaac9658a3a353017 aac84.pdf.
[64] Silbey, R., and Harris, R. A.; 1984; Variational calculation of the dynamics of a two level system interacting with a bath; Journal of Chemical Physics; 80; 2615.

[65] Arnold, W & Kohn, H.I.; 1934; The chlorophyll unit in photosynthesis. J. gen. Physiol. 18, 109–112.
[66] Hu, Thorton Ritz Schulten 2002.
[67] Duysen, L. N. M. (1952). Transfer of excitation energyin photosynthesis. PhD thesis, Utrecht.
[68] Duysen L.N.M. 1964; Photosynthesis. Progr. Biophys. molec. Biol. 14, 1–104.
[69] Miller K.R. (1982); Three-dimensional structure of a photosynthetic membrane. Nature 300, 53–55.
[70] Stark Monger, T. & Parson, W. (1977); Singlet-triplet fusion in Rhodopseudomonas sphaeroides chromatophores. A probe of the organization of the photosynthetic apparatus. Biochim. Biophys. Acta 460, 393–407.
[71] Waltz, T. & Ghosh, R. (1997); Two-dimensional crystallization of the light-harvesting I – reaction centre photounit from Rhodospirillum rubrum. J. molec. Biol. 265, 107–111.
[72] Scholes, G.D., Flemming G R, Olaya Castro A, grondelle R van; *et al.*; 2011; Nature-Chemistry; 3(10); 76–774.
[73] Green B.R. & Parson W.W.; (2003); Light Harvesting Antennas in Photosynthesis; Ed, Green, W.W (Eds).
[74] Scholes, G. D.; 2003; Long-range resonance energy transfer in molecular systems. Annual Review of Physical Chemistry, 54, 57.
[75] Sauer K., Cogdell R. J., Prince, S. M., Freer, A., Isaacs, N. W. & Sheer S, H.; 1996; Structure-based calculations of the optical spectra of the LH2 bacteriochlorophyll-protein complex from Rhodopseudomonas acidophila. Photochem. Photobiol. 64, 564–576.
[76] Renger; T; 2009; Light Absorption and Energy Transfer in the Antenna Complexes of Photosynthetic; Photosynth. Res.; 102; p. 471.
[77] Thornber, J. Philip, Richard J., Cogdell, Richard j., Beverly K. Pierson, Beverly K., Richard, E. B. Seftor, E.B., Richard E.B.; (1983); Pigment-protein complexes of purple photosynthetic bacteria: An overview; https://onlinelibrary.wiley.com/doi/abs/10.1002/jcb.240230113.
[78] Rene A Brunisholtz; & Herbert Zuber; 1992; Structure, function and orgarnization of antenna polypeptides and complexes from the three families of Rhodospirillaneae: Journal of photochemistry and photobiology B: Biology; 113–140.
[79] Hawthornthwaite A.M. & Cogdell R.J.; 1991; Bacteriochlorophyll binding proteins. In: H.Scheer (ed.); The Chlorophylls, pp 493–528. CRC Press, Baton Rouge, by JN Sturgis - 1996.

[80] Stark, W. *et al.*; 1984; The structure of the photoreceptor unit of Rhodopseudomonas viridis.; EMBO J. 3, 777–783.

[81] Monger, T. & Parson, W; 1977; Singlet-triplet fusion in Rhodopseudomonas sphaeroides chromatophores. A probe of the organization of the photosynthetic apparatus. Biochim. Biophys. Acta 460, 393–407.

[82] Sundström, V. & Van Grondelle, R.; (1991); Dynamics of excitation energy transfer in photosynthetic bacteria. In Chlorophylls (ed. H. Scheer), pp. 627–704. Boca Raton: CRC Press.

[83] Sundström, V. & Van Grondelle, R.; (1995); Kinetics of excitation transfer and traping in purple bacteria. In Anoxygenic Photosynthetic Bacteria (eds. M. R. Blankenship, M. T. Madigan & C. E. Bauer), pp. 349–372. Dordrecht: Kluwer Academic Publishers.

[84] Germeroth, A., Lottspeich, F. & Robert, B & Michel, H.; (1993); Unexpected similarities of the B800–850 light-harvesting complex from Rhodospirillum molischianum to the B870 light-harvesting complexes from other purple photosynthetic bacteria. Biochemistry 32, 5615–5621.

[85] Aagaard, J. & Sistrom, W.; 1972; 1972; Control of synthesis of reaction center bacteriochlorophyll in photosynthetic bacteria. Photochem. Photobiol. 15, 209–225.

[86] Borisov, A.Y. & Godic, V.I.; 1973; Excitation energy transfer in photosynthesis. Biochim. Biophys. Acta 301, 227–248.

[87] Francke, C. & Amesz, J.; 1995; The size of the photosynthetic unit in purple bacteria. Photosyn. Res.46, 347–352.

[88] Freiberg, A.; 1995; Coupling of antennas to reaction centers. In Anoxygenic Photosynthetic Bacteria (eds. R. E. Blankenship, M. T. Madigan & C. E. Bauer), pp. 385–398. Dordrecht: Kluwer Academic Publishers.

[89] Cogdell *et al.*; 1996; The purple bacterial photosynthetic unit. Photosyn. Res. 48, 55–63.

[90] Mauzerall, D. & Greenbaum, N.; 1989; The absolute size of a photosynthetic unit. Biochim. Biophys. Acta; 974, 119–140.

[91] Grossman, A. R., Bhaya, D., Apt, K. E. & Kehoe, D. M.; 1995; Light-harvesting complexes in oxygenic photosynthesis: diversity, control, and evolution. Annu. Rev. Genet. 29, 231–288.

[92] Fromme, P.; 1996; Structure and function of photosystem I. Curr. Opin. struct. Biol. 6, 473–484.

[93] Gantt, E.; 1996; Pigment protein complexes and the concept of the photosynthetic unit: chlorophyll complexes and phycobilisomes. Photosyn. Res. 48, 47–53.

[94] Hankamer *et al.* 1997; Hankamer, B., Barber, J. & Boekema, E. J.; (1997); Structure and membrane organization of photosystem II in green plants. Annu. Rev. Plant. Physiol. 48, 641–671.
[95] Borisov, A.Y. & Godic, V.I.; 1973; Excitation energy transfer in photosynthesis. Biochim. Biophys. Acta 301, 227–248.
[96] Zuber, H. & Brunisholtz, R.; 1981; Structure and function of antenna polypeptides and chlorophyll–protein complexes: principles and variability. In Chlorophylls (ed. H. Scheer), pp. 627–692. Boca Raton: CRC Press.
[97] Sundström, V. & Grondelle, R.; 1995; Kinetics of excitation transfer and traping in purple bacteria. In Anoxygenic Photosynthetic Bacteria (eds. M. R.
[98] Pullerits, T. & Sunstrom, V.; 1996; Photosynthetic light-harvesting pigment–protein complexes: toward understanding how and why. Acc. Chem. Res. 29, 381–389.
[99] Flemming, G.R. & Van Grondelle; 1997; Femtosecond spectroscopy of photosynthetic light-harvesting systems. Curr. Opin. struct. Biol. 7, 738–748.
[100] Karrasch, S. *et al.*; 1995; 8 5 AI projection map of the light-harvesting complex I from Rhodospirillum rubrum reveals a ring composed of 16 subunits. EMBO J. 14, 631–638.
[101] Clayton, R. K.; 1973; Primary processes in bacterial photosynthesis. Annu. Rev. biophys. Bioeng. 2, 131–156.
[102] Govindjee, (Ed.); 1982; Photosynthesis. New York: Academic Press.
[103] Scheer, H. (Ed.); 1991; Chlorophylls. Boca Raton: CRC Press.
[104] Deisehofer, J. & Norris, J.R. Eds; 1993; The Photosynthetic Reaction Center. Volumes I and II. San Diego: Academic Press.
[105] Flemming, G. & Van Grondelle (1994); The primary steps of photosynthesis. Physics Today 47, 48–55.
[106] Fyfe, P.K. & Cogdell, R.J.; 1996; Purple bacterial antenna complexes. Curr. Opin. struct. Biol. 6, 467–472.
[107] Sundström *et al.*; 1999; Photosynthetic light-harvesting: reconciling dynamics and structure of purple bacterial LH2 reveals function of photosynthetic unit; J. phys. Chem. (B), 103, 2327–2346.
[108] Cogdell R.J. *et al.*; 1999; How photosynthetic bacteria harvest solar energy. J. Bacteriol. 181, 3869–3879.
[109] Krueger, B. *et al.* 1999a; The light harvesting process in purple bacteria. Acta. Phys. Pol. (A), 95, 63–83.
[110] Rodrick K. Clayton; 1966; The bacterial photosynthetic reaction center. Brookhaven Symp. Biol. 19, 62–70.

[111] Andrew John Young; 1991; https://doi.org/10.1111/j.1399-3054.1991.tb02490.x;]
[112] Dexter, D.; (1953); A theory of sensitized luminescence in solids. J. Chem. Phys. 21, 836–850.
[113] Damjanovic *et al.*, A., Ritz, T. & Schulten, K.; (1999); Energy transfer between carotenoids and bacterio chlorophylls in a light harvesting protein. Phys. Rev.(E), 59, 3293–3311.
[114] Kosztin, I.; Schulten, K. Biophysical Techniques in Photosynthesis II, volume 26 of Advances in Photosynthesis and Respiration. Aartsma, Thijs; Matysik, Joerg, editors. Springer; Dordrecht: 2008. p. 445–464.
[115] Ritz T, Park S, Schulten K. J Phys Chem B. 2001; 105:8259–8267.
[116] Engel GS, Calhoun TR, Read EL, Ahn T-K, Mancal T, Cheng Y-C, Blankenship RE, Fleming GR.; 2007; Nature.; 446(7137):782–786.
[117] Panitchayangkoon G, Hayes D, Fransted KA, Caram JR, Harel E, Wen J, Blankenship RE, Engel GS.; 2010; Proc Natl Acad Sci USA.; 107(29):12766–12770.
[118] Sener MK, Hsin J, Trabuco Leonardo G, Villa E, Qian P, Hunter CN, Schulten K.; 2009; Chem Phys.; 357:188–197. [PubMed: 20161332].
[119] Gobets B, van Stokkum IHM, Rögner M, Kruip J, Schlodder E, Karapetyan NV, Dekker JP, van Grondelle R. Biophys J. 2001; 81:407–424. [PubMed: 11423424].
[120] Şener M, Schulten K. Phys Rev E. 2002; 65:031916.
[121] May, V.; OKühn, J. Wiley, S.; 2004; Inc. Charge and energy transfer dynamics in molecular system; Wiley-VCH.
[122] Weiss, U.; 2008; Quantum dissipative systems. World Scientific Publishing Company.
[123] Cardeira AO, Leggett AJ.; 1983; J Ann Phys (NY).; 149:374–456.
[124] Janusonis J, Valkunas L, Rutkauskas D, van Grondelle R.; 2008; Biophys J.; 94(4):1348–1358.
[125] (a) Redfield, A.G.; 1957; On the theory of relaxation processes, IBM Journal of Research and Development; 1; 19–31. (b) Redfield, A.G.; 1965; "The Theory of Relaxation Processes", Adv. Magn. Reson. 1, 1–32.
[126] (a) Sumi, H.; 1999; J. Phys. Chem. B, 103, 252. (b) Sumi, H.; 2002; Chem. Rec., 1, 480.
[127] (a) Scholes, G.D., G.R. Fleming, G.R.; 2000; J. Phys. Chem. B, 104, 1854. (b) Scholes, G.D., Jordanides, X., Fleming, G.R.; 2002; J. Phys. Chem. B, 105, 1640.
[128] (a) Mukamel, S.J.; 1995; Priciples of Nonlinear Optical Spectroscopy. (b) Mukamel, S.J.; 1998; Chem.Phys.; 108; 7763.

[129] Zhang, W. M.; Meier, T., Chernyak; V. *et al.*; 1998; Exciton-migration and three-pulse femtosecond optical spectroscopies of photosynthetic antenna complexes; J. Chem. Phys., 108(18), 7763–7774 (1998).
[130] Jang, S.; Newton, M.D.; Silbey, R.J.; 2004; Phys. Rev. Lett., 92, 218301.
[131] Vladimir I.; Novoderezhkin, Miguel A.; Palacios, Herbert van Amerongen & Rienk van Grondelle; 2004; Energy-Transfer Dynamics in the LHCII Complex of Higher Plants: Modified Redfield Approach.
[132] Yang, M., and Fleming, G. R., 2002; Influence of phonons on exciton transfer dynamics: comparison of the Redfield, Förster, and modified Redfield equations; Chem. Phys., 275, 355–372.
[133] Kühn O, Sundström V. J Chem Phys. 1997; 107(11):4154–4164.
[134] Renger T, May V, Kühn O.; 2001; Phys Rep.; 343:137–254.
[135] Tanimura Y, Kubo R.; 1989; J Phys Soc Jpn.; 58(4):1199–1206.
[136] Ishizaki, Akihito; Fleming, Graham R. J.; 2009; Chem Phys.; 130(23):234111–10.
[137] Chen L, Zheng R, Shi Q, Yan Y.; 2009; J Chem Phys.; 131(9):094502–11.
[138] Hsin J, Strumpfer J, Sener M, Qian Pu, Hunter CN, Schulten K.; 2010; New J Phys.; 12:085005.
[139] Strumpfer J, Schulten K.; 2010; The effect of correlated bath fluctuations on exciton transfer.
[140] Dijkstra AG, Tanimura Y.; 2010; New J Phys. 2010; 12(5):055005.
[141] Yang M, Damjanović A, Vaswani HM & Fleming GR.; 2003; Energy transfer in photosystem I of cyanobacteria Synechococcus elongatus: model study with structure-based semi-empirical Hamiltonian and experimental spectral Density. Biophys. J.; 85:140–158.
[142] Ouyang Y, Andersson CR, Kondo T, Golden SS & Johnson CH.; 1998; Resonating circadian clocks enhance fitness in cyanobacteria; Proc. Nat.Acad. Sci. USA; 95:8660–8664.
[143] Damjanovic A, Vaswani HM, Fromme P, Fleming GR.; 2002; J Phys Chem B.; 106.
[144] Vasilev S, Bruce D.; 2004; Plant Cell.; 16:3059–3068.
[145] Wu J, Liu Fan, Shen Young, Cao J, Silbey RJ.; 2010; New J Phys.; 12:105012.
[146] Noy D, Moser CC, Leslie Dutton P.; 2006; Biochim Biophys Acta.; 1757:90–105. [PubMed: 16457774].
[147] Noy D.; 2008; Photosyn Res.; 95:23–35. [PubMed: 17968671].

Chapter 3

Possible Role of Quantum Coherence in Photosynthesis

"The important thing in science is not so much to obtain new facts as to discover new ways of thinking about them". —Sir William Henry Bragg; In lecture during Nobel Prize For Physics, 1951

3.1 Introduction: Role of Quantum Coherence in Photosynthesis

The role of quantum mechanics has traced a long history in the history of the biological research, because of their biological significance, especially related to the photosynthesis which appears indisputable. Not only that among many of necessary factors for nearly all form of life on the earth, it has been found to be an absolutely one, because after all, biological molecules are nothing but the conglomeration of molecules, atoms of different size and categories, bound together with the forces which has been defined as of a 'quantum origin'.[1–3] To be quite precise, the process in Photosynthesis is responsible for providing chemical energy for plants, algae and bacteria, whereas a heterotrophic kind of organisms depend on these species only through these biological process, taking them as their ultimate food source. The solar energy, here, acts as a storage of fuel, playing the vital link in between the supplier of energy, i.e., of the Sun and life on Earth, acting as a receiver.

Initial biological step in photosynthetic process involves the capture of solar energy from sunlight acting as a fuel. Starting with the process of light harvesting, it then follows transfer of the excitation energy, possessing very high efficiency, i.e., the process of excitation trapping, as a next, followed by charge separation at the reaction center (RC). There, specialized pigment–protein complexes takes the vital role in transforming sunlight into electronic excitation. After being initiated by light absorption, the excited state dynamics initiated by light absorption, plays the central role in the primary reactions of photosynthesis. The highly efficient solar-energy collection, followed by the transfer, ultimately converts into photosynthesis, but, all of them remain within the excitation lifetime which, is accomplished, precisely, with the help of a special kind of membrane-bound pigment-protein complexes, called light-harvesting antenna. In particular, with this purpose, all of these have evolved, in order to fulfill ultra fast energy transfer occurring in light harvesting. If not successfully transferred from the excited chromophore, the excitation energy relaxes back again to the electronic ground state. This could happen either of two ways, i.e., via emission of a photon (radiative decay) or through processes of various nonradiative origin.

Recently, i.e., almost from the beginning of last century, a new idea started blossoming, called Quantum biology which proposes deployment of some notoriously counter intuitive behavior, regarding life's molecular mechanisms. There, the researchers reported evidence that photosynthesis — the process with the help of which, green plants and some other bacteria turn sunlight into chemical energy — gains light-harvesting efficiency through the exploitation of the phenomenon, called "quantum coherence". Even though, researchers, attached to these problems, have been in continuous discussion, i.e., what it means by the recently or very recently measured experiments which could be associated with the complexity of the processes, after the examination being completed, — the associated problems, still remained. That is, even though the biological molecules are constructed of atoms, the associated forces, being of a 'quantum origin', bound them together. But, still the role of quantum coherence (and more recently, entanglement), being dependent on time or energy scales, have a direct and dictating role to play in biological function, even if keeping their roles still controversial.

After lots of suggestions and propositions being placed, quantum superpositions and/or quantum transport phenomena has/have been suggested as a possible cause, responsible for the efficiency and robustness of energy transport phenomena, present in biological systems. Critical observations in these experiments, comprised the observation of coherent oscillations or "quantum beats" via femtosecond laser spectroscopy which can be observed in many of different light harvesting systems. Truly, the results, obtained from recent experiments[4–9] on photosynthetic 'Light Harvesting Complexes' (LHC) together with their constituents, e.g., the Fenna–Matthews–Olson (FMO) pigment-protein complex in green sulfur bacteria, have put on the strong evidences which suggest, especially, the presence of quantum coherence playing an unavoidable and a crucial role, not only, in one of the very fundamental but also an important biological processes, i.e., the energy transport together with energy conversion, this way, persuading the proper investigation further on. In this experiment, critical experimental observations have been performed in order to measure coherent oscillations or "quantum beats", observed via femtosecond laser spectroscopy. This comprise the observation of coherent oscillations, observable in many different light harvesting systems. Consequently, substantial number of important experimental and theoretical studies have been followed recently, with the aim of observing the way which could have been followed during ultrafast energy transfer in light harvesting, particularly, with the aim of investigating the role of coherence.[136] As has already been known, — during the early steps of photosynthesis, it is the light-harvesting complexes which acts the vital and primary role, in absorbing and transferring solar excitation energy to the reaction centre (RC), i.e., the site of energy conversion. The next step is then the conversion of the excitation energy into a trans-membrane electrochemical potential with near-unity quantum efficiency where almost every absorbed photon is converted into a charge-separated state.

Even though, presently, we have the substantial knowledge about the pathways and timescales of charge separation,[10] the precise mechanism in order to have high efficiency of this process, is still not well established. In elucidating the design principle of natural light-harvesting antenna,[10,12,159] numerous investigations have been devoted by now. The result obtained currently, has concluded that a key element of their efficient functioning is related to the spatial arrangement of their pigment molecules and also with their electronic interactions. However, the answer to the question of electronic coherence, presumably present in these systems, depends on,

(1) Firstly, on physiological temperatures
(2) Secondly, this could influence the energy transfer process, for long enough.

In 1990s, after the famous and successful experiments,[159] producing the remarkable results, related to the direct visualization of coherent nuclear motion in the excited state of the primary electron donor in the bacterial RC,[159] there have conclusive suggestions about the functional role for these motions, in the primary electron-transfer reaction. This view, also, has been supported later on by theoretical models.[12,13] But, in a practical sense, at that time, **no** coherence has been observed between the reactant and the product of that reaction. This could be the reason why very recently, the observation of long-lived coherence in photosynthetic complexes light-harvesting[4,14–17,19,248] and oxidized bacterial RCs which have been found unable to perform charge separation,[5,21] has triggered an **intense debate** regarding the role of quantum coherence in promoting the efficiency of photosynthesis. In fact, no clear correlation between coherence and efficiency of energy or electron transfer has been presented. In this context, quantum coherence between the electronic states involved in energy or electron transfer introduces correlations between the wave functions of these states,[25,26] enabling the excitation to move rapidly and coherently through simple multiple pathways in space.[26] Therefore, the quantum coherence effect may render the process of energy and electron transfer less sensitive to the intrinsic disorder of pigment-protein complexes and in this way, allow these systems in reaching successfully to their final state avoiding energy losses.

These evidences of the quantum coherence between electronic excited states play a role akin to energy transfer involving the superposition of electronic quantum states, which seems, at once, enable in exploring many energy-transmitting pathways. If so, quantum mechanics could claim assisting the fundamental energetic process that drives all life on the surface of the Earth in one way or other. But a controversy still remains regarding the precise knowledge about the pathways and timescales of charge separation.[10] Other way, it means, the precise mechanism responsible for such high efficiency of this process, still remains possessed by many controversies, yet to be finalized. Anyway, out of various applications of theoretical developments, there after followed by corresponding physical applications, this mechanism has been advanced vastly in developmental aspects, placing ample of experimental evidences obtained where the presence of the physical phenomena has already firmly been established, thus establishing the fact that during energy transfer in between electronic excited states, one of the most important, unavoidable role, is played by quantum coherence and entanglement.

In the present chapter, a glimpse of the role of quantum coherence has been presented, consequently followed by that of entanglement phenomena, manifested during the process related to photosynthetic light harvesting. Followed through these aspects, while trying to establish their implications, the first question, appears as: **IS, in Biological processes, can there be really a functional role for quantum mechanics or coherent quantum effects?** Even though, this question has been as old as quantum theory, only recently, a few of positive outcome in this situations have been reached, i.e., after applying the advanced techniques in the measurements on biological systems of ultra-fast time-scales, thus shedding light on the possible answers of the related queries. Here, by using the term "functional", to the concerned process, we imply a typical role played by the presence of coherent quantum dynamics by which something, either more efficiently or otherwise impossible could be achieved in comparison to a classical mechanism alone. Thus, even though we adhere to the concept of "nature taking advantage of quantum mechanics" as an inspiring notion, we need to be careful enough in putting the evidence for and against it which, again, must be examined quite carefully, thus leading us to try discussing only experimentally-verifiable systems in this review.

Thus the subject, we try to deal here presently, has been growing at a phenomenal pace, out of which, we can only have a try in summarizing the main points for each system. For this, we consider, as in this topic, a few of recently raised crucial controversies. For example, recent experiments[4–9] on photosynthetic 'Light Harvesting Complexes' (LHC), together with their constituents (e.g., the Fenna–Matthews–Olson (FMO) pigment-protein complex in green sulfur bacteria), have suggested that both the processes of quantum coherence and entanglement might have a conclusive and decisive roles in the functioning of most fundamental and important biological processes: energy transport and energy conversion. For example, quantum coherence, manifested in the delocalized eigenstates of photo-excitations in photosynthetic complexes have been found playing a few of fundamental roles in spectral properties, i.e., energy tuning, and energy transfer dynamics of photosynthetic light harvesting. Added to this, the effects of excitonic coherence has been observed to be more difficult for the proper analysis. That is why these have been the subject of intensive research in the past few years. Spectroscopic measurements (2DES) on the FMO complex at liquid-nitrogen (77 K)[4] at room temperatures,[7,8] have produced results showing time-dependent oscillations, presumed to be quantum beating in the amplitudes of the spectral signals. Inspired by the observed results, matching the predictions of quantum theory, it has been proposed that a intricate interplay of quantum coherent excitation transference and environmental dephasing assisted transport[27] processes are there in helping the increase in the efficiency of the energy transport process.

In 1997, after the experiment of Savikhin, Buck and Struve[14] saw unusual oscillations in their pump–probe data, the first hint of the existence of coherence among excited states in a photosynthetic light harvesting complex appeared. Even though the oscillations in pump–probe data were quite common in displaying its existence in other experiments also, but interestingly, these oscillations, on the contrary, appeared in anisotropy measurements, thus giving the indication that a simple vibrational wave packet might not be the real cause. The authors[14] modeled the experiment by creating grouping of the seven accessible electronic states present in the system, into two bins, as a result of which the presence of oscillation has been manifested as the resultant effect of electronic coherence among the binned states. This insightful model has also been verified later on by using 2D electronic spectroscopy.[15,28,29] However, importantly, the first direct observations of quantum coherence in biological systems, has been occurred in 2007, when the same FMO-complex[31,32,41] was interrogated with 2D spectroscopy.[4] Two-dimensional Fourier transform electronic spectroscopy,[4,28,29] i.e., 2DES has mapped these energy levels and their coupling, present in the Fenna–Matthews–Olson (FMO) bacteriochlorophyll complex[41] which can be found in green sulphur bacteria. In fact, acting as an energy 'wire' to the reaction center (RC), this connects a large peripheral light-harvesting antenna, i.e., the chlorosome. The documented spectroscopic data clearly demonstrated the dependence of the dominant energy transport pathways on the spatial properties of the excited-state wave functions of the whole bacteriochlorophyll complex.[15,33]

But the intricate dynamics of quantum coherence, which has no classical analogue, was largely neglected in the analyses — even though electronic energy transfer involving oscillatory populations of donors and acceptors was first discussed more than 70 years ago[34] and electronic quantum beats arising from quantum coherence in photosynthetic complexes have already been predicted[35,36] and indirectly observed.[14] But, it was Engel *et al.*[4] who extended previously developed investigations of the FMO bacteriochlorophyll complex, as a result of which there have been the direct evidence for remarkably long-lived electronic quantum coherence, playing an crucially important part in energy transfer processes within this system.

In fact, in photosynthetic light harvesting, quantum coherence could have different meaning depending on different circumstances. For example, quantum coherence can be referred to as the quantum superpositions of localized molecular excitations occurring naturally because electronic couplings between molecular excitations lead to delocalized eigenstates (a.k.a. excitons) which, being present at all times, is continuously created, destroyed, and recreated by the interaction of the electronic system with the surrounding nuclear degrees of freedom. By the term stationary eigenstates, the associated picture can be conceptualized as the result of coarse-graining via a convenient mean-field approximation. Thus, in a photosynthetic complex, the wave function associated with an electronic excitation is never stationary, perpetually, always evolving under the influence of the fluctuations in its condensed-phase environment. These environmental fluctuations modulate the energy and couplings, associated with the collective molecular system, thus leading to dynamical transitions of excitation energy transfer. As a result, the associated physical picture underpins all quantum dynamical processes, only in the condensed phase.

However, sometimes, coherence in light harvesting alludes to the coherent wave-like dynamics of energy transfer which actually reflects the superposition of excitonic eigenstates. The subsequent energy transfer to the reaction center is commonly rationalized in terms of excitons, moving on a grid of biomolecular chromophores following a typical timescales, i.e., <100 fs. However, the way we understand the energy transfer today, includes the fact, i.e., the excitons are being delocalized over a few neighboring sites, but, on the contrary, the role of quantum coherence is considered as irrelevant for the transfer dynamics because it decays, typically within a few tens of femtoseconds. However, very recently, this idea, i.e., giving an orthodox picture of incoherent energy transfer between clusters of a few pigments, sharing delocalized excitons, has been challenged after the successful development of ultrafast optical spectroscopy experiments with the Fenna–Matthews–Olson protein where, the interference oscillatory signals, obtained, up to 1.5 ps, were reported and interpreted as direct evidence of exceptionally long-lived electronic quantum coherence.

To be more precise, the coherence is represented in the molecular site basis in the former case, whereas, in the latter case, the representation of coherence is on the delocalized exciton basis. It is quite important to note that exciton basis are special in the sense that they are related to the spatial arrangement of chromophores and also the energy eigenstates of the Hamiltonian, respectively. Among both types of coherence effects, quantum coherence is manifested in the delocalized eigenstates of photoexcitations in photosynthetic complexes which play a fundamental role in spectral properties, energy tuning, and also the energy transfer dynamics of photosynthetic light harvesting.[35,37] But, the effects of excitonic coherence are more difficult to analyze and have been the subject of intensive research since the past decade. Spectroscopic measurements (2DES) on the FMO complex at liquid-nitrogen (77 K)[28] and room temperatures[41] have shown time-dependent oscillations, presumed to be quantum beating, in the amplitudes of the spectral signals, matching the predictions of quantum theory. Inspired by all such kind of observations, it has been proposed that a intricate interplay of quantum coherent excitation transfer and environmental dephasing help to increase the efficiency of the energy transport process.[26,27,37–39]

Thus, to start with, there should be precise meaning of 'quantum coherence' by defining this phenomena explicitly, following the famous statement of Flemming *et al.*,[136] i.e., "coherence is present at all times and is continuously being created, destroyed, and recreated by the interaction of the electronic system with the surrounding nuclear degrees of freedom". Here, the picture referring to stationary eigenstates means nothing but, in essence, of coarse-graining via a convenient mean-field approximation. Characteristically, in a photosynthetic complex, the wave function of an electronic excitation is never stationary but evolve perpetually, under the influence of the fluctuations, present within its condensed-phase environment. As a result of such kind of environmental fluctuations, modulation in the energy and couplings of the collective molecular system occurs, leading to dynamical transitions of excitation energy transfer in which, the asssociated physical picture underpins all quantum dynamical processes only in the condensed phase.

As per the definition given by Mohseni *et al.*,[37] quantum coherence happens due to the existence of off-diagonal elements of the density matrix representing the ensemble. However, in their particular approach, this group[37] set aside the questions of 'quantum' versus 'classical' coherences controversy; instead, simply assumed that the system in question is best described quantum mechanically, presuming that the coherences in such a system are also quantum in nature. But the basic difficulty still remains there. Insufficiency in definition, still remains because of the fact that the density matrix as well as the magnitudes of its off-diagonal elements, in particular, depend on the 'basis set', while writing down the matrix. But, quite importantly, they considered the density matrix only as their basis, in the Hamiltonian eigen-basis. Thus, if the arguments of Lambert[40] are to be followed in order to explain the functionality, we should imply the role in such a way so that the presence of coherent quantum dynamics achieves something which could be more efficient or otherwise impossible than it could have been achieved by a classical mechanism alone. Hence, though the concept, i.e., considering that the nature taking utmost advantage of quantum mechanics, controlled by the quantum mechanics only, appears to be an inspiring concept, we should take the points both for and against, examining both of them carefully.

3.2 Fenna–Matthews–Olson Photosynthetic Complex: FMO

In light harvesting structures, most theoretical studies of non-classical phenomena to date have been dependent on the FMO complex of green sulfur bacterium so far the experimental results are concerned because this is an extremely well characterized pigment-protein structure. The FMO complex,[41,42] a crucial part of the photosynthetic system of green sulfur bacteria, is only one component, present in many types of light-harvesting complexes, found in nature. By structure, this complex is a trimer in the bacterial species "prosthecochloris aestuarii". This complex, perhaps, has been by far, one of the most well studied and characterized, available only in extremely low light environments, for example, under microbial mats or in the deep ocean at 77K.

The FMO (Fig. 3.1 & Fig. 3.2), is an example of a pigment–protein complex (PPC) network through which electronic excitations can migrate via excitonic couplings on individual pigments. The model associated with this, assumes that just after the photon impinges on the peripheral antenna of the light harvesting complex, absorption of the photon immediately occurs, in order to produce an exciton through electronic excitation which then traverses a network of seven chromophores or sites in one unit of the trimer. This exciton can either recombine, representing a loss of the excitation or function as a type of molecular 'wire', this way,- funneling light energy, captured in the chlorosome antennae, to a reaction centre (RC). There, finally, the energy is used to initiate chemical reactions.[42] The related experiments have already demonstrated the existence of strong quantum coherences in between multiple pigments. Not only that, the results have demonstrated highly efficient energy relaxation in this system which proceeds via coherently delocalized exciton states.

In addition, importantly, 'wave-like beating' between these excitons has also been observed to persist on timescales >550 fs, i.e., a significant fraction

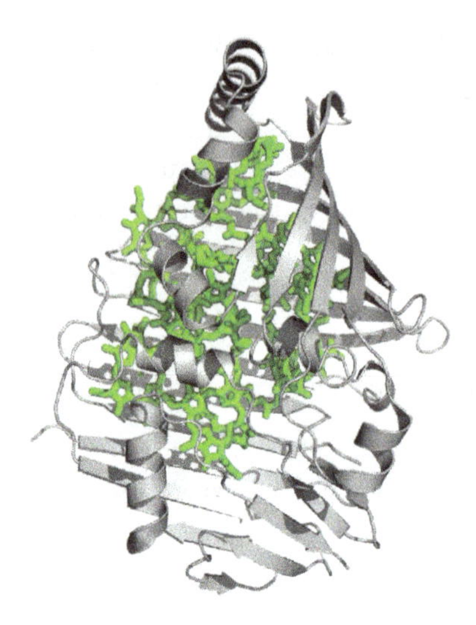

Figure 3.1. The Fenna–Matthews–Olson complex consists of a protein backbone (grey) containing a hydrophobic pocket that holds seven strongly coupled bacteriochlorophyll molecules (green).

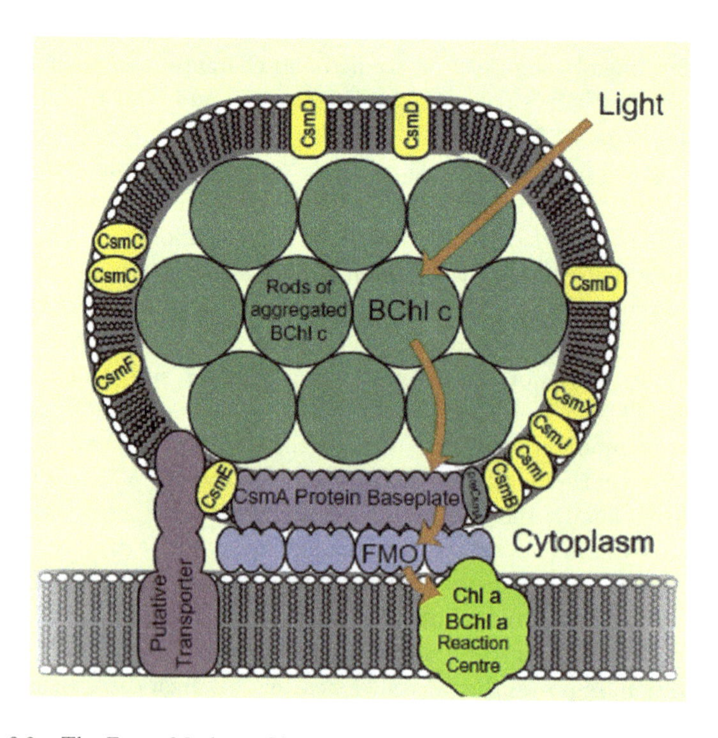

Figure 3.2. The Fenna-Mathews-Olson complex serves as an excitonic wire linking the chlorosome to the reaction centre.

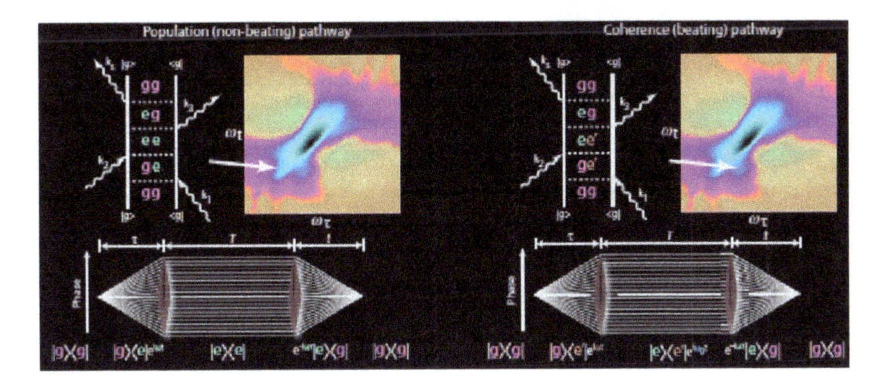

Figure 3.3. The left panel shows a constant (population) pathway that does not beat during the waiting time, T, The pathway is expressed both as a Feynman diagram and also as a lens diagram showing how phase spreads and rephrases. The time evolution is shown bellow the lens diagram in the form of a time-dependent factor in front of the density matrix element. The right panel shows a pathway that beats during the waiting time because the pathway involves coherence between excited states.

of the typical transport time in FMO.[4] This, i.e., FMO is a trimer of three identical units, each being composed of seven bacteriochlorophyll molecules, embedded in a scaffolding of protein molecules. In fact, this apparatus functions in a similar way across a broad range of photosynthetic organisms. Though the size and configuration of the antenna and component systems can differ greatly, as such, they have evolved themselves to a large, extremely broadband antennae, operating at near perfect quantum efficiency. In the ultrafast optical spectroscopy experiments with the Fenna–Matthews–Olson protein, interference in oscillatory signals, up to 1.5 ps, were reported and interpreted as direct evidence of exceptionally long-lived electronic quantum coherence.

In the seminal work of Engel *et al.*[4]

Illustration of the excitation energy transfer in the FMO protein of green sulfur bacteria. The eight BChl a pigments of the monomeric subunit of the trimeric FMO protein are oriented as depicted. The excitation energy enters from the base plate at the top and is transferred to the reaction center complex at the bottom. The blue, green, and red surroundings of the pigments indicate high-, intermediate-, and low-energy exciton states, respectively, to which the respective pigments contribute.

The strong quantum beating have been revealed in the amplitudes and shapes of the diagonal peaks. However, the thermodynamic efficiency of

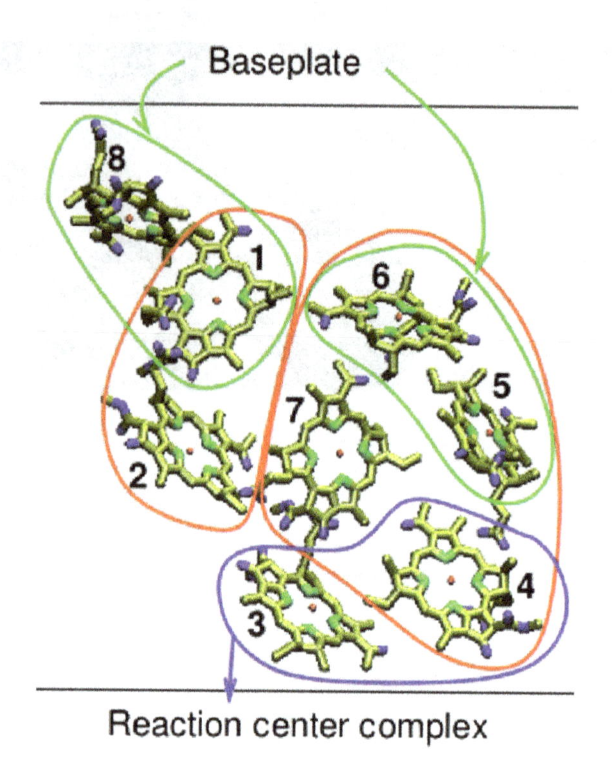

Figure 3.4. Excitation energy transfer and decay of coherences in the FMO protein.

this complex has been observed nowhere near unity. Quite interestingly, the typical characteristics possessed by this process, even after having such kind of thermodynamic efficiency, arising from a series of 'downhill' energy transfer steps, becomes fast and irreversible, — however, happens at the cost of thermodynamic efficiency. This typical archaea provides a fabulous model organism for studying strategies for light-harvesting and energy transfer, where, effectively, we are exploiting 2.4 billion years of evolutionary research and development.

The Fenna–Matthews–Olson (FMO) complex has proved itself to be an ideal model system for photosynthetic light-harvesting because:

(1) By now, the X-ray structure of FMO is known to within 2.2 A;
(2) Though sparingly, the complex has been found to be water soluble;
(3) The complex has been observed as highly asymmetric, yielding not optically dark states;

(4) Even after having only seven strongly coupled bacteriochlorophyll pigments, each monomer has been observed to be computationally tractable;
(5) The three monomeric units, found in the trimer, are extremely weakly coupled;

Figures 3.1 & 3.2, as shown, is the crystal structures which has been solved by Tronrud *et al.*[32] This protein acts as a 'spacer' i.e., it separates the chlorosome antenna from the reaction centre, being a gigantic, almost crystalline, rod-like structure (Figures 3.1 & 3.2) containing 250000 (bacteriochlorophylls), with the purpose of permitting reductant to diffuse to the reaction center and refill the hole after charge separation. Thus, for example, the Fenna–Matthews–Olson complex in green sulpher bacteria, effectively serves as an excitonic wire, linking the chlorosome to the reaction centre (RC).

Even though the complex itself is trimeric, the monomeric units have been found as very weakly coupled. Spectroscopically, we can think ourselves as blind to everything but active only to the optically active modes. There are seven bacteriochlorophyll molecules in the FMO complex which, being estimated, each found having an electronic transition near 800 nm (formally a Qy transition), coupling to one another electrostatically, following typically dipole–dipole interactions. A 7×7 Hamiltonian, can be constructed out of them which method has been followed by Vulto *et al.* (1999),[43] Renger *et al.* (2001)[44] and also by Hayes & Engel (2011).[46] Thus, as an ultimate goal, seven delocalized excitonic states are possible to construct, with the purpose of probing diagonalization of the matrix.

Especially, in the case of green sulpher bacteria, it has been observed that basic function of the FMO complex is to transport excitations from the sunlight-harvesting LHC antenna (chlorosome) to RC, initiating a charge-separation process, in order to generate chemical potential needed for underlying biochemical processes to be active. Quite recently, an additional eighth bacteriochlorophyll molecule (BChl) has been discovered by Tronrud *et al.*[32] (Fig. 3.5(b)). This sits in between the base plate and the rest of the FMO complex. After a photon being caught by the chlorosome antenna, it is transferred through the base plate to the FMO complex. (see Figs. 3.1 & 3.2). It should be noted that until recently, as the 8th BChl was not discovered in structural models of the FMO complexes, most of the studies, based on the FMO complex, up till that period, considered only a model with 'seven' BChls. However, considering the simplified picture, each BChl

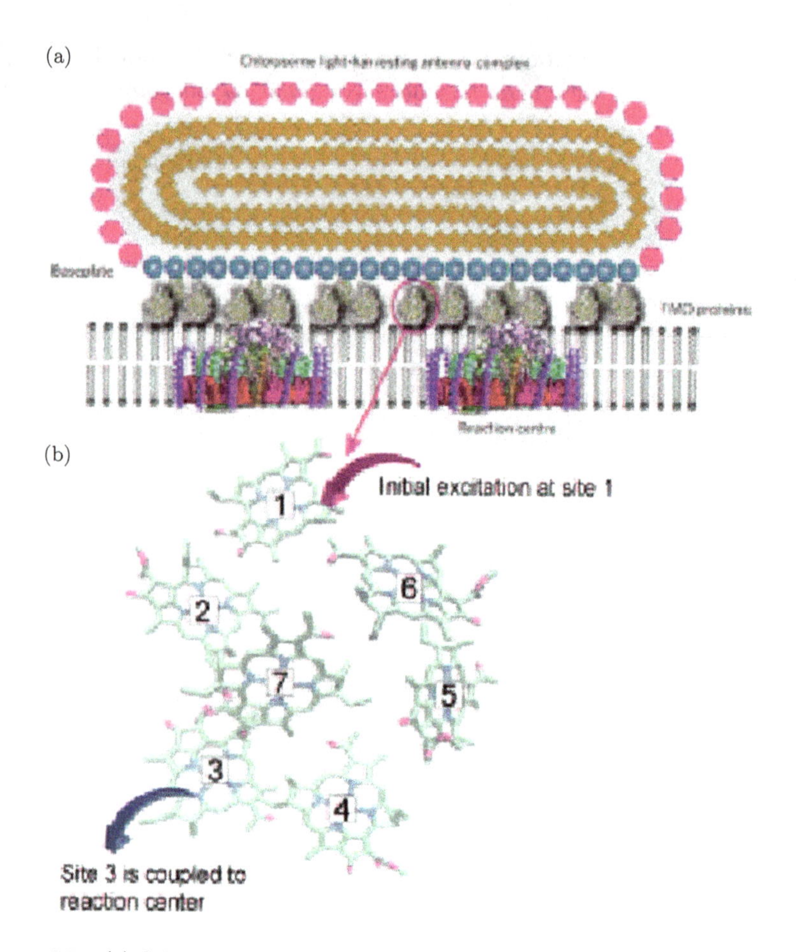

Figure 3.5. (a) Schematic diagram of roughly 200,000 BChl-c molecules encased in a proteinlipid structure from the chlorosome antenna. (b) There is a rendering of the structure of one of the FMO pigment protein complexes. The function of FMO in green sulfur bacteria is to transport excitations from the sunlight harvesting LHC antenna (chlorosome) to the RC where it serves to initiate a charge-separation process.

molecule, being excited from its ground state into its first singlet excited state, form a molecular exciton. As a next, after a photon being caught by the chlorosome antenna, it is transferred through the base plate to the FMO complex, irreversibly entering the RC, ultimately igniting the charge-separation process. After the start of ignition in the charge-separation process, the excitation keeps on passing from one BChl molecule to the next, until it reaches the molecule or site, closest to RC.

After entering the RC, it starts the process of charge-separation in a irreversible way so that electronic couplings between molecular excitons enables excitation energy transfer between BChl molecules. Here, the protein acts like a 'spacer', separating the chlorosome antenna, a gigantic, almost crystalline with a more or less rod-like structure having (250,000 bacteriochlorophylls) from the reaction centre (RC), permitting reductant to diffuse to the RC, refilling it again, i.e., the hole, after charge separation. In this way, FMO complex, effectively acts as an excitonic wire linking the chlorosome to the RC, (as shown in Figure above). For the FMO protein, the calculated dephasing times of inter-exciton and optical coherences are in the range of 50 and 75 fs, respectively which, being significantly shorter than the lifetimes of the exciton states, establishes the dominance of pure dephasing processes. This result, significantly, reflects the partial localization of excited states and their modest spatial overlap in this system.

During the continuing debate about the importance of probable role of quantum coherences in photosynthesis, the FMO complex has come into the central stage of discussion due to its absolutely important status in quantum biology. A number of exemplary studies have advocated regarding this aspect after observing certain form of "environmental protection of excitonic coherence" which has been observed in 2D spectra[5,47,143] as the source of long-lived oscillations. Very recently, ultrafast nonlinear spectroscopy has been successful in probing the dynamics related to energy transfer in the Fenna-Matthew-Olson (FMO) and other photosynthetic aggregates.[4,5,7,15] This complex can also be cited as an example of a pigment-protein complex (PPC), a network through which electronic excitations can migrate on individual pigment via excitonic couplings. Not only this, these experiments also have provided evidence for the existence of significant quantum coherences in between 'multiple pigments' caused by the presence of wave-like beating between excitons, having timescales >550 fs which has been observed to persist a significant fraction of the typical transport time in FMO. Thus, recently, considerable interest has been growing, in order to explore the possible existence of functional role for the quantum coherence in the remarkably efficient excitation energy transfer in FMO and other PPCs.

Within the Chlorobium tepidum FMO complex at 77k, direct observation of quantum beating signals among the excitons gives the direct manifestation of the quantum coherence itself, directly observable through quantum beating signals among the excitons. This kind of wave-like characteristics of the energy transfer can explain its extreme efficiency within the photosynthetic complex by allowing the sampling in the vast areas of phase space due to its requirement in finding the most efficient path for the energy transfer. However, no quantum/molecular mechanics,[52,226] based on dynamic studies of the FMO protein could identify correlations in site energy fluctuations, even though, in a normal mode analysis of the FMO spectral density, correlations have been observed, 'but' only at very low vibrational frequencies.[44,45] However, these kind of calculated correlations, practically showed no influence on the populations of exciton states.

In this respect, by using proper approximation (justified for the FMO protein), the evolution of inter-exciton coherences can be found as independent of the evolution of populations, even after excluding a direct functional influence of these coherences. If considered from a structural point of view, it can be opined as the inhomogeneous charge distribution in the FMO protein so that, on one hand, it leads to varying site energies of the pigments and, on the other hand, gives rise to coupling constants of different local exciton-vibrational origin, suppressing correlations in site energy fluctuations. Interestingly, the first effect is meant for the use in directing the excitation energy toward the reaction center (RC) whereas the second, for leading to an efficient dissipation of the excess energy of excitons. Interestingly, both these effects has been observed leading to a fast decay of inter-exciton coherences in femtosecond spectroscopic experiments.

For the FMO protein, the dephasing times have been calculated for the inter-exciton and optical coherences which have been found in the range of 50 and 75 fs, respectively, significantly shorter than the lifetimes of the exciton states. However, it points out to the dominance of pure dephasing processes, reflecting the partial localization of excited states and their modest spatial overlap in this system. At this point, with the expectation of having extremely fast decoherence, it is appropriate to consider an alternative mechanism (beyond the spatial overlap of excitons) so that it could lead to long-lived inter-exciton coherences, availing the correlations in site energy fluctuations among different pigments. A number of studies have advocated such kind of "environmental protection of excitonic coherence" as being the source of long-lived oscillations in the case of 2D spectra.[5,47,143] However, no quantum mechanics/molecular mechanics based dynamic studies of the FMO protein could identify such kind of correlations in site energy fluctuations.[52,226]

In a normal mode analysis of the FMO spectral density, correlations were observed only at very low vibrational frequencies,[45] but, when calculated, have been found, in a practical sense, having no influence on the populations of exciton states. Added to it, while these correlations, artificially been introduced for higher-frequency components of the spectral density, it has been observed that this can lead to protect the inter-exciton coherences, at the same time, effectively hampering exciton relaxation, thereby also helping the spatial transfer of excitation energy. Hence, it could be concluded that correlations in site energy fluctuations, allowing for long lived inter-exciton coherences, are basically detrimental for the light-harvesting function.[44,45]

Whether coherence is actually generated under natural excitation conditions (i.e., by sunlight) is still a heavily debated topic,[53,55] with some works dismissing the idea of coherence under sunlight,[53] while others following an early suggestion of representing sunlight by a series of ultra-short bursts.[56] We would like to point out that, in a secular approximation, i.e., justified for the FMO protein (Fig. 3.2), the evolution of inter-exciton coherences is independent of the evolution of populations, obviously excluding a direct functional influence of these coherences.

From a structural point of view, it is the inhomogeneous charge distribution in the FMO protein which, on one hand, leads to varying site energies of the pigments and, on the other hand, gives rise to different local exciton-vibrational coupling constants, suppressing correlations in site energy fluctuations. The first effect is used to direct the excitation energy toward the RC, the second effect leads to an efficient dissipation of the excess energy of excitons whereas both of them leading to the observation of a fast decay of inter-exciton coherences in femtosecond spectroscopy experiments. Thus, as the electrostatic tuning of site energies by the protein environment is used by many photosynthetic PPCs, e.g., those of higher plants,[57,58] we think that the mechanisms analyzed for the FMO protein are quite general for the above mentioned cases.

In general terms, this apparatus (FMO) (Figs. 3.1 & 3.2) can be described, functioning in a similar way across a broad range of photosynthetic organisms, having the size and configuration of the antenna and component systems, different greatly.[59] For example, as has been observed, in principle, we should eliminate all kind of classical models[243] and instead take into consideration of purely thermal environmental effects,[226] producing the similar efficiency so that we could come to a conclusion unambiguously, stating that high efficiency transport, itself is an example of functional quantum biology. An experimentalist is more likely able, only to affect a "noisy" or "fuzzy" measurement of the sites in the FMO complex. Along these lines, Kofler *et al.*[61,62] have shown that "fuzzy" measurements may lead to our observations of a classical world even in the presence of quantum dynamics. Further future works should consider more realistic models of noise in the FMO complex, including, potentially correlated noise[52,221] and non-Markovian effects.[218] In addition, the very question raised, i.e., whether the dynamics of observed in-vitro (in experiment) also occur in-vivo (nature) is not yet settled. This fact, however, makes the question of certainty quite vulnerable about the nature of the energy transport, especially, from the antenna (chlorosome and base plate) to the FMO complex itself. *At this point, one might find necessary in considering an application of Bell's inequality to study the ability of the FMO complex to preserve entanglement.*

3.2.1 *Exciton Transfer Process*

By now, we have been more or less well conversed with the idea about the most well-known way for describing the excitation-transfer process, i.e., the Förster model which applies the dipole-transition coupling in between two chromophores (sites) giving rise, as a result, to rates for excitation (exciton) hopping between interacting sites. However, even after being quite successful one, the Förster model neglects quantum coherence in between different sites but experiments have already shown that the exciton can move coherently among several chromophores which ultimately lead to the existence of electronic quantum beating[14] and perhaps, also coherent collective phenomena, the well-known results obtained in case of optical example of superradiance.[66–69] In addition, some other components of LHCs in other species, also suggest the existence of coherence and long-range entanglement in larger structures. As an example, the LHC of certain types of purple bacteria can be mentioned which contain two types of ring-like antenna, LH1 and LH2, substantially larger than the FMO complex. Associated spectroscopic studies on these systems have suggested that photo-excitations may delocalize among[4,5] BChl sites within the strongly coupled rings.[66,67,69]

Another interesting issue to be discussed is that the energy transport through the FMO complex has been observed to be quite fast (100's of femtoseconds), with the efficiency of many LHC systems being very high at the same time. But it remains quite non-trivial in finding the proper reason and explanation as well, why the quantum yield of light-to-charge conversion in some photosynthetic units[10] to be so high, i.e., up to 95%, and explaining how the excitations navigate the energy landscape of the LHC so successfully. In addition, the need to navigate this landscape quickly, must be understood in terms of losses, occurring during the transport through the LHC system and its components, not in terms of the overall energy processing rate of the entire PSU (LHC and RC systems) which has been observed to be relatively slow in comparison. For example, the 'reset time' or 'turnover' rate of the RC in purple bacteria, is normally found to be of the order of 1 kHz but the energy absorption rate of a single BChl in an antenna, in bright conditions, is only 10 Hz. The reason behind is thought that, in totality, the large number of BChl in the antenna provide enough energy to optimally use this 1 kHz turnover rate of the RC, indicating a large separation of time-scales. The fast transport through the FMO complex, for example, is needed to "beat" the rate at which excitations are lost due to fluorescence relaxation, not because of the overall "clock rate" of the light harvesting complex.

Also, one should note that the "energy" efficiency of photosynthetic biomass, e.g., a fuel source, is relatively low, even compared with photovoltaic efficiency,[70,71] as the down-stream biochemical reactions being able to turn chemical potentials into biomass, have extremely low energy efficiency. These are among few vital cases, needed to have a deeper understanding of all such kind of quantum phenomena. To avail this, powerful spectroscopic tools have been and being developed, improvised as far as possible to provide opportunities in studying the dynamics of the excitations, specially, in the pigment-protein complexes at the femto-seconds[61,69] time scale. In this front, two-dimensional electronic spectroscopy (2DES) can be categorized as a powerful technique capable of probing the couplings and the dynamics of energy flow on a two-dimensional (2D) map in the frequency domain allowing direct observation of "coherence" between electronic excitations. However, a full understanding of exactly how much information could be extracted with such techniques is still being discussed and researched.[72,73]

3.2.2 *Experimental Signature of Quantumness: Coherence & Beating: Information Through 2DES*

Recently, the many of the discoveries achieved by scientific research communities, took the advantage in using various kind of advanced techniques of powerful experimental tools, possessing unprecedented ability and sensitivity. They availed great opportunities, to probe some of most sensitive quantum effects, obtained from the related theoretical research related to photosynthetic systems. In order to demonstrate the importance of such kind of provision in these new insights, specially, in the case of 2DES, primarily, we need to describe how this information is manifested in a 2-D spectrum for a model system. Particularly, it has already been an established fact that 2D electronic spectroscopy have been successful in providing a unique tool, specifically, to electronic coherence in excitonic systems.[26,74,75,159]

Basically, a 2D experiment is a four-wave mixing process, within which there are three laser pulses. These pulses interact with the sample, resulting in the creation of an electronic polarization, thus generating the corresponding signal (Fig. 3.3). On the other hand, the 2DES experiment uses femto-second laser pulses for the excitation of the electronic absorption bands of the system, to be studied (e.g., a protein dispersed in solution) and this way, monitor the change of these excited states with time. As a result, this delivers information on how energy has been redistributed among the exited states; for example, how the energy transfer process causes excitation energy to flow from one molecule to another.

In the above experiment, two periods having time delay, can be controlled in the apparatus, i.e.,

(1) The time delay between the first and second pulses (τ, coherence time)
(2) The time delay between the second and third pulses (T, i.e., population time)

At fixed τ and T, the signaling field emitted in the photon-echo phase-matching direction, $k_s = -k_1 + k_2 + k_3$, is combined with respect to the time delay attenuated local-oscillator pulse for heterodyne-detection and frequency which is resolved for obtaining a spectrum. This spectrum can be regarded as the Fourier transform of the oscillating signal fields, in between the third pulse and the signal t (i.e., re-phasing time). Thus, a sequence consisting of three pump pulses is allowed to interact with the sample. As a result, this causes stimulation of a third-order coherent response, after which, the frequency and phase are being resolved using frequency domain heterodyne detection. With the help of the results from interferogram, both the magnitude and phase of the signal can be obtained. As a result, the response evolved, during the time period between the first two pulses, can be considered as a coherence between the ground and resonant excited states and defined as a 'one-quantum coherence', whereas, the initial time delay can therefore be referred to as the coherence time (τ).

In a series, the second interaction promotes the evolution to populations of the excited states and that of the ground state together, including coherences in between excited states and zero-quantum coherence. Here, the time delay in between the second and third pulses is referred to as the waiting time (t). Next to this the final pulse appears stimulating the sample into a radiative coherence, thus emitting the third-order signal. Here, the rephasing time τ is referred as the delay in between the third pulse and the emitted signal. As a next step, i.e., the two-dimensional spectra at fixed waiting times, is then generated by taking a two-dimensional Fourier transform of the properly apodized signal over the τ and t dimensions. Now, in order to obtain complementary information, the relative ordering of the pulse in the first two beams can be varied with the purpose of accessing the additional non-rephasing pathways.

This way, one can obtain the total signal by separating this into rephasing and non-rephasing contributions. This measurement can be repeated by varying τ, and then obtaining the Fourier, transformed with respect to τ to obtain a 2D spectrum in frequency domain (ω_τ & ω_t) at a fixed population time T.[74, 159] Thus, in practical sense, 2DES is quite similar to pump–probe spectroscopy where, a probe pulse arrives at various times after the pump. Next, it is then used for examining the way the excited-state population has redistributed among the excited states. Another typical characteristics of 2D electronic spectroscopy is that it records the 'signal' at the level of the field, rather than the 'intensity', indicating its sensitivity towards the quantum phase evolution of the electronic system, during the population time. Thus, the broadband pulse technique is capable of interacting with multiple exciton states whenever, it becomes necessary for the production of their superpositions (coherences), offering an important consequence, i.e., as a result of this, the induced coherence in the exciton basis, undergoes oscillatory phase evolution which as a consequence, finally leads to beating signals, being a function of time, associated with pulse-induced density-matrix dynamics of the system, by showing the Liouville pathways.

In the Figure 3.6(b) the Liouville pathways very clearly depict the contributions to the beating signals showing pulse-induced density-matrix dynamics of the system contributing to the beating signals. Briefly stating, in this experiment, a sequence of three pump pulses interacts with the sample, causing stimulation of a third-order coherent response. By applying technique of the frequency domain heterodyne detection, a stage is reached when the frequency and phase get resolved, giving the results in the form

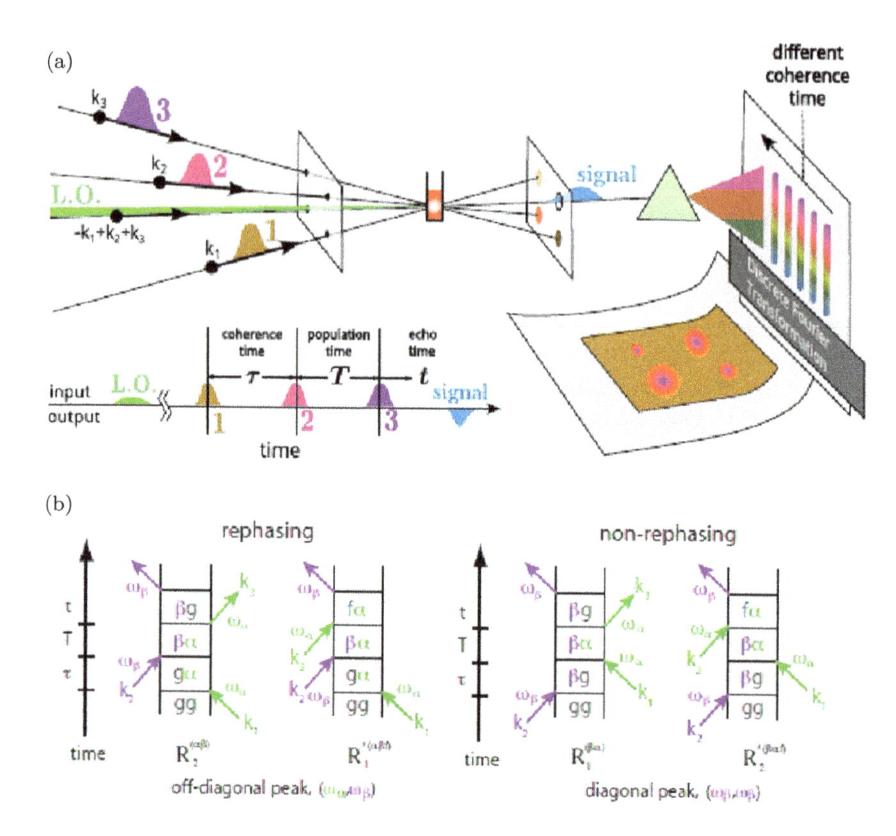

Figure 3.6. (a) Schematic of a 2DES setup. (b) Liouville pathways that contribute to the beating signals in the 2DES spectra. Contributions to 2DES signals are often depicted as double-sided Feynman diagrams as shown here. These diagrams represent the pulse-induced time evolution of the density matrix of the system. Light-matter interactions are depicted as arrows. An incoming arrow indicates an absorption by the system, whereas an outgoing arrow represents a stimulated emission event. Density-matrix elements are depicted in the center.

of an interferogram, providing both the magnitude and phase of the signal. The time evolution of the coherence during the population time T possess an oscillating phase factor which appears in the $2D$ spectra as a resulting quantum beats, being a function of T.

The above experiment provides an incisive tool in probing quantumness of an electronic process, i.e., the corresponding quantum beating signals indicating the presence of quantum coherence. Here, the resulting signal is visible during the waiting time delay T. Importantly, these signals manifest oscillations having both the absolute amplitude and the real portion of the response which characteristics should be emphasized as the main way by which 2DES differs from techniques in pump–probe spectroscopy. The pump pulse is spectrally resolved in a 2D electronic spectrum. For this, one can think of the pump pulse as effectively labeling the system according to the electronic absorption bands and reporting this labeled state along the excitation axis, wexcite. After some time, i.e., the waiting time (t_2), within which the system is free to evolve, the probe pulse interacts to effectively "read out" the current state of the previously labeled system, and finally, that current state is recorded along the detection axis, i.e., wexcite.

The above experiment can be described as follows: The response evolved as a coherence in between the ground and resonant excited states, during the time period between the first two pulses can be defined as a 'one-quantum coherence' and, the initial time delay, thereafter, referred as the coherence time (τ). Importantly, the second interaction is responsible for the promotion in evoluting the populations of the excited states, populations of the ground state, coherences between excited states and zero-quantum coherences, the corresponding time delay in between the second and third pulses referred to as the waiting time (t_2). The final pulse stimulates the sample into a radiative coherence, emitting the third-order signal. The delay in between the third pulse and the emitted signal is referred to as the rephasing time (τ). Finally, the two-dimensional spectra at fixed waiting times, are generated by taking a two-dimensional Fourier transform of the properly apodized signal, over the τ and t dimensions. The relative pulse ordering of the first two beams can be varied to access additional non-rephasing pathways, which can provide complementary information. The total signal, in this way, can be separated into rephasing and non-rephasing contributions.

The frequency–frequency correlation map is generated following this technique where each excitation frequency can be traced, correlated to each detection frequency.[28,76] In order to access additional non-rephasing pathays, the relative pulse ordering of the first two beams can be varied so that complementary information can be provided. In addition, the total signal can be separated into rephasing and non-rephasing contributions. However, for multi-chromophoric systems, the spectroscopic description given above, remains incomplete. The reason behind this is the presence of broad bandwidth of the ultra fast pulses, generally spanning multiple transitions. As a result, the second pulse may or may not interact with the same excited state, after the initial excitation of a coherence between the ground and excited state. Thus, if interaction of second pulse, having different states occurs, as a result, we are left in a coherence in between excited states rather than a population, evolving phase with the energy difference between the excitonic excited states. However, the period of this phase is slow enough to be observed (i.e., beating pathway) so far as the waiting time dimension of the 2D spectra is concerned.

Importantly, prior to the data published[4] in 2007, electronic coherence had never been observed in $2D$ spectra. The beating signals observed, were reproducible, both in phase and frequency, indicating that the signal was of molecular origin but also, were fully consistent, found with the expected beating pattern which has already been predicted by looking at excitonic energy differences and weighting the intensity with the appropriate excitonic transition dipole moments. The precise origin of this beating pattern itself has been explained only later on, by carefully dissecting the 'Feynman diagrams', associated with quantum coherence.[21] This observation led directly to a detailed understanding, relating also the line shape of the beating signals. Importantly, it should be noted that the signal appeared in the absolute magnitude because, the signal itself interferes with a constant, non-oscillating signal. Firstly, there is constructive interference, followed by a destructive interference, and so on.

In the present case, however, as the experiment has been based on theoretical predictions,[77] this first signal observed, appeared as a surprise, showing beating for 70 fs only, even though the sampling times were not uniformly taken. (The sampling started with 10 fs steps, moving to 15 fs steps, then 20 fs steps until finally sheer panic cased 30 fs step sizes, as the beating persisted.) Such non-uniform steps required adaptation of the non-uniform fast Fourier transform (NFFT),[78] where the associated algorithm normally identifies a large family of solutions, consistent with more complete data, thus achieved, so that thorough analysis would confirm this assignment.[48] The quantum beating in FMO has been found capable of being used in interrogating many of new arena, associated with of the Hamiltonian thus formed which govern energy transfer and relaxation. This has been now a quite established fact that the beating signals arise from the difference of energy, present among the states and hence the beating spectrum comes into help in understanding the Hamiltonian, where, related Fourier transformation along the waiting time axis shows many new features remaining still within the congested 2D spectrum.[46]

For example, these signals, arising from quantum beating on the peaks, forming the beating spectrum, arise from energetic differences among states, found to be present in the main diagonal or rephasing spetra, but not expected showing beating[5] at all. Not only that, the beating signals have been observed to be remarkably robust[46] and any attempt to scramble vibrational modes or to shift resonances with isotopic substitution miserably failed to affect the beating signals and found to constrain significantly any kind of microscopic model, related to the role of the protein bath. Because, for the most simple model invoked, i.e., a single, finely tuned mode (out of over 100000) to drive the coherence — the isotopic data shows that this attempt simply fails to satisfy the query. Ultimately, becoming the source of understanding the Hamiltonian, **such beating indicates that not only the populations are oscillating, the probability of finding the excitation in a given state is also oscillating in time**.

But importantly, such kind of non-secular effects require populations to be driven by coherence only. Because, when an oscillatory coherence drives a population, the population oscillates 90 degrees out of phase from the coherence which is exactly the signal, observed. Not only that Fourier transforms along the waiting time axis shows many new features within the still congested 2D spectrum.[46] There have been still active discussions about the mode or the ways regarding how vibrational modes may mediate this long-lived coherence. Regardless of the microscopic explanation, such robustness implies that the quantum coherence presents a tractable engineering target. After suggestions by Voronine *et al.*,[79] signals were isolated that showed direct coupling between populations and coherences.[8] Not only that following the suggestions by Palmieri *et al.*,[79] signals were isolated that show direct coupling between populations and coherences,[7] arising from quantum beating on peaks in the main diagonal, present in rephasing spectra.

Even though these results were not expected to show beating,[5] it established the oscillating nature of the populations. Not only that these results also indicate the probability of finding the excitation in a given state, oscillating in time. But, interestingly, such kind of non-secular effects demands the related populations to be driven by coherence. Because, an oscillatory coherence driving a population, makes the population oscillating 90^0 out of phase from the coherence and that will be the exact signal to be observed. Interestingly, this observation raises many new queries. For example, FMO is a non-degenerate system such that population oscillation implies energetic oscillations. But, again, it is quite important to recall that energy oscillations within a subsystem are not forbidden — and hence, energy can still be conserved, but it must be traded with the environment.

But, this interpretation significantly changes our view regarding the role played by the related protein bath.

It is an well known fact that protein, being strongly coupled to the bacteriochlorophylls, trade actively energy with the corresponding system. Interestingly, this has been a surprising finding for giving the answer to an important query which scientists had long wondered i.e., how chlorophyll could sit in a soft, polarizable environment, yet not dissipate energy. Here we are, capable of giving the answer that query, i.e., coherence solves this problem — the coupling exists, but all of those prior measurements (hole burning, transfer efficiency, Stokes shifts) were blind regarding the coupling aspect. The crucial reason for happening so, is the interpretation of that data, implying incoherent, dissipative coupling. Added to that, interestingly this coupling, being responsible for coherences and populations, also explains why populations oscillate, including the answer to the question, i.e., why coherence persist for so long. It rather becomes crucial to report that, — in fact, the long-lived coherence borrows lifetime from the populations to which itself couples.

Finally, another handle to analyze quantum beating has been provided by the dephasing rates of coherences. Early on, Hayes *et al.*[46] recognized that different coherences dephase with rather different rates. The pattern does not match any simple scheme, for example, higher frequency beats found dephasing faster, nor are resonances have been found to be evident with any particular bath modes. After the emergence of more complete data sets, a more formal approach for the analysis was introduced that exploits a modified Fourier transform (z-transform), involving complex frequencies — thus have been able, in isolating both dephasing rates and beat frequencies at the same time.[80,81]

3.2.3 *Vibrionic Nature of 2DES Spectrum*

After the vibronic nature of the spectroscopy of molecules being established, vibrational wave packets, coherent superpositions between vibrational levels, and possible electronic superpositions can be prepared,[85–88] based on the excitation, with the help of a short optical laser pulse (with a broad spectral bandwidth), the vibrational frequencies being often quite similar in magnitude to the difference in energy between the exciton states. However, in such cases, one of the prerequisite challenges, demanding in 2DES, is assigning the 'waiting-time-dependent oscillations' to vibrational and electronic coherences. Anyway, it turns out that[24,86] more-detailed analysis of 2DES is capable of discriminating electronic and vibrational coherences. Based on this, however, it has been concluded that a long-lived electronic coherence (having a dephasing time of 170fs) contributes to the oscillatory amplitude of the cross-peak in PC645. Although 170 fs is incredibly short on the time scale of biology, it has been found to be comparable with the time scale of the most rapid energy transfer processes, as could be observed in the photosystem I, for example.

3.2.4 *Limitations Associated with 2DES Techniques*

Although we do not discuss the experimental techniques here, we would like to comment on some of the limitations associated with 2DES. 2DES is based on Fourier transform techniques and, therefore, requires at to-second (10–18-second) timing precision and mechanisms for phase stabilization.[87,88] Therefore, although 2DES can be used to obtain more direct, unambiguous information regarding the system, one of the drawbacks of employing this technique is that it is harder to implement experimentally than pump–probe spectroscopies. Another limitation involves in the difficulty faced in extracting kinetics from the rich wealth of information contained in the $2D$ spectra. However, the Ogilvie group[89] and the Scholes group[180] have made a huge progress in this regard by extending well-established techniques for extracting kinetics from $1D$ pump–probe spectra to $2D$ electronic spectra.

3.2.5 *Debate Over the Interpretations of 2DES Experiments*

There have been lots of debate, so far as the interpretation of the 2DES experiments is related, in particular, to the observed facts showing substantial similarity in the oscillations, observed in the data of the vibrational coherence and those appearing in electronic coherence of 2DES data.[91,92] It has already been an established fact by now which states that in a typical photosynthetic protein, several vibrational modes are accessible by stimulated Raman scattering, leading to oscillations in pump probe and 2DES measurements, having frequencies, quite similar to those, reported to be exciton coherences. These results have placed a great challenge as a research subject with the aim of finding the interpretation of the ultrafast measurements as observed in the related experiment. For clarity, differences in between two types of coherence, i.e., vibrational and electronic coherence can be distinguished as follows: Vibrational coherence refers to pure nuclear motion of wave packet induced by photoexcitation. This characteristics of resonant Raman active normal modes present in the molecule, could be present either on the ground or excited electronic state. But on the contrary, electronic coherence at the present case refers to the coherent excitation of two electronic states, coupled by electronic interactions but uncoupled from any kind of vibrational modes in the system. As a result, the nature of the modes appears to be quite critical whenever their meaningful interpretations are needed. Observing the presence of distinct nature in their 2DES data, Turner *et al.*[24] made a systematic approach for the observations of the signatures related to different types of coherences, in order to identify and differentiate the observations of the corresponding types of coherences, present in the three- versus- four-level system.

The three-level system of 2DES system, was modeled with the aim of more detailed studies, by using a ground state, $|g\rangle$ and two electronic excited states, $|\alpha\rangle$ and $|\beta\rangle$ which, for the present, has been assumed to be excitonic states of the system (e.g. model dimer). A coherent superposition between these two electronic states is signaled by the off-diagonal density elements. In the present case, Inline formula has been considered with the aim of giving the signatures associated with purely electronic coherence in between electronic eigenstates. The four-level system has been assumed with the aim of being representative of a single chromophore (two-level system) with an additional vibrational level, $|1\rangle$, present, both in the ground and excited states. In such a case, vibrational coherence becomes possible as being either in a ground-state vibrational wave packet Inline Formula or as an excited state vibrational wave packet Inline Formula. Here, comparison of the expected dynamics, present in a three- and a four-level system provides a good test case for the simplest model of electronic or vibrational coherence. But, in reality, this procedure of discrimination in between electronic and vibrational coherences has been found to be valid only in the limit of strong electronic coupling. Three distinct methods have been applied in 2DES experiments with the purpose of distinguishing oscillations, occurring in between three- and a four-level system and that also, together with corresponding identification, occurring due to the presence of the extra $|g_1\rangle$ state.

But it is important to mention about the significant differences existing in between NMR and optical 2DES. In the optical case, both the optical excitation beams and signal beams have well-defined directions. Hence, for a given spatial geometry, choice of different pulse sequences can be done just by alternating the time ordering of the pulses in the different beams.

This way, in one pulse sequence, a photon echo sequence is achieved, referred as the 'rephasing' spectrum and corresponding 'non-rephasing' spectrum is meant whenever the time ordering of the first and second pulses is reversed. Again, the evolution of the cross peaks in the rephasing and non-rephasing spectra were predicted to behave differently for a three- and a four-level system. The reason behind this is the presence of an additional vibrational level in the excited state of the four-level system which eventually opens several additional pathways, capable of creating a coherence even on the ground vibrational level.

But, as a result, the observation of oscillations becomes practically insensitive to the choice of pulse sequence, because, the motion in the ground-state vibrational wave packet becomes insensitive to the choice of pulse sequence. Moreover, for the three-level systems, there exist a limited number of excitation pathways, creating a coherence in the excited states. Hence, the corresponding observation of oscillations becomes sensitive to the pulse sequence. Now, if in the cross peak, the origin of the oscillations arise from a three-level system, they will occur only in rephasing spectrum. But in contrast, oscillations owing to stimulated Raman scattering will be observed in both rephasing and non-rephasing spectra.[92] However, in reality, this kind of signature is quite difficult to achieve, as high-quality data is essential, as the non-rephasing signal is quite often significantly weaker than the rephasing signal. For example, a mode, meant for the magnitude of non-rephasing signal, might appear above the noise level in the Fourier transform but in case of the real non-rephasing signal, this does not appear in the Fourier transform, might be, due to even minor errors in phasing.

Conversely, due to the presence of broad dispersive features, there might be an artificial appearance of a mode in the magnitude trace, centered on the diagonal, overlapping the cross peak of interest. Again, a mode may be present in the oscillations of a trace but hidden in the Fourier transformed data due to the spectral overlap with features, owing to a baseline or population decay which was not properly subtracted, prior to performing the Fourier transform. These are among few examples which illustrates some of the difficulties which could arise during assigning an oscillatory feature, based on the absence of a mode in a noisy, weak signal. A careful statistical analysis of data is important when drawing conclusions from such kind of analysis. For instance, in the measured 2DES spectra for PC645, the oscillations of a cross peak, observed below the diagonal were carefully monitored. Not only that the observed oscillations were found fit to a sum of eight damped cosine functions, identifying several oscillating components, showing the contributions to the overall signal. Then the rephasing and non-rephasing contributions were separated and examined independently. Both components showed clear signs of oscillations, i.e., albeit showing similarity in general, a Fourier transform of the separate components indicated two modes to be very close in frequency: 21 THz (700 cm^{-1}) and 26 THz (870 cm^{-1}). Interestingly, one of those frequency components was clearly visible in the rephasing trace while absent in the non-rephasing trace.

Two-color technique has also been followed for further investigations on $PC645$. In these experiments, two pump beams excited the sample, by a time-delayed probe beam, for monitoring the dynamics together with similar geometry, adopted in typical 2DES measurements. In the present case, three narrow band pulses were used to excite and probe all possible pathways instead of using three broadband pulses, in order to excite and probe a limited number of specific pathways. This approach reduces the number of contributions to the overall signal and identify more clearly the signatures of coherence in PC645. In the reported results, the separated, especially, non-overlapping spectra for the two excitation pulses were chosen, as this choice of non-overlapping spectra ensures that signal measured in the chosen phase-matched direction is radiated by molecules, left in a coherence, only after interacting with the two excitation pulses, but having no contribution from molecules, left in a population. This technique, providing a direct measurement of the lifetime for a single coherence, have been applied by Richards *et al.*[153] who observed a coherence with a lifetime of 500 fs in PC645 at cryogenic temperatures. In its presented form, however, the experiment did not distinguish between vibrational or electronic coherences, so it is difficult to identify clearly which coherence lifetime was measured. However, similar to transient absorption spectroscopy with a narrow band pump, this technique also has the potential in clarifying the dynamics of individual coherence, as the number of pathways contributing to the overall signal is significantly reduced.

Very recently, proposal has been been made with the aim of distinguishing coherences in between electro-oscillations.[156] Here, in a system, the comparison has been done in between the spectrally integrated pump-probe response of a vibrational coherence with a single excited electronic state and that of a coupled electronic dimer in the limit of an infinite bandwidth pump — infinite bandwidth probe experiment. In ultrafast pump-probe spectroscopy, excitation of a two-level electronic system, with accessible vibrational levels on the ground and also excited states, creates a vibrational wave packet which oscillates on either the ground or excited state manifold. For a single probe-pulse wavelength, as a result, the wave packet oscillations increase or decrease the probability of absorption or emission, leading to oscillations in the pump-probe spectrum.[157, 158]

However, for a spectrally integrated measurement, the total probability for absorption or emission does not change and the oscillations in the pump-probe signal average to zero amplitude. By contrast, in a system with two excited electronic states, being in a coherent superposition, the evolution of two wave packets on the different electronic states depends on the relative accumulated phase of the excitation, present in each of the electronic states and does not spectrally average to zero. Through a careful analysis of the two cases, i.e., vibrational and excitonic coherence, conclusion can be arrived demonstrating that as the bandwidth of both pump and probe pulses are increased in a spectrally integrated pump-probe experiment, the oscillations due to a vibrational coherence should decay to zero, linearly with the pump and probe pulse temporal durations, whereas the excitonic coherence should have residual oscillations. In principle, this technique can provide a relatively simple alternative approach for the identification of different modes, present in the ultrafast spectroscopic measurements.

3.3 Non-trivial Quantum Features in Photosynthetic Phenomena: Coherence and Decoherence

3.3.1 *What is Quantum Coherence?*

With the aim of having a precise meaning of 'quantum coherence', following Engel,[4] quantum coherences can be defined as the off-diagonal elements of the density matrix, representing the ensemble. Particularly, both the density matrix and the magnitudes of its off-diagonal elements are dependent on the basis set used to construct the matrix. As defined by Engel,[4] this constrain does not allow this definition to be termed as sufficiently defined in a specific way. In reality, this definition also lacks sufficiency because, using the Hamiltonian eigen basis, in order to construct the definition of quantum coherence, it demands for the enormous experimental and interpretive simplifications.

However, this definition is very well suited to spectroscopic methods in measuring the transition energy. As the Hamiltonian operator corresponds with energy observations, governing temporal evolution, the population elements along the main diagonal of the density matrix represent the probability of finding the system in a given energy level. Therefore, unitary dynamics of both the populations (diagonal) and coherences (off-diagonal) density matrix elements in the Hamiltonian eigen basis are also quite simple. Coherences evolve phase based on the energy difference in between the two associated Hamiltonian eigenstates only. Hence, the equation of motion for the density matrix in any basis set can be written as,

$$\frac{\partial \rho}{\partial t} = -\frac{i}{\hbar}[\widehat{H}, \rho]. \tag{3.1}$$

The above equation can be written in a more simplified way by expressing this in the Hamiltonin form, i.e.,

$$\frac{\partial \rho_{ij}}{\partial t} = -\frac{i}{\hbar}(\epsilon_i - \epsilon_j)\rho_{ij}, \tag{3.2}$$

Here, ϵ_i represents the energy of the ith eigenstate whereas the corresponding populations ($i = j$) remain constant and coherences ($i \neq j$) evolve phase with time following the relation

$$\rho_{ij}(t) = e^{\frac{-i(\epsilon_i - \epsilon_j)t}{\hbar}} \rho_{ij}(0) \tag{3.3}$$

for $i \neq j$. The above equation shows that the population of Hamiltonian eigenstates remains constant. But, importantly, there will be oscillation due to coherence. This fact appears to point out to a clear strategy for observing quantum coherence which, however, is not that much easy in practical sense.

Now, the next vital as well as crucial question arise stating how does quantum manifest while observing coherence phenomena in that situation when the Hamiltonian eigenstates remain constant but coherences oscillate. In light harvesting, we mean coherence in between exciton states to imply the wave-like character of energy flow. In the site basis, the presence of long-time or steady state coherence simply means that the eigenstates of the system are delocalized (exciton) states.

In natural harvesting, the fact that electronic coherence is observed via an ultra fast spectroscopic method, creating initial coherence and recording its loss as a function of time, has created a certain amount of confusion regarding the significance of coherence. This arises due to to possibility of a typical question which might be: "*Does the coherence only matter during the first few hundred femtoseconds of the energy transfer, following the absorption of light?*" But the basic premise of this question might be misleading. Because, the ultrashort pulse excitation simply serves to coordinate the ensemble in time only, allowing the related observation of dynamical processes through theoretical modeling, enabling deductions about the system's Hamiltonian. There are two mechanisms which mainly dictate, the observed coherence in the ensemble decays. Among them, the first one

(i) Ensemble dephasing, maintaining the coherence in individual complexes, but disrupts the correlation between the oscillatory behavior of individual members of the ensemble. This process leads to the decay of observable oscillations in an experiment.
(ii) The second is microscopic dephasing, or decoherence, which, in an individual complex destroys the coherent superpositions between excitons

During full quantum-mechanical treatment of an open quantum system, the environment (or commonly called the 'bath') is modeled as a second quantum system, coupled to the primary system of interest. The ideal model of the bath, as if the world minus the system is mathematically intractable, the bath is usually modeled as an effective set of external DOF, for instance, a set of harmonic oscillators[96] or a set of spins.[97] Then, reasonably, we can suggest that the system-bath interactions create and recreate coherence throughout the energy flow in the process. But, in case of an ensemble measurement, different members of the ensemble (in a few hundred femtoseconds) become rapidly uncorrelated as a result of which the associated coherence is not directly observable. In such case, both the dephasing processes contribute to the decay of oscillations which is measurable in two dimensional electronic spectroscopy (2DES).

As a next step, for a moment, let us even set aside the fact, i.e., the temporal evolution need not be unitary in describing a subsystem having a reduced dimensional description which practically, imposes this condition in order to ignore the states of the system in a bath, so that, as a result, the energy can flow into those states, giving rise to a dissipative dynamics. In such a situation, our aim is to focus on the oscillatory beating signals, as a marker of quantum coherence in the Hamiltonian eigen basis. But in practice, generally, such kind of oscillatory beating signals do not arise from the coherence, present in other basis sets and we lack both experimental and theoretical methods to observe such characterization of microscopic dephasing in the condensed phase. Rather, we expect the fluctuations within an individual complex to be uncorrelated with those in other members of the ensemble, so that the microscopic dephasing becomes significantly slower than the ensemble dephasing. Indeed, reasonably, it could be suggested that the system-bath interactions create and recreate coherence throughout the energy flow process.

But, in practice, different members of the ensemble rapidly (in a few hundred femtoseconds) become uncorrelated in an ensemble measurement and the coherence can not directly be observed. In fact, both of the dephasing processes contribute to the decay of oscillations which can be measured in two dimensional electronic spectroscopy (2DES). Among many of important points, the crucial condition for effective coherence is that the relevant process must be followed in a short time scale, otherwise microscopic dephasing (decoherence) would be completed before the occurrence of population transfer, all of which considerations have been found to be directly relevant, for example, to the nature of photosynthetic energy transfer. At the first glance, it could appear that a simple rule has been crafted out for the identification of quantum, i.e, coherence oscillation. But we must observe this oscillation in actual sense, i.e., we need to make observations A of the system which, within the density matrix formalism, corresponding to an operator, $\widehat{A}$, can simply be written as,

$$\langle A \rangle = Tr(\widehat{A}\rho)$$

Interestingly, at this point, if instead of Hamiltonian operator, we want to measure energy of any system evolving under unitary dynamics, we need to be blind to Hamiltonian operator, i.e., to coherences (it is not surprising as unitary dynamics conserve energy and hence no oscillation are possible). In fact, as the Hamiltonian matrix is diagonal when expressed in its eigen basis, the populations are constant under unitary dynamics so that, finally, we get

$$\langle E \rangle = Tr(\widehat{H}\rho) = \sum_i \epsilon_i(\rho_{ii}) \tag{3.4}$$

In the above equation, as the populations are constant under unitery dynamics, it establishes the fact that the observed energy remains independent of the effect of coherences. Not only that, — for any kind of associated phase, we must select a measurement which is not commutable with the Hamiltonian. But as the non-commuting operators also means not (in general) sharing eigenfunctions, this condition rather forces us to ensure that the operator is not diagonal in the Hamiltonian eigen basis. Even though, in principle, we could select any operator spectroscopically, the dipole operator can be selected as a most convenient operator. Adopting a simple form in such case, the light of a proper colour, couples the ground and excited states through the dipole operator. After the interaction, these have some probability of being found in the excited state. But, as this operator does not commute with the Hamiltonian, it must contain off-diagonal elements when written in the Hamiltonian eigenbasis. Thus, as shown in Equation (3.4), these off-diagonal elements will make possible the contribution from coherences to the observable quantity. In a spectroscopic measurement, the observable associated with the dipole operator is manifested as the amplitude (and therefore intensity). That is, when we set out to measure quantum coherence, we will look for periodic oscillations in the amplitude of the signal fluctuations, called 'quantum beats.' which is an ubiquitous feature of quantum coherence.

Practically, when the dipole operator is examined, the same beating would occur with any operator, not commuting with the Hamiltonian. This makes us able in thinking this beating as probability, washing back and forth across sites, or as oscillations of the electrodynamic polarization via the dipole operator. In reality, we can state it as 'all of these things'. By summing up all the conjectures, the following statement could be placed: *coherence is present at all times, continuously being created, destroyed, recreated as a result of the interaction of the electronic system with the surrounding nuclear degrees of freedom.*

Stationary eigenstates, produced in this way, can be stated as a result of coarse-graining via a convenient mean-field approximation. Another crucial point to be mentioned here, is that the wave function, associated with an electronic excitation within a photosynthetic complex is never stationary, perpetually evolving under the influence of the fluctuations and that also, only in its condensed-phase environment. These environmental fluctuations, as a result, modulate the corresponding energy and couplings of the collective molecular system which finally leads to dynamical transitions of excitation energy transfer. The physical picture underpins that all of quantum dynamical processes, even though appearing stationary, perpetually evolve under the influence of the fluctuations in its condensed-phase environment.

Finally, question of time scale is a very important factor, related to coherence, in playing a role so that the relevant process occurs in a short time scale. Otherwise, the process of microscopic dephasing (decoherence) would be complete before the occurrence of population transfer. Importantly, these considerations have been found to be directly relevant to recent discussions which relates the nature of photosynthetic energy transfer and, for example, other ultrafast optical biological processes such as vision under solar radiation.

3.4 Interference Phenomena in Photosynthesis: Coherence & Decoherence

3.4.1 *Interference Phenomena in Photosynthesis*

In quantum system, experimental observables having the pathways of coupled degrees of freedom, may become entangled, thus leading to interesting interference effects. It can be demonstrated that the interferences of quantum pathways of matter are related to the entanglement of excitons in chromophore aggregates which can be manipulated through the interactions with quantum optical fields. But even though common dynamical observables can be obtained from the expectation values of operators, but, it contains incomplete information carried by the related wave function which can be represented by Liouville space pathways, describable by the joint evolution of the bra and ket containing the system's density matrix. Their pathways become entangled, in the presence of two or more systems, leading to a new and interesting quantum effects. Focusing on long-lasting excitonic coherence and coherent wave-like dynamics, — in the present case, the central point should be the effects of quantum interference. But the famous quintessential question immediately arises, stating *to what extent are quantum mechanical effects involved in light harvesting which should be posed with proper emphasis on quantum interference.*

Coherence allows energy transfer pathways to interfere with each other and produce results that are not describable by the laws of classical probability. For example, in case of photosynthesis, laws of quantum probability are formulated by summing amplitudes associated with each energy transfer pathway through a multichromophoric complex. In order to yield a probability versus time for the evolution of population densities, the modulus is squared to the cross-terms which arise naturally, following this procedure (e.g., after correcting the classical rate law for quantum mechanical interference, *amplitude of pathway, multiplied by that of another pathway*); This is why experiments is so critical while probing quantum interference between explicit pathways, i.e., it is meant for the understanding and control of quantum coherence effects in photosynthesis. Also, in order to discuss the possible roles of different quantum aspects in light harvesting, it is rather essential in giving the importance to proper importance on the problems related to long-lasting excitonic coherence and coherent wave-like dynamics. In that case, it appears that the central point should be referred to as a problem, essentially related to "effects of quantum interference".

All of quantum effects related to dynamics, are observed ultimately to be associated with interference among pathways. Feynman, in his famous works related to path formulation, clearly pointed out this, i.e., the evolution of the wave function in space-time.[183] Usually, the wave function is not possible to be experimentally observable (with the exception in quantum tomography and quantum information algorithms).[104–109] In fact, when two or more systems are coupled, their pathways become entangled and this leads to another kind of new and interesting quantum effects. As common dynamical observables are given by the expectation values of operators the results contain less than the complete information carried by the wave function. In such case, these has found to possible for representation by Liouville space pathways which describe the joint evolution of the bra and ket, thus representing the system's density matrix.[110,111] But, when two or more systems become coupled, their pathways get entangled, leading to another kind of new and interesting quantum effects. As raised, famously, by Flemming and their group and others, the question of "*to what extent are quantum mechanical effects involved in light harvesting?*" should be placed putting the emphasis on quantum interference.

Quite often, discussions on quantum effects in light harvesting focus on long-lasting excitonic coherence and coherent wave-like dynamics. But, in fact, the central point should be the effects of quantum interference and hence, the above quintessential question should be posed with an strong emphasis on quantum interference. Already being an well known fact, coherence allows energy transfer pathways to interfere with each other, producing results, non-describable by the classical probability laws. Quantum probability laws are formulated by summing amplitudes associated with each energy transfer pathway through a multi-chromophoric complex, the modulus of which quantity, when squared, yields a probability versus time in assessing the evolution of population densities. Finally, the cross-terms arising out of the probability from the procedure (e.g. amplitude of pathway 1 multiplied by amplitude of pathway 2) correct the classical rate law for quantum-mechanical interferences. Experiments, probing quantum interference in between explicit pathways, are quite critical for the understanding as well as control of quantum coherence effects in photosynthesis.

For example, we can state Pearlstein's random-walk model in considering the idea about i.e., what does quantum mechanics do for energy transfer? A short answer can easily be offered, i.e., it changes the way we think about the energy jumps in. Let us now consider that there exists two pathways for transferring excitation from molecule 'A' to molecule 'B' in a light-harvesting complex: directly from A to B (P_{AB}) and there could also be way of a third molecule, C (P_{ACB}). Classically, the probability of the energy transfer is simply the sum of the probability of taking each path:

$$P_{total} = P_{AB} + P_{ACB}$$

But, in quantum mechanics that probability law is modified, resulting in the common explanation that both paths are taken simultaneously. What happens here, is that the probability of energy transfer from A to B is calculated differently: We assign a probability amplitude to each path, sum those amplitudes, then convert the sum to a probability by taking the modulus squared, i.e.,

$$\Pi_{total} = |A_{AB} + A_{ACB}|^2$$

The result of this procedure points to the fact that the pathways can interfere, as if, we have added those waves after representing each of them as a wave. If the crests of the waves are prepared to be lined up for the two paths, it will be the constructive interference for boosting up the energy transfer rate relative to the classical calculation. Here, we should be aware that the quantum law reduces to the familiar probability law when facing to a particular situation. This means when we lose the ability to discriminate the waves, a process called decoherence operates and the system behaves in such a intrinsically complex way that all the constructive and destructive interferences cancel on average, an idea, exploited by Miller (2012)[112] in semiclassical simulations of dynamics. In case of photosynthesis, quantum probability laws are formulated by summing amplitudes associated with each energy transfer pathway through a multichromophoric complex. The squared modulus of that quantity yield a probability versus time for the evolution of population densities. The cross-terms that arise naturally from this procedure (e.g. amplitude of pathway 1 multiplied by amplitude of pathway 2) correct the classical rate law for quantum-mechanical interferences.

For example, consider a system that exhibits two dominant pathways (e.g. the FMO complex). If the measurements of energy transfer efficiency depends on the conditions that either one of the two experiments probing quantum interference between explicit pathways, i.e., one pathway is blocked, then the results obtained, will be critical for the understanding and control of quantum coherence effects in photosynthesis. Not only that the results, thus obtained, can be compared to that of the unmodified system, which will bring us to have a real truth behind the idea regarding the contributions of excitonic quantum coherence so far as the quantum efficiency of light harvesting is concerned. Now, from the present case, for example, consider a system that exhibits two dominant pathways (e.g. the FMO complex). If the measurements of energy transfer efficiency, on the conditions that either one of the two pathway is blocked, can be performed and compared to the results of the unmodified system then we can truly bring to light the contributions of excitonic quantum coherence to the quantum efficiency of light harvesting. One of the most important application of coherence is the constructive or destructive interference of amplitude transferred along two (or more) distinct paths in an energy transfer network, which effect has been discussed by Cao, Silbey and coworkers[113] in detail. It is interesting to note that in a recent theoretical analysis, Beratan and coworkers[114] demonstrated that the two views of electron tunneling in proteins can be put in a unified description based on a simple criterion that depends on the dynamical fluctuations of the effective coupling. However, experimental evidence that can be used to distinguish the two limits by measuring interference effects is still lacking.

3.4.2 *Quantum Coherence*

As has already been stated earlier, quantum super-positions of the states of a composite system, give rise to non-trivial quantum features, i.e., coherence and entanglement. While quantum coherence is generally associated to interference effects, entanglement refers to non classical correlations between distinguishable modes or subsystems of a multipartite complex.[98,99] In photosynthetic antenna, the typical characteristics of electronic coherence, observed during energy transfer in the complexes, has reinvigorated some of crucial queries: for these molecular systems, whether there exists any practical functionality, so far as the coherence and/or entanglement are concerned. Olaya and the following group investigated[101,118] quantitative relationships in between the quantum yield of a light-harvesting complex and the corresponding distribution of entanglement among its components. Their study emphasized on the measure of entanglement yield or average entanglement, surviving a time scale, comparable to the average excitation trapping time. Importantly, they considered the Fenna-Matthews-Olson (FMO) protein of green sulphur bacteria as a prototype, following which they showed that there is an **inverse relationship** in between the quantum efficiency and the average entanglement in the distant donor sites suggesting that long lasting electronic coherence among distant donors might help modulation of the light harvesting function.

In fact, applying the framework of entanglement, Fassioli *et al.*,[102] quantified the strength of electronic coherence in a light harvesting system for different initial conditions by investigating the distribution of such coherences among the molecular sites and bath parameters. Through detailed studies, they assessed the effects of the coherences which survive a time scale of possible biological relevance. For example, both the time for average excitation trapping and that of long-lasting electronic coherence are quantified by the entanglement yield, defined as the average value of entanglement at times of possible trapping events. The obtained results suggests that being a long-lived electronic coherence in between well-separated pigments, it may modulate the efficiency profile by being function of the system-bath coupling.

This modulation is manifested as an inverse relationship in between quantum efficiency and the entanglement yield among distant pigments, acting as excitation donors. In their challenging approach, the theory of photosynthetic energy transfer, i.e., possible quantitative relations in between electronic coherence and the light-harvesting function, have also been developed as well. Assuming the FMO complex under an weak system-bath coupling, the relationships between the transfer efficiency and the entanglement present in the system, has been investigated at the time of possible occurrence of trapping events out of which valuable results obtained. The long-lived coherence, thus developed, is spatially distributed in such a way that the average entanglement among donor dimers have been found to be in inverse relationship with the quantum yield.

In light harvesting, coherence is implied as the wave-like character of energy flow in between exited states. This, in the site basis, is stated as due to the presence of long-time or steady electronic coherence. These can be observed with the help of an ultrafast spectroscopic method which is able of creating initial coherence, recording simultaneously its loss, as a function of time. But this very characteristics has created a certain amount of confusion regarding the significance of coherence in natural light harvesting. A typical question might be: *Does the coherence only matter during the first few hundred femtoseconds of the energy transfer following the absorption of light?* But, may be, the basic premise of this question is not incorrect. It is a known fact that the ultra-short pulse excitation simply serves to coordinate the ensemble in time and allow observation of dynamical processes which, importantly, through theoretical modeling, enable deductions of the Hamiltonian of the related system. The observed coherence in an ensemble have been found to decay by two mechanisms:

(1) The first is called as ensemble dephasing that does not destroy coherence in individual complexes, but disrupts the correlation between the oscillatory behavior of individual members of the ensemble, thus leading to the decay of observable oscillations in an experiment.
(2) The second, known as microscopic dephasing, or decoherence, destroys the coherent superpositions between excitons in an individual complex.

At the present state, we do not have either experimental or any theoretical method of such grade so that we would have been able to observe and characterize the condensed phase, known to be related for the characterization of microscopic dephasing. Hence, at this point, we expect the fluctuations within an individual complex to be uncorrelated with those, present in other members of the ensemble and also expect the microscopic dephasing to be significantly slower than the ensemble dephasing. Indeed it could be suggested that through out the energy flow process, system-bath interactions create and recreate coherence. But, in an ensemble measurement, different members of the ensemble have been observed to become uncorrelated quite rapidly (nearly, in a few hundred femtoseconds) making the possibility of direct observation of coherence impossible. In such cases, both the dephasing processes contribute to the decay of oscillations that is measurable with the help of two dimensional electronic spectroscopy (2DES).

Romero *et al.*[103] while discussing their results of 2DES, revealed about the presence of coherent effects in photosynthetic complexes. For this, they employed three ultra short and spectrally broad laser pulses, separated by controlled time delays, so that, as a result, the broadband excitation creates coherent superpositions of electronic/vibrational states, showing specific features in the 2D spectra. In explicit way, this can be stated as the representation of coherence wavelength in the initial excitation, whereas the rephasing wavelength can be taken as the subsequent emission. Not only that, in the absence of coupling, contributions from excited-state absorption and emission cancel each other. Thus, there will be no yielding off-diagonal peaks in the spectrum, considered as essential for obtaining the proper signal of such coupling. But, due to the presence of coupling, the cancellation is no longer complete and a so-called cross-peak emerges[136] as a result. In such cases, in order to explain coherence phenomena in a consistent way, two-dimensional spectroscopy (2DES) offers an excellent technology in probing the coupling between energy levels.

In quantum systems, the type of coherence produced, is dependent on the basis set of states, harboring the coherence which can be over any state—i.e., electronic, vibrational or vibronic. For the purpose of discussion, let us distinguish these two types on either the ground or excited electronic state of coherence — vibrational coherence and electronic coherence. Vibrational coherence refers to pure nuclear wave packet motion, induced by photoexcitation which is a typical characteristic of resonance in the molecule of Raman active normal modes whereas electronic coherence refers to the coherent excitation of two electronic states within the system, but typically, coupled by electronic interactions but uncoupled from any kind of vibrational modes in that system. Assignment of such nature of the modes has been found quite critical for a meaningful interpretation of the results to be obtained. In fact, coherence in the site/chromophore basis, refers to exciton states, i.e., the delocalized states that are stationary when evolved by the Hamiltonian. These excitons (site coherences) are however, ubiquitous in light harvesting systems, produced normally in the presence of strong coupling in between chromophores, this way, generating splittings in the energy levels.

As such, originally, below-diagonal cross-peaks oscillating as a function of T [population time] were assigned to electronic coherences between excitons. Keeping that interpretation in mind, the vibrational coherences are considered as modulated diagonal amplitude[26] whereas energy transfer appearing as non-oscillating cross-peaks. More recently,[102] based on specific vibrational modes, present in the electronic or vibrational (vibronic) modes, have been proposed. Based on displaced potential energy surfaces and considering both diagonal and off-diagonal contributions together with those models, are considered and these modes are taken to be resonant with the excitonic manifold energy gaps. These models also consider the role of non-equilibrium vibrations,[102] including the role of ground-state vibrational coherence.[91]

Not only that, coherence in between excitons can also be stated as 'the superpositions' in the "energy basis" (basis of stationary states — i.e., excitons or vibronic states), the signature of which can be observed through quantum beats in spectroscopic experiments. These can be represented as the oscillations in between two energy basis states (excitons), having different energies. If a system is considered as being excited by laser via its excitons, obviously, the observed features of the system will also be in between excitons. Lastly, states undergoing coherent processes (when dephasing from the environment is smaller than the interaction energy of the said process) harbor the process 'coherence' also. Thus, as the outcome of all such processes are quantum mechanical, this result produces 'quantum superpositions' which can be evolved by themselves. All such types of coherence, thus, can be represented as off-diagonal elements of a density matrix, in some basis. Importantly, if we were to represent the density matrix of the system in the site basis, off diagonal elements would refer to the coherence only in between sites (excitons).

It should be pointed out that in this context, some statements of a few works done, offers a rather controversial views regarding the role of quantum coherence. Even though serious questions have been raised about the feasibility and essentiality for the attainment of high efficiency in light harvesting, Fassioli-Scholles group opined that due to the present, definite and certain ample of evidences, the coherence effects involved in the dynamics of photosynthetic process, has already been established. The key to this efficiency can better be understood, for example, due to the mechanism, present in ultrafast light-harvesting processes. In such situation, the coherence effects involved in the associated dynamics, can be observed as the process of balance in between the ultrafast (femtosecond to picosecond timescale) transfer of electronic excitation within and between antenna complexes. Not only that, this high grade of technically balanced phenomena can very well be compared in between the time for relaxation to the ground state and that for the scaling of the excitation diffusion with the size of the antenna system.

3.4.3 *Decoherence*

Two flavors could be noticed while discussing environmental interactions with the chromophoric system where both of them cause fluctuations and heterogeneity in their electronic structure (chromophores), the substantial implications of which can be observed in case of the coherence scales of the excitons and their interactions. These are

(1) Dynamic Disorder: The first effect can be defined as the disorder, capable of changing in environment of the phase relationships, already present in between excitons. In the sense of Dynamicity, this disorder localizes the state to a smaller region if the energy of environmental interactions becomes larger than that in between chromophores or excitons, in effect, meaning a separability of timescales in between

energy transfer and localization timescales. The exciton remains at least partially delocalized in this regime, meaning thermal equilibrium with the associated environment. On the other hand, under strong localization conditions, the electronic states are dressed with environmental fluctuations. As a result of this, phase relationships in between any states are lost much faster than the transfer time. Finally, states become localized, placing the system into an equilibrium state, determined by the Boltzmann distribution under thermal conditions.

(2) Static Disorder: The second effect, called static disorder, arises from heterogeneity of chromophore excitation energies which alters the Hamiltonian and thus the stationary states. The corresponding localization effect takes place whenever the difference in between chromophoric excitation energies becomes larger than the coupling strength — i.e., when the electronic state transition remains away from being in resonance. In fact, chromophores, normally, do not mix with different energies except those having the same energy, already resonant and hence capable of mixing fully, provided, the interaction energy is larger than the environmental perturbations. Quite interestingly, disorder caused by the environment, effectively, can bring prior non-resonant chromophores into resonance through the shifts in the chromophore excitation energies.

This kind of differences in chromophore excitation energies can be, either due to presence of different molecular species (chromophores) in the system or via different local perturbations of the electrostatic environment caused by the protein environment, surrounding identical chromophores. Not only that, whenever the local electrostatic environment slowly changes, i.e., very slow (picosecond — estimating speed of sound in a protein, i.e., $\approx$2 nm/ps),[115] associated conformational dynamics can also lead to a similar effect, i.e., a shifting in the chromophore excitation energy. In many light harvesting systems, the environmental interaction energy is observed to be of the same order as that of electronic coupling. This means that the Förster regime can not be applied because of the inability in separating time and energy scales in the system. The effect produced is such that the interactions between chromophores and environment place the system in a regime that allow just enough quantumness for producing some non-trivial effects over the energy transfer time scale. Interestingly, in particular cases where long range transfer is needed in between weakly coupled chromophores, the Förster regime is perfectly applicable. Also, even within a single system, a subset of chromophores may be weakly coupled in the Förster regime, compared to a subset of more strongly coupled chromophores, thus harboring quantum exotica., i.e., indeed, possibility might indeed be there in having examples of this kind, existing in nature; (namely the eighth chromophore of the FMO complex).

But, in the so-called perfect quantum systems i.e., without noise, the excitons may be found in a superposition, hence, may suffer destructive interference at the trap site/chromophore, although this depends on the initial superposition, thereby making 'energy funneling' impossible. Furthermore, considering a complex energetic landscape through space, an excitation might not go up (too far) in energy, if lying at a local minimum. However, exception could be in a specific case when the exciton is already higher in energy, lying in a superposition with the energetically lower lying states, thus funneling to a global trap. In energetically disordered systems, for example, light harvesting antennae, having a fluctuating environment and capable of inducing a robustly tuned level of dephasing, can also be beneficial in a certain way. Rapid localization can be obtained out of perfectly coherent transport, producing rapid localization due to the lack of order and correlations in energies of the system (known as Anderson localisation) and an infinitely incoherent system produced what is called the quantum Zeno effect (where the excitation state cannot travel anywhere as it is constantly being measured/localised by the environment).[116]

The fluctuating environment can also be thought of as bringing chromophore excitation energies into resonance (de-trapping) in a transiant fashion. Also, partial spatial and energetic order can enhance transport by providing structure, allowing transport of excitations to converge to a key acceptor chromophore. But, even after that still continuing considerable debate, i.e., how quantum coherence can add efficiency to the transport of exciton. The argument that a light harvesting system will sample many pathways simultaneously and select the most energy efficient one, depending on the state of the environment, may not need to be invoked. Instead, starting at a high energy state, the system may stochastically select the global energetic minimum away from a local energetic minimum, may be in an incoherent manner with greater probability instead of being trapped in a local minimum and fluorescing. Of course, having a coherent superposition of a local trap and non-trap state would mean that the system has a probability of moving away from the local trap instead of the 'classical' case where it is stuck until it fluoresces or FRETs. Adding all such kind of complicated associated physical criteria, even then the role of quantum coherence in increasing efficiency still remains one of the greatest open questions.

3.4.4 *Quantum Superposition*

In a composite system, quantum superposition of the states give rise to non-trivial quantum features, for example, coherence and entanglement. While quantum coherence is generally associated to interference effects, entanglement refers to non classical correlations between distinguishable modes or subsystems of a multipartite complex.[118,218] Recent evidence of electronic coherence during energy transfer in photosynthetic antenna complexes has again reinvigorated the critical review discussing whether coherence and/or entanglement has any practical functionality for such kind of molecular systems. Olaya and the following group investigated[118] quantitative relationships between the quantum yield of a light-harvesting complex and corresponding distribution of entanglement among its components. Their study emphasized on the entanglement yield or average entanglement surviving a time scale, comparable to the average excitation trapping time. Importantly, considering the Fenna-Matthews-Olson (FMO) protein of green sulfur bacteria as a prototype, they found an "inverse relationship" between the quantum efficiency and the average entanglement between distant donor sites which suggested that long lasting electronic coherence among distant donors, might help modulation of the light harvesting function.

In fact, using the framework of entanglement, Fassioli *et al.* quantified the strength of electronic coherences in a light harvesting system, investigating also the distribution of such coherences among the molecular sites for different initial conditions and bath parameters. Also, they studied the effects of the coherences surviving a time scale so that it becomes biologically relevant. The average excitation trapping time and the long-lasting electronic coherences are known to be quantified by the entanglement yield, defined as the average value of entanglement at times when there is possibility trapping events to occur. Thus the obtained results suggest that being a long-lived electronic coherence in between well-separated pigments, it may modulate the efficiency profile which is a function of the system-bath coupling. This modulation manifests itself by showing an inverse relationship in between quantum efficiency and the entanglement yield among distant pigments, thus acting as excitation donors.

In their challenging approach,[101] the theory of photosynthetic energy transfer, i.e., possible quantitative relations between electronic coherence and the associated light-harvesting function, have been developed. Assuming the FMO complex under weak system-bath coupling, they investigated the relationships between the transfer efficiency and the entanglement, present in the system at the time of possible occurrence of trapping events and obtained valuable results.

The long-lived coherence, developed in this way, is spatially distributed in such a way that the average entanglement among donor dimers exhibits an inverse relationship with the quantum yield. In light harvesting, we imply coherence between exciton states as the wave-like character of energy flow which, in the site basis, is the presence of long-time or steady electronic coherence, observed via an ultrafast spectroscopic method, creating initial coherence. Also, at the same time, loss is recorded as a function of time. But, this has created a certain amount of confusion so far as the significance of coherence in natural light harvesting is concerned. A typical question might be: "*Does the coherence only matter during the first few hundred femtoseconds of the energy transfer, following the absorption of light?*".

But, the basic premise of this question might be incorrect. The ultrashort pulse excitation simply serves to coordinate the ensemble in time and allow observation of dynamical processes which, through theoretical modeling, enable deductions about the Hamiltonian of the related system. The observed coherence in an ensemble decays by two mechanisms:

(1) First is ensemble dephasing, that does not destroy coherence in individual complexes, but disrupts the correlation between the oscillatory behavior of individual members of the ensemble, leading to the decay of observable oscillations in an experiment.
(2) Second is microscopic dephasing, or decoherence, that destroys the coherent superpositions between excitons in an individual complex.

At the present stage, we currently lack both the experimental and theoretical methods in observing and characterizing the condensed phase, related to the characterization of microscopic dephasing. Nonetheless, as we expect the fluctuations within an individual complex to be uncorrelated with those in other members of the ensemble, at the same time, we also expect the microscopic dephasing to be significantly slower than the ensemble dephasing. Indeed it appears reasonable to suggest that the system-bath interactions create and recreate coherence throughout the energy flow process. However, in an ensemble measurement, different members of the ensemble rapidly (in a few hundred femtoseconds) become uncorrelated and the coherence is not directly observable. In such cases, both the dephasing processes contribute to the decay of oscillations that can be measured in two dimensional electronic spectroscopy (2DES).

2DES results reveal the presence of coherent effects in photosynthetic complexes using three ultrashort and spectrally broad laser pulses, separated by controlled time delays. Following this process, the broadband excitation creates coherent superpositions of electronic/vibrational states giving rise to specific features in the 2D spectra. Putting it in explicit way, the coherence wavelength represents the initial excitation, while the rephasing wavelength can be thought of as the subsequent emission. Without coupling, contributions from excited-state absorption and emission cancel each other, yielding no off-diagonal peaks in the spectrum which is essential for having the proper signal of such coupling. But in the presence of coupling, the cancellation is no longer complete and a so-called cross-peak emerges.[136]

In such cases, two-dimensional spectroscopy (2DES) provides an excellent probe of the coupling between energy levels, thus explaining coherence phenomena in a consist ent way. In quantum systems, the type of coherence produced, is dependent on the basis set of states harbouring the coherence. These coherences can be over any state—electronic, vibrational or vibronic. For the purpose of discussion, we distinguish these two types of coherence—vibrational coherence and electronic coherence. Vibrational coherence refers to pure nuclear wave packet motion induced by photoexcitation and is characteristic of resonance Raman active normal modes in the molecule on either the ground or excited electronic state. Electronic coherence here refers to the coherent excitation of two electronic states coupled by electronic interactions but uncoupled from any vibrational modes in the system. Assignment of the nature of the modes is critical to a meaningful interpretation of the results.

Firstly, coherence in the site/chromophore basis refers to exciton states, i.e., the delocalized states that remains stationary when evolved by the Hamiltonian. Excitons (site coherences) are ubiquitous in light harvesting systems and are generally produced when strong coupling between chromophores is present and generates splittings in the energy levels. In fact, as stated earlier, originally, below-diagonal cross-peaks oscillating as a function of T [population time] were assigned to electronic coherences between excitons. Within that interpretation, the vibrational coherences modulated the diagonal amplitude,[65] whereas energy transfer appeared as non-oscillating cross-peaks. More recently, electronic or vibrational (vibronic) models, with both diagonal and off-diagonal contributions, have been proposed which are based on specific vibrational modes, resonant with the excitonic manifold energy gaps. Not only that, these models are based on the role of non-equilibrium vibrations[12,108] also, including the role of ground-state vibrational coherence[133] as well as models based on displaced potential energy surfaces.

Secondly, coherence between excitons can be stated as the superpositions in the "energy basis" (basis of stationary states — i.e., excitons or vibronic states). These coherences can be observed through quantum beats in spectroscopic experiments, being represented as the oscillations between two energy basis states (excitons) with different energies. As laser excites the system via its excitons, the observed features of the system will also be between excitons. Lastly, states undergoing coherent processes (when dephasing from the environment is smaller than the interaction energy of the said process) harbor the process 'coherence'. Outcomes of such processes are quantum mechanical, producing quantum superpositions which can be evolved by themselves. All of these types of coherence can be represented as off-diagonal elements of a density matrix in some basis. Not only that, if we were to represent the density matrix of the system in the site basis, off diagonal elements would refer to the coherence between sites (excitons).

Though some statements of a few works done in these context offers a rather controversial views regarding the role of quantum coherence, i.e., whether it is essential for attaining the highly efficient light harvesting in photosynthesis, Fassioli-Scholles group[92] opined that because of the presence of definite and certain ample of evidences, the fact that coherence effects are involved in the dynamics of photosynthetic process, has already been established. Therefore, the key to this efficiency can better be understood, for example, in the presence of the mechanism of ultrafast light-harvesting processes. In such situation, the coherence effects involved in the associated dynamics, i.e., looking from another angle of consideration, can be observed as the process of balance between the ultrafast (femtosecond to picosecond timescale) transfer of electronic excitation within and between antenna complexes. Not only that, this high grade of technically balanced phenomena can very well be compared, also with the time for relaxation to the ground state and the scaling of the excitation diffusion time with the size of the antenna system.

3.5 Quantum Coherence: Electronic and Vibrational Coherence

Plants, algae, and photosynthetic bacteria use surprisingly sophisticated optimizations at the quantum mechanical level to harvest the sun's energy. The observation of coherence phenomena within light-harvesting complexes after short laser-pulse excitation has inspired advances in our understanding of light-harvesting optimization, highlighting the interplay of electronic excitations and vibrations. In quantum systems, the type of coherence produced, is dependent on the basic set of states that harbor the coherence. These coherences can be over any state — electronic, vibrational or vibronic. Firstly, coherence in the site/chromophore basis refers to exciton states — the delocalised states that are stationary when evolved by the Hamiltonian. It is an well established fact now that excitons (site coherences) are ubiquitous in light harvesting systems, generally produced when strong oupling between chromophores is present, generating & splittings in the energy levels. Secondly, coherence between excitons are states that are superpositions in the energy basis (basis of stationary states—excitons or vibronic states). Signature of such kind of coherences can be observed through quantum beats in spectroscopic experiments, as they represent oscillations between two energy basis states (excitons) with different energies. A laser excites the system via its excitons and hence observed features of the system will also be between excitons.

Lastly, it can be stated that undergoing coherent processes (when dephasing from the environment is smaller than the interaction energy of the said process) harbor process coherence itself. This is the case where the outcomes are quantum mechanical, producing quantum superpositions and can evolve superposition themselves at the same time. It has already been an established fact that a separation of electronic and nuclear coherences in a (realistic) molecular aggregate, based on the experimental data alone is questionable. On the basis of a molecular aggregate's Raman spectra and excitonic structures, it is experimentally more feasible, in a more definite way, to distinguish coherences evolving in ground or excited states. It is to be noted that all of these types of coherence can be represented as off-diagonal elements of a density matrix in some basis. Furthermore, if we were to represent the density matrix of the system in the site basis, off diagonal elements would refer coherences between sites (excitons).

3.5.1 *Electronic Coherence or Exciton Delocalization*

Remarkably, recent studies, performed by Fassioli *et al.*[102] have established convincing evidences, stating that, indeed, some of these fascinating molecular aggregates are designed to sustain quantum coherent transfer for longer than expected duration, and that also can be achieved at temperatures as high as in biological conditions. A direct implication of such studies, is related to the underlying process, i.e., the effect of quantum coherence in light-harvesting processes in vivo.[119] After the observation of coherent excited state dynamics in light-harvesting systems, a new impetus did put the end of long-standing debate and discussion regarding relationships between energy transfer efficiency and quantum features, for example, exciton delocalization (electronic coherence) together with its coherent oscillatory behavior.[101] So far as the practical functionality of quantum coherent phenomena is concerned, it appears to be subtler than efficiency maximization. For example, both of the structure and arrangement of antenna complexes appear, not for the optimization of the fast transfer solely but quite efficient in possessing the ability in modulating their function under different environmental conditions[101] (and references there in). All of these facts, may raise a vital question, i.e., how electronic coherence, together with its possible coherent evolution, may satisfy the requirement for having such an built-in mechanisms, so that, this process, successfully modulates transfer efficiency, even under different environmental parameters.

Firstly, coherence in the site/chromophore basis refers to exciton states — the delocalised states that are stationary when evolved by the Hamiltonian. All of such types of coherence can be represented as off-diagonal elements of a density matrix in some basis. Furthermore, if we were to represent the density matrix of the system in the site basis, off diagonal elements in the spectral data and associated quantum beats must be interpreted by building a theoretical model that supports the observations. However, multiple models could often be invoked to reproduce and fit the data which practically brings us facing with the obvious difficult task, i.e., trying to understand the fundamental source of coherence. To the end, the interpretations of the results of spectroscopic studies have ebbed and flowed and there are rather evidence of some form of coherent interference phenomena underlying the spectra. For this, the spectral data and associated quantum beats must be interpreted by building a theoretical model which can often be invoked to reproduce and fit the observed data. Gaining an understanding of the engineering principles behind the possible maintenance of such a delicate 'quantum concert' is paramount to many applications. Many theoretical studies highlight that having a "tuned amount of quantum coherence" in disordered light harvesting systems, can enhance energy transfer efficiency and robustness far more than would be the case of energy incoherently hopping between chromophores (Förster limit).

Indeed interestingly, it may be that light-harvesting systems are able to take full advantage of unique quantum "interference effects" which posses the possible efficiency for the regulation of initial state conditions[49,101] or, might be, through accumulative quantum phases in closed transfer pathways.[113] Another quite outstanding point of the recent experimental results to be mentioned here, is that the electronic coherence has been observed to expand several molecules across the whole antenna complex, including distant as well as weakly interacting pigments.[4,6] Recently, many of motivated studies have been in this field of research, i.e., long-range electronic coherence, exclusively in order to study the electronic energy transfer from the perspective of quantum entanglement.[25,38,101] However, still an intriguing question remains, i.e., whether "*correlations or interference between distant pigments play a part in the efficient functioning of a light-harvesting complex or they are just a consequence of exciton delocalization with no particular relation to transfer efficiency*".

Finally, all these attempts, involved in the search of the actual processes going on, have reached to a suggestion stating that electronic energy transfer in complex biological and chemical systems can involve quantum coherence, even at ambient temperature conditions which phenomenon has been found in some of the photosynthetic proteins too. But, even though the role of quantum coherence in purple bacteria light harvesting was first established by Hu *et al.*[122] in 1997, the intricate dynamics of quantum coherence, which has no classical analogue, was largely neglected in the analyses, even though electronic energy transfer involving oscillatory populations of donors and acceptors, was first discussed more than 70 years ago by Perrin.[34] There, electronic quantum beats arising from quantum coherence in photosynthetic complexes have been predicted and indirectly observed from spectroscopic data which clearly have documented the dependence of the dominant energy transport pathways on the spatial properties of the excited-state wave functions of the whole bacterio-chlorophyll complex.

However, the process by which these vibronic effects change or optimize the function of light-harvesting complexes has remained still unclear. In other words, we should learn what kind of the design principle while making advancement of two-dimensional electronic spectroscopy (2DES), in order to quantify the vibronic mixing, among the light-absorbing molecules of a light-harvesting complex, for example, from cryptophyte algae.[184] Out of extensive studies, the related phenomena has been revealed stating that, a striking reallocation of absorption strength, in turn, provides a robust increase in the rate of energy transfer of up to 3.5–fold. It has been realized that absorption-strength redistribution, induced by vibronic coupling, provides a 'multiplicative' increase in the rate of energy funneling. This establishes a bio-inspired design principle for optimal light-harvesting systems.

3.5.2 *Vibrational Coherence & Vibronic States*

It has already been established that both the energy transfer as well as charge separation in photosynthesis are rapid events with high quantum efficiency, caused by production of collective excitations (excitons) forming a coherent superposition of electronic and vibrational states of the individual pigments. Recently, the problem of the 'photo-induced coherence' observed in two-dimensional electronic spectra (2DES) of light harvesting complexes[4, 19] has attracted a lot of attention suggesting the origination of observed coherence from a superposition of electronic eigenstates, its long lifetime being of central importance for the high efficiency of light harvesting systems.[4] The detailed analysis by Cheng and Fleming[61] showed that electronic coherence beatings in an electronic dimer lead to an oscillating diagonal peak in the non-rephasing spectrum, while the corresponding rephasing peak remains static. Turner *et al.*[24] have applied this argument on his experimental data for showing the distinction in between electronic and vibrational coherence. But, during simultaneous presence of electronic and vibrational DOFs (Degrees of Freedom) in a coupled system, the energy levels in the excited state acquires a mixed character,[34] i.e., the excited state energy level structure will then be similar to the electronic case, but certain levels may possess also a significant vibrational character.[4, 123]

Recently, in the nonlinear spectroscopic experiments, long-lived coherences have been observed in photosynthetic antenna complexes, indicating that the vibrations, resonant with the exciton splittings, is able to modify the delocalization of the exciton states. Not only that this produces additional states, thereby promoting directed energy transfer, allowing a switching in between the two charge separation pathways. Importantly, this apparent persistence of electronic coherence well beyond optical dephasing times, for example, brings to light the role of nuclear motion in sustaining electronic correlations. This happens whenever the delocalization timescale matches that of vibrational motion.[4,56] The observation of such kind of short laser-pulse coherence phenomena within light-harvesting complexes has inspired advances in our understanding of light-harvesting optimization, highlighting, importantly, the interplay of electronic excitations and vibrations.

Based on quantitative modeling, the exciton-vibrational coherences observed in 2D photon-echo of the photosystem II reaction center (PSII-RC) suggested that the theoretical results, reflecting underlying electronic–vibrational resonances, may play a functional role in enhancing energy transfer. Following extensive research works, involving two-dimensional electron spectroscopy consensus among recent theoretical studies and 2DES predictions, the ultimate conclusion states that, in reality, light-gathering macromolecules in plant cells, play the role of transferring energy by taking advantage of **molecular vibrations** whose physical descriptions have no equivalents in classical physics. Here, the vibronic coupling in between donor and acceptor states is the prevailing mechanism giving rise to long-lived coherent oscillations, having signatures of electronic interactions.[6,61,123]

Recently, similar vibronic coherence was identified in charge separation also in photosystem II of plants.[23] Henceforth, the following conclusion can be drawn, i.e., the coincidence of the frequencies of the most intense vibrations with that of the splittings within the manifold of exciton and charge-transfer states in the PSII-RC, is not occurring **by chance**, but reflects a fundamental principle, i.e., how energy conversion in photosynthesis can be activated. But still, there remain key questions:i.e., how vibronic coherence affects the energy-transfer mechanism and whether it is harnessed for improvement of the function of light-harvesting complexes. Also, it should be noted that in vibronic states, appearing frequently in light harvesting systems, information about the quantum state depends on the whole system (i.e. the vibrational states plus the electronic states). In such cases, where the vibrational mode has energy equal to the energy gap between electronic states, the vibronic coupling can particularly be strong, having some significant effects for light harvesting.

Firstly, the electronic states is capable of borrowing coherence lifetime from the vibrational states with which they are entangled. This leads to vibronic coherences (that is, a coherent superposition between vibronic states which can be either more vibrational or more electronic in character), capable of remaining coherent, longer than purely electronic coherences. Secondly, vibronic states can effectively enhance spectral overlapping between electronic (excitonic) states, thereby enhancing resonant energy transfer. This effect can be enhanced by the possible phase locking of the vibrational coherences on different but identical chromophores, ultimately, leading to synchronised fluctuations. Such kind of phase locked behavior can be generated either by mechanical contact (closely packed chromophores) or by electromagnetic interactions or oscillations in the electric field of an effector molecule, caused by changes in charge density from the vibrations, thereby forcing oscillations in a detector molecule.

Here, the vibrational states of molecules play the substantially important role in describing the quantum state of the system. Just like the coherence between electronic states play an crucial role in the dynamic evolution of the system, in a similar fashion, coherent superpositions of vibrational modes do so. Among various kind of involvement to be mentioned, for example, a singly excited vibrational mode in the ground or excited electronic state, might be involved with different vibrational modes, that also, on the same molecule. Vibrations on different chromophores can also exist in tandem, being coupled and phase locked (coherent) or exist across many chromophores. More to it, the role of vibrations and vibrational coherence becomes more complex, caused by the superposition of different vibrational modes across different electronic states of the molecule and also in between them. These effects can be observed spectroscopically through Raman transitions and particularly important, — in electronic energy transfer which happens whenever their energy matches the energy gaps in between excitons. Thus, the additional vibrational energy helps in bringing excited chromophore states into resonance that would not happen otherwise.

In some cases, the coupling between electronic and vibrational states can be so strong that they are no longer separable. In fact, under such instances, alterations to the nuclear structure, distort the electronic orbitals, affecting the vibrational states due to which it will be very difficult in making approximation, i.e., to make electronic states as separate from vibrational states. In such cases, the mixed, inseparable states are referred to as "vibronic" states.

By Now, we are aware with the fact that in quantum system, the type of coherence produced, is dependent on the basic set of states, harboring the coherence. Many of the mechanisms underpinning coherent dynamics in photosynthesis can be over any state — electronic, vibrational or vibronic, i.e., it appears to be vibronic in nature, i.e., the transfer of excitation energy between excited electronic states harbors coherent, coupled states comprising a mixture of both electronic and vibrational degrees of freedom. Prediction of the quantum theory, related to photosynthetic phenomena, also predicts that electron transfer pathways could exhibit interference effects.[125] Regan and Onuchic,[126] in their detailed theoretical analysis of electron transfer through the azurin protein, suggested that quantum interferences between multiple distinct pathways play important roles in this protein.[125,128–130] However, definite experimental evidence for the quantum interference effects in long-range electron transfer in proteins still remains elusive.[124]

3.5.3 *Quantum Interference: Beating & Role of Chromophore*

Quantum superposition of molecular electronic states is characterized as possessing very fragile nature because of thermal energy fluctuations and the static conformational disorder, induced by the intimate surrounding of constituents in the system's molecules. In reality, the manifestation of the observable, expected to be manifested, associated with the dipole operator, is amplitude (and therefore intensity). In practical sense, in fact, whenever we set out to measure quantum coherence, we look for periodic oscillations in the amplitude of the signal called 'quantum beats' — a ubiquitous feature of quantum coherence. These beats were ascribed to coherent nuclear motions but more modern work focuses on optical studies related to these systems. For practicality, besides this kind of dipole operator, the same beating would occur with any operator, not commuting with the Hamiltonian which brings us in concluding this 'beating' as probability, walking back and forth across sites, or equally well, as oscillations of the electro-dynamic polarization via the dipole operator.

In a ultrafast laser spectroscopy of several light harvesting proteins, the manifestation of such kind of observable, associated with the dipole operator, is amplitude (and therefore intensity). This means, while measuring quantum coherence, we will look for periodic oscillations in the amplitude of the signal, dubbed as 'quantum beats', persisting for hundreds of femtoseconds, as putative signatures for quantum transport phenomena. For practical reason, when the dipole operator is examined, the same beating would occur with any operator that does not commute with the Hamiltonian. Thus, this 'beating' can be thought as *probability*, walking back and forth across sites or equally well, as *oscillations* of the electro-dynamic polarization via the dipole operator. The associated studies look at the response of an unpaired electron to a magnetic field, over time and a coherent resonance pattern (quantum beats between spins),[218] i.e., quantum beats have been found, ascribed to coherent nuclear motions.

But, more recent works focuses into these systems, through optical studies, the manifestation of such observables, associated with the dipole operator, i.e., amplitude (and therefore intensity). This means, while measuring quantum coherence, we need to look for periodic oscillations in the amplitude of the signal, i.e.,,so called 'quantum beats'. Even though the dipole operator is examined for practicality, the same beating would occur with any operator, not commuting with the Hamiltonian, i.e., we can think of this 'beating' as the probability, walking back and forth across sites or, as oscillations of the electro-dynamic polarization via the dipole operator. In reality, as there are evidence of some form of coherent interference phenomena underlying any spectrum,[218] the spectral data and associated quantum beats must be interpreted by building a theoretical model that supports these observations. However, multiple models could often be invoked to reproduce and fit the data. But this brings us, practically,- facing the core of obvious difficult task, i.e., *trying to understand the fundamental source of coherence.*

To the end, the interpretations of the results of spectroscopic studies have ebbed and flowed and there are rather evidence of some form of coherent interference phenomena underlying the spectra. Specially, gaining an understanding of the engineering principles behind the possible maintenance of such a delicate 'quantum concert' is paramount for many kind of practical applications. In most light harvesting systems, coherent quantum beats are observed only in those cases of coherences which are stable over a time scale, in proportion with the relevant energy transfer. However, the advancement of ultrafast, femtosecond laser technology have facilitated the development of spectroscopic techniques for probing the energy transfer pathways through light harvesting in photosynthetic systems as well as that of proteins. In fact, the presence of beats is a signature of underlying coherent phenomena, observed in most of biological light harvesting and photosynthetic systems/proteins with lifetime ranging from tens of femtoseconds through the picoseconds. But the difficulty lies in determining the nature of these coherences and to establish the fact whether they play a substantial role in photosynthesis and/or light harvesting. Many theoretical studies highlight that a "tuned amount of quantum coherence" in disordered light harvesting systems, can enhance energy transfer efficiency and robustness far more than it would be for the case of energy, incoherently hopping between chromophores (the Förster limit).

Since the initial discovery of quantum beats in the FMO complex, there has been considerable debate, related to the origin of coherence. Two dominant coherent oscillations are observed on the cross peaks between excitons one and two (160 cm^{-1}) and excitons one and three (200 cm^{-1}) with coherence times ranging from 100 fs to 1.1 ps, depending on the method and author.[4, 147, 148] Not only the oscillatory features are observed in peak widths, quantum beats have also been seen to grow in amplitude at early times indicating possible transfer of coherence from higher energy excitonic states relaxing to lower lying ones. Under such circumstances, coherence transfer could then support the sampling of multiple relaxation pathways to the sink, i.e., exciton.

Not only that, transfer of coherence via quantum transport is also supported by the changes in the oscillatory phase in between populations and coherences (π phase flip). However, the question remains still unclear whether coherence arises from vibrations or electronic superpositions. In fact, growing evidence suggests that it may not be possible to separate vibrational and electronic dynamics of the system. In many cases, the gaps in exciton energy and the energies of vibrations, resident on chlorophylls — from Raman studies, are found resonant. These facts could result in strong coherent coupling and vibronic states having long coherence times. Interestingly, the strongest quantum beats, observed are found to be resilient to alterations in the excitonic structure showing similar frequencies and dephasing times. Both observations of robustness to changes in the vibrational environment and exciton structure suggest that quantum beats may arise from vibronic coherence or at least, the strongest vibrational and electronic coherence are energetically resonant.

3.5.4 *Resonance in Quantum Beating: Vibrionic Model*

Fuller[149] showed resonance among many of the quantum beats having energy gap involving special pair exciton states, with similar frequency in type II reaction centers and found in good agreement when simulated applying vibronic model. This resonance appears to speed up energy relaxation, energy transport and charge separation, coming in bursts of population (probability) increase of the charge separated state, suggesting the coherences as of a mixed electronic–vibrational (vibrionic) nature which may enhance the rate of charge separation in oxygenic photosynthesis. Interestingly, following their views in photosynthesis, when the energy of a collective vibration of two chromphores have been observed to match the energy difference between the electronic transitions of these chromophores, a resonance occurs and efficient energy exchange taking place between electronic and vibrational degrees of freedom, proving that as the energy associated to the vibration is higher than the temperature scale, only a discrete unit or quantum of energy is exchanged. This points to the fact that during energy transference from one chromophore to another, the collective vibration displays properties, having no classical counterpart.

At this juncture, it should be pointed out that other bio molecular processes, for example, the transfer of electrons within macromolecules (like RCs in photosynthetic systems), the structural change of a chromophore upon absorption of photons (like in vision processes) or the recognition of a molecule by another (as in olfaction processes), have already been noted to be influenced by specific vibrational motions. These results suggest a closer examination of the vibrational dynamics involved in the processes of other biological prototypes, exploiting truly non-classical phenomena, if any. For example, Huang's group[150] followed the technique of super-cooling, the photosynthetic bacteria and succeeded in observing a single photon, appeared to excite different chromophores [pigments] simultaneously which they explained as most likely, due to electronic coupling between the co-factors (pigment), whereas, precisely positioned proteins specify the coupling.[150]

For practicality, though the dipole operator is examined, the same beating would occur with any operator that does not commute with the Hamiltonian. We can think of this 'beating' as probability, walking back and forth across sites or equally well, as oscillations of the electro-dynamic polarization via the dipole operator. In reality, it could be depicted as all of these things- in a broad sense, as the evidence related to some form of coherent interference phenomena underlying any spectrum which can be availed. Hence, the spectral data and associated quantum beats must be interpreted by building a theoretical model supporting the observations. However, multiple models could often be invoked to reproduce and fit the data which practically brings us facing with the obvious difficult task, i.e., trying to understand the fundamental source of coherence. Coherence between excitons can thus produce path coherence in the energy transfer processes throughout the complex down to the trap site, i.e., funneling to the RC. Since the initial discovery of quantum beats in the FMO complex, there has been considerable debate, related to the origin of such coherence.

Two dominant coherent oscillations are observed on the cross peaks between excitons one and two (160 cm^{-1}) and excitons one and three (200 cm^{-1}) with coherence times ranging from 100 fs to 1.1 ps, depending on the method and author.[4,7,8,48(a)&(b),147(a)&(b)] Not only the oscillatory features are observed in peak widths, quantum 'beats' are also seen to grow in amplitude at early times indicating possible transfer of coherence from higher energy excitonic states relaxing to lower lying ones. Under such circumstances, coherence transfer could then support the sampling of multiple relaxation pathways to the sink, i.e., exciton. Also, transfer of coherence via quantum transport is also supported by the changes in the oscillatory phase in between populations and coherences (π phase flip). However, even after that, the question remains still unclear whether coherence arises from vibrations or electronic superpositions.

Even though interpretations have still been mired with controversies, when chromophore complexes form excitons, delocalised across the chromophore array, beats often arise from the superpositons of excitons states. In such case, the beat frequencies and decoherence times commensurate with that of vibrational and vibrionic states. Hence it does not seem likely that vibrational coherences are to play key roles in light harvesting. However, it appears that they might have an important role in the function of reaction centers (RC) and bacteriorthodopsin. Another important point to mention is that time scales of coherence and that of energy transfer does not appear separable and not are the energy scales of couplings and environmental effects in most of the light harvesting systems. Strictly speaking, *energy transfer lies in an intermediate region which neither strictly classical nor is it purely a quantum mechanical.* It still remains with some degree of unresolved issues, i.e., whether it enhances or diminishes transfer. Recent theoretical studies indicates the energy transfer occurring in this regime to be enhanced beyond that of classical hopping and purely quantum transport.[144,145]

Thus, present growing consensus is that most of observed quantum beat signals, relevant to energy transport, are of mixed vibrionic nature.

Hence, vibrionic coherences are likely to play contributing role during the energy transportation within the protein/pigment complexes i.e., when the vibrational energy is found to be iso-energetic with the excitonic energy gaps. Finally, it appears that there exists a nontrivial quantum effect within the individual light harvesting and photosynthetic proteins. The presence of strongly coupled chromophores leads to the exciton states, having broader spectral coverage, and most probably coherent dynamics. The role of vibrionic coherence, in this picture, probably lies in exciton transfer through the protein's chromophore system, depositing the excitation in some form of final acceptor or trap state. But, in contrast, there exists very little evidence for quantum effects being involved in energy transfer between individual proteins.

Finally, up till now, there exists substantial amount of debatable evidence for the process coherence in light harvesting systems, namely, coherence occurring between different pathways of a reaction centers (RC) where several initiating states are coherent before a pathway is selected. All major diagonal peaks in 2DES spectra, are found to be connected by cross peaks, indicating transfer between excitons. Even though specific path ways are preferred there are many possible pathways existing for the transference of energy from initially excited states to the lowest energy state. Some evidences suggest this to occur during transport through the FMO complex. For example, in case of purple bacteria, the presence of coherence in between the exciton states of B800 ring, appear to have quantum control via a phase shift when it is being transported down to the B850 ring. This phase shift would suggest a selection of pathways from an initial superposition based on the state of the environment. However, this fact is a potential example of manifestation regarding robustness in energy transfer. The fastest and thus most prevalent energy transfer steps tend to involve transfer between exciton states with some spatial overlap, quite evident in the putative composition of chromophore sites in each exciton.

In fact, growing evidence suggests that it may not be possible to separate vibrational and electronic dynamics of the system. In many cases, the gaps in exciton energy and that caused by vibrations, resident on chlorophylls, from Raman studies, are found resonant. This fact could result in strong coherent coupling and vibronic states with long coherence times. Many of researchers have attempted to tackle this problem, i.e., the biological relevance and robustness of coherence to perturbations in the FMO system, but the strongest quantum beats, observed, are resilient to alterations in the excitonic structure showing similar frequencies and dephasing times.

Both the observations of robustness to changes in the vibrational environment and that of exciton structure leads to the suggestion that quantum beats may arise from vibronic coherence or at least the strongest vibrational and electronic coherence are energetically resonant. Many of the quantum beats are resonant with the energy gap involving special pair of exciton states having similar frequency to that of type II reaction centres.[149]

According to Fassioli and his group's view,[63] this kind of comparison of the expected dynamics of a three- and a four-level system provides a good test case for the simplest model of electronic or vibrational coherence. However, they emphasized that this procedure for discriminating between electronic and vibrational coherence is validated in reality, only in the strong electronic coupling limit. Simulations of a vibronic model have been found in good agreement with the data. Thus this resonance appears to speed up energy relaxation, energy transport and charge separation, coming in bursts of population (probability), increase of the charge in separated state in simulations. The spectral data and associated quantum beats must be interpreted by building a theoretical model that supports the observations. However, multiple models could often be invoked to reproduce and fit the data, practically bringing us facing with the obvious difficult task, i.e., understanding the fundamental source of coherence.

To the end, the interpretations of the results of spectroscopic studies have ebbed and flowed and there are rather evidence of some form of "coherent interference" phenomena underlying the spectra. For this, the spectral data and associated quantum beats must be interpreted by building a theoretical model which can often be invoked to reproduce and fit the observed data. For this, gaining an understanding of the engineering principles behind the possible maintenance of such a delicate 'quantum concert' is paramount to many applications. Many theoretical studies highlight that having a "tuned amount of quantum coherence" in disordered light harvesting systems, can enhance energy transfer efficiency and robustness far more than would be the case of energy incoherently hopping between chromophores (the Förster limit). In most light harvesting systems, coherent quantum beats have been observed only in the cases having coherence stable over a time scale being in proportion with relevant energy transfer times. The advancement of ultrafast, femtosecond laser technology, have facilitated in developing spectroscopic techniques for probing the energy transfer pathways through light harvesting, photosynthetic systems including that of proteins.

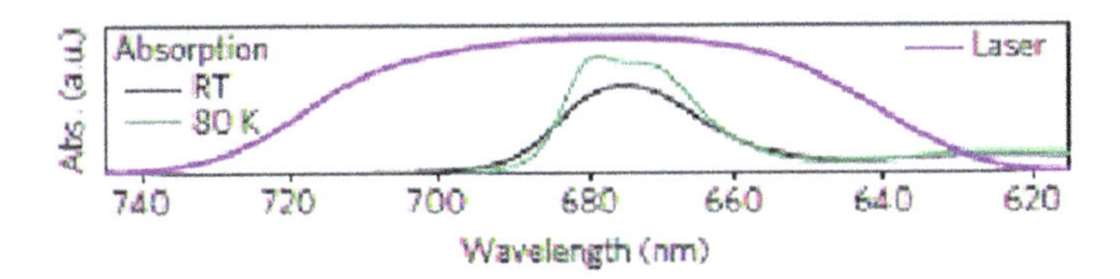

Figure 3.7. The electronic structure of the PSII RC.: Absorption spectra and laser spectral profile. RT, room temperature.

The presence of beats is a signature of underlying coherent phenomena, observed in most of biological light harvesting and photosynthetic systems/proteins having lifetime ranging from tens of femtoseconds through the picoseconds. This is evident in the putative composition of chromophore sites in each exciton and also the overlap between each exciton participating. throughout the complex, coherence between excitons produce path coherence in the energy transfer processes down to the trap site, finally, funneling to the RC. Since the initial discovery of quantum beats in the FMO complex, **there has been considerable debate, related to the origin of coherence**. Two dominant coherent oscillations are observed on the cross peaks between excitons one and two (160 cm^{-1}) and excitons one and three (200 cm^{-1}) with coherence times ranging from 100 fs to 1.1 ps, depending on the method and author.[4,7,8,46,147,148] Not only the oscillatory features are observed in peak widths, quantum beats are also seen to grow in amplitude at early times indicating possible transfer of coherence from higher energy excitonic states relaxing to lower lying ones. Under such circumstances, coherence transfer could then support the sampling of multiple relaxation pathways to the sink, i.e., exciton. Not only that, transfer of coherence via quantum transport is also supported by the changes in the oscillatory phase in between populations and coherences (π phase flip). However, the question remains **still unclear** whether coherence arises from vibrations or electronic superpositions.

At this juncture, we comprehensively can combine experimental (2DES) and theoretical (standard Redfield theory) methods to demonstrate the presence of electronic-vibrational (vibronic) coherences in the PSII RC and that whether these coherences strongly correlate with efficient and ultrafast charge separation. Importantly, 2DES reveals the presence of coherent effects in photosynthetic complexes through the use of three ultrashort and spectrally broad laser pulses, separated by controlled time delays

Fourier transform with respect to the coherence time τ i.e., (time between the first and second pulses) and with respect to the rephasing time t (time between the third pulse and the signal) yields the two-dimensional (2D) electronic spectrum in the frequency domain, correlating the absorption, ω_τ, and emission, ω_t — frequencies for a fixed population time T (time between the second and third pulses). Here, the broadband excitation creates coherent superpositions of electronic/vibrational states giving rise to specific features in the 2D spectra. Originally, below-diagonal cross-peaks oscillating as a function of T were assigned to electronic coherences between excitons. Within this interpretation, the vibrational coherence has been stated to modulate the diagonal amplitude,[26] the corresponding energy transfer appeared as non-oscillating cross-peaks. More recently, electronic-vibrational (vibronic) models, with both of their diagonal and off-diagonal contributions, have been proposed which is specifically based on the vibrational modes, resonant with the excitonic manifold energy gaps,[23,86,91] focusing on energy transfer efficiency. This approach included the role of non-equilibrium and also the ground-state vibrational coherence,[91] as models based on displaced potential energy surfaces.[86] The authors, Chin *et al.* (2013)[192] showed that the resonant electronic — vibrational configuration sustains, regenerates, or even creates coherence between electronic states during the timescale of energy and electron transfer. This mechanism does not require coherent laser excitation, and is also valid for incoherent sunlight excitation.[92,191]

However, debate has been still there, in determining the nature of the long lived quantum beats, specially, at physiological conditions, observed in time-resolved spectra of molecular aggregates. However, it has been accepted now as an established fact that quantum beats, as an ubiquitous feature of quantum coherence, are the oscillatory patterns observed in spectroscopic measurements whenever experiments are related in studying, especially that of transport phenomena, like, photosynthesis, thus putting confident evidence of some form of coherent interference phenomena underlying the spectra. But it is quite important to note that in most light harvesting systems, coherent quantum beats have been observed 'only' in those cases where the coherences are sufficiently stable over a time scale. Also, this must be in proportion with the relevant energy transfer times. The advancement of ultrafast, femtosecond laser technology have facilitated the development of spectroscopic techniques for probing the energy transfer pathways through light harvesting photosynthetic systems as well as that of proteins.

But the difficulty lies in determining the nature of these coherences. Even though interpretations are still mired with controversies, when chromophore complexes form excitons, delocalised across the chromophore array, beats often arise from the superpositons of excitons states. In such case, the beat frequencies and decoherence times commensurate with that of vibrational and vibronic states. Hence it does not seem likely that vibrational coherences are to play key roles in light harvesting. However, it appears that they might have an important role in the function of reaction centers (RC) and bacteriorthodopsin. Another important point to mention is that time scales of coherence and that of energy transfer does not appear separable and not are the energy scales of couplings and environmental effects in most of the light harvesting systems. Strictly speaking, *energy transfer lies in an intermediate region* which neither strictly classical nor is it purely a quantum mechanical. It still remains with *some degree of unresolved issues*, i.e., whether it enhances or diminishes transfer. Recent theoretical studies indicates the energy transfer occurring in this regime to be enhanced beyond that of classical hopping and purely quantum transport.[144,145]

However, present growing consensus opines that most of observed quantum beat signals, relevant to energy transport, are of mixed vibrionic nature. Then, it appears that vibrionic coherences are quite likely playing contributing role at the time of the energy transportation within the protein/pigment complexes, i.e., when the vibrational energy is found to be iso-energetic with the excitonic energy gaps. Thus, in a inclusive way, we can arrive at some point of conclusion stating that there exists a nontrivial quantum effect within the individual light harvesting and photosynthetic proteins. Most probably, the presence of strongly coupled chromophores leads to the exciton states, having broader spectral coverage and coherent dynamics. Here, probably, the role of vibrionic coherence, lies in exciton transfer through the protein's chromophoral system, depositing the excitation in some form of final acceptor or trap state.

But, in contrast to above scenario, there exists very little evidence for quantum effects being involved in energy transfer between individual proteins. Up till now, there exists debatable evidence for process coherence in light harvesting systems, namely, coherence occuring between different pathways of a RC where several initiating states are coherent before a pathway is selected. In 2DES spectra, all major diagonal peaks are observed to be connected by cross peaks indicating transfer between excitons. There are many possible pathways for energy transfer from initially excited states to the lowest energy state, although specific processes are preferred. Some of the evidences among them, suggest this to occur during transport through the FMO complex.

For example, in purple bacteria, the presence of coherence in between the exciton states of B800 ring, appear to have quantum control via a phase shift when transported down to the B850 ring. This kind of shifting would suggest a selection of pathways from an initial superposition based ton the state of the environment. However, this fact puts an potential example, for the manifestation of robustness in the process of energy transfer. The fastest and thus most prevalent energy transfer steps might involve transfer in between exciton states with some spatial overlap. In each exciton, this puts an evidence for the presence of the putative composition of chromophore sites in each exciton and eventually leading to the fact, i.e., the overlap between each exciton participating in energy transfer. Coherence between excitons can thus produce 'path coherence' also, in the energy transfer processes throughout the complex down to the trap site, thus, funneling to the reaction centre (RC).

In order to analyze the spectral data, associated quantum beats must be interpreted by building a theoretical model that supports the observations. Even after widely exploring the nature of the quantum interference (beats), many ambiguities still remains. For example, largely due to the fact that when energies are (near) resonant, the difficulties arise in accurately teasing apart vibrational and electronic quantum coherent mechanisms, originating from noise caused by the presence of environment (may be, due to the presence of bath etc). To tackle this problem, most researchers introduced the idea of vibronic states in light harvesting systems, exhibiting quantum beats. In such a condition, the transfer of excitation energy between excited electronic states harbor 'coherent entangled states' comprising a mixture of both electronic and vibrational degrees of freedom. Interestingly, among almost all of these light harvesting systems, these kind of quantum beats of various forms have already been observed.

Butkus *et al.*[131] discussed in detail, developing the conditions when and how long-lived electronic quantum coherences, originating from recently proposed inhomogeneous broadening mechanism, are enhanced and reflected in the 2DES-spectra of the excitonically coupled molecular dimer. They established the mechanism, i.e., how, caused by a disordered sub-ensemble, depending on the amount of inhomogeneous broadening, the excitonically coupled molecular system can establish long-lived electronic coherences, as a result of which the dephasing caused by static energy disorder becomes significantly reduced. Also, it has also been established that depending on the amount of inhomogeneous broadening, the excitonically coupled molecular system can establish long-lived electronic coherences for which the dephasing due to static energy disorder becomes significantly reduced.

Finally, it is quite interesting to note that, this controversy, i.e., whether coherence is caused by vibrations or electronic superpositions, has remained still unsolved. Indeed, growing evidences have been suggesting that it might be a quite difficult job to take the decision i.e., how to separate vibrational and electronic dynamics of the associated system. Because, the gaps in exciton energy and the energies of vibrations residing on chlorophylls, observed from Raman studies, are often found in resonance. This fact could result in strong coherent coupling of vibronic states having long coherence times. Also, the quantum beating frequencies between specific excitons remain unchanged upon these perturbations suggesting that quantum beats may arise from vibronic coherence only or that the strongest vibrational and electronic coherences are energetically resonant. Electronic or vibronic coherence is potentially manifest as a high-frequency mode, dephasing over 60 fs and associated coherent domain can potentially complete a sampling of the possible energy transfer pathways, under the high-frequency coherences and also found to be effectively able in sustaining and regenerating electronic coherence on energy transfer time scales.

Following Heitler — London approximation, this kind of collective excitation ability can be reflected as the coherent superposition of molecular excited states, thus establishing the excitonic spectrum.[131] Another established fact, worth to be mentioned here is that coherent nature of electronic excitations also play a very important role during the functioning of various molecular systems, having range from macromolecules of biological origin to artificial nanostructures used for design of specific molecular electronic devices. In fact, the local environment, inducing static and dynamic disorder, caused by the variability of the local environment, disturb the coherent relationship between the molecular wave functions. Distribution of such molecular transition energies and coupling strengths, not being changed during the time interval of the processes, are referred to as the diagonal static disorder, the off-diagonal being termed as static disorder. Here, the dynamic disorder arises from the time-dependent fluctuations of a system's electronic and nuclear parameters on the time scales of the relevant ultrafast processes.

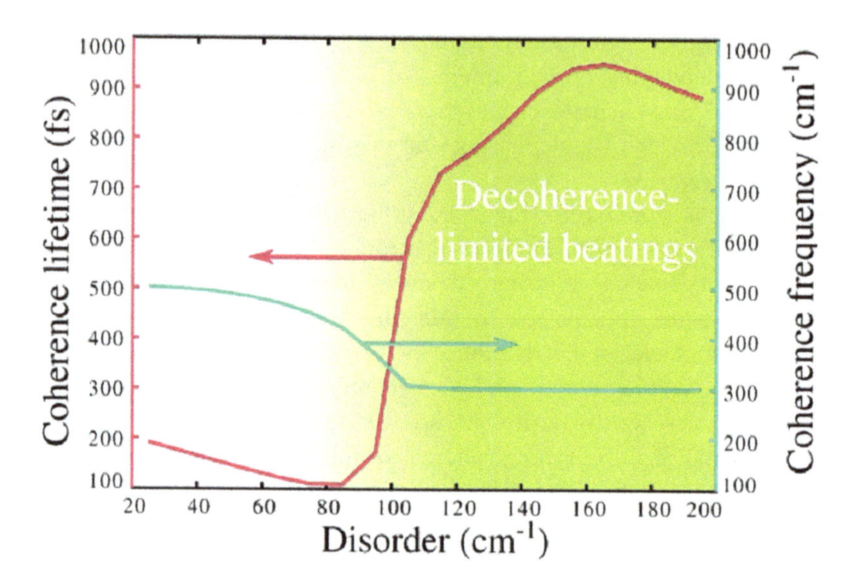

Figure 3.8. Disorder vs. Coherence lifetime and Coherence frequency.

In fact, both the static and dynamic disorder influence the exciton dynamics, to the extent of exciton delocalization, thus influencing significantly on excitation dynamics including photochemical properties of large molecular crystals and aggregates.[33,133] For example, in photosynthetic complexes,[134] a substantial exciton delocalization has been found to be quite crucial so far as the robust, efficient, and untrapped excitation transfer through a network of coupled molecules concerned. Along with many advantages related to excitation selectivity or spectral and temporal resolution, offered by the femtosecond laser excitation-detection, these methods also provide the possibility of observing the long-lived quantum beats, thus establishing different aspects, related to quantum mechanical superposition of electronic, vibrational, or mixed states.[128]

Very recently, it has been revealed that the non-adiabatic effects of coupling to the nuclear degrees of freedom cause the speed-up of excitation dynamics or charge transfer[127,128] and references there in. Though these have been revealed from the analysis of such quantum beats, however, the influence of the disorder on the coherent beats, has still not clearly been understood in convinced way.

Ongoing debate, related to the nature of the long-lived quantum beats, observed in time-resolved spectra of molecular aggregates at physiological conditions, still have been continuing. Butkus *et al.* studied in detail the conditions related to the origin of coherences and also, from recently proposed inhomogeneous broadening mechanism, are enhanced and reflected in the 2DES of the excitonically coupled molecular dimer. They presented the conditions for long-lived electronic quantum coherences originating from recently proposed inhomogeneous broadening mechanism, for being enhanced and reflected in the 2DES, after studying electronic spectra of the excitonically coupled molecular dimer. They established that depending on the amount of inhomogeneous broadening, the excitonically coupled molecular system can establish long-lived electronic coherences, caused by a disordered sub-ensemble, for which the dephasing due to static energy disorder becomes significantly reduced. On the basis of such considerations, it can efficiently be explained why the electronic and vibrational coherences were or were not observed in a range of recent experiments, in which, depending on the amount of inhomogeneous broadening, the excitonically coupled molecular system can be established. However, the nature of the long lived quantum beats, observed in time-resolved spectra of molecular aggregates, at physiological conditions, is still being debated. On the basis of these results, they[127,128] explained why the electronic or vibrational coherence were or were not observed in a range of recent experiments, caused by a disordered sub-ensemble, for which the dephasing due to static energy disorder becomes significantly reduced.

A number of quantum beats have been observed in spectroscopic measurements of the chlorosome where a set of long-lived modes appear with low frequency dephasing over 1 ps, in Chloroflexus and 1.5–2 ps in Chlorobium.[136] These modes have been ascribed to either interfering ground state vibrations following resonant impulsive Raman scattering or intra/inter-molecular vibronic modes[137,140] because, on the basis of a molecular aggregate's Raman spectra and excitonic structures, it is experimentally more feasible to distinguish coherences evolving in ground or excited states. Discussing coherences in coupled molecular aggregates involving both electronic and nuclear degrees of freedom, Mancal *et al.*[22] concluded that a general distinguishing criterion based on the experimental data alone, cannot be devised, implying that the coherences are purely electronic in nature, because 'vibronic exciton coherences' will oscillate with frequencies, different from those of the vibrational mode. However, in such case, the dephasing time of the coherences could provide crucially important information on the relative contribution of vibrational and electronic degrees of freedom (DOF).

In fact, these coherent beatings have been observed independent of the spectral line shape evolution, different parts of the emission spectrum showing different vibrational modes and phase flips.[136] Hence, the diffusion of excitons through the chlorosome probably does not produce electronic coherence; however, electronic coherences may be enhanced by intensity, borrowing from vibronic coupling.[140, 141] Detailed studies[140, 141] found the vibrational coherences dephasing too quickly while aiding in transferring coherence between coherent domains, given that the time scale of the random energy diffusion process between coherent domains remains 4–5 ps between rods which, however, may aid in intra-domain relaxation.[140, 141] Through observation and measurement of the changes in the line-shapes, the energy transfer in the single coherent domains, has been found, taking place in around 20–30 fs.

Again, while studying the movement of exciton peaks in the spectra (spectral motions), it shows the sign of excitons experiencing correlated changes in different regions of the spectrum, with the frequencies, agreeing with that of excited state vibrations.[142] The correlation have been observed to last for much of the first picosecond of the spectral evolution. Through their interesting observations, they[142] explained this kind of correlations, i.e., the residence of vibrations on individual chlorophylls, comprising each different excitons where local motions on one chromophore affect multiple excitons, also including the global exciton energy. Or, might be, excitons inheriting 'vibrational phase information' from each other, enhance quantum 'beating' lifetimes. Another important point here is that the coupling of vibrational modes in different chromophores, at distant ends of the protein, cannot be a mechanical effect. On the contrary, it might be caused due to the presence of the 'too slow' speed of sound in a protein. So, rather a necessity arises for the electromagnetic interactions to posses a correlating vibrations in distinct chromophores, seperated by protein. Interestingly, synchronized oscillations may stem from 'embedding multiple identical chromophores' with same internal vibrational modes, in close proximity, residing in an organized protein matrix.

But, it should be noted that, at the same time, the response of the chromophores ensures the correlations to be still valid among the spectral motions in the excited state so that, at the end, there appear growing consensus that most observed quantum beat signals, relevant to energy transport, are of a mixed vibronic nature. Vibronic states result from mixing of vibrational states with electronic states in a non-separable manner. Vibronic coherences are likely to be important in energy transport within pigment/protein complexes when the vibrational energy is iso-energetic with the exciton energy gaps. Taken together, there is considerable evidence for *non-trivial quantum effects* within individual light harvesting and photosynthetic proteins. The presence of strongly coupled chromophores leads to exciton states, with broader spectral coverage and possibly coherent dynamics. On the other hand, vibronic coherences may play a role in excitation transfer through the protein's chromophore system, depositing the excitation in some final acceptor or trap state.

3.6 Debatable Role of Vibrionic & Electronic Dynamics in Coherence

Growing evidence points to the impossibility in separating vibrational and electronic dynamics of the system. From Raman studies, in many such cases, the gaps in exciton energy and the energies of vibrations, resident on chlorophylls, are found resonant which fact could result basically, in strong coherent coupling and vibronic states with long coherence times. Many have attempted to tackle the problem of biological relevance and robustness of coherence to perturbations in the FMO system. Interestingly, the strongest quantum beats that have been observed are resilient to alterations in the excitonic structure showing similar frequencies and dephasing times. Both observations of robustness to changes in the vibrational environment and exciton structure suggest that quantum beats may arise from vibronic coherence or at least the strongest vibrational and electronic coherence are energetically resonant. Among them, many of the quantum beats are resonant with the energy gap involving special pair exciton states with similar frequency to that of type II reaction centres.[148] Simulations of a vibronic model are in good agreement with the data. This resonance appears to speed up energy relaxation, energy transport and charge separation, which comes in bursts of population (probability) increase of the charge separated state in simulations.

Fuller and his group reported the observation of coherent dynamics persisting on a picosecond timescale at 77 K in the photosystem II RC, using two-dimensional electronic spectroscopy [2DES]. Supporting simulations suggest the coherences as of a mixed electronic–vibrational (vibrionic) nature which may enhance the rate of charge separation in oxygenic photosynthesis. According to their views,[147] in photosynthesis, when the energy of a collective vibration of two chromphores matches the energy difference between the electronic transitions of these chromophores, a resonance occurs and efficient energy exchange between electronic and vibrational degrees of freedom takes place. Providing that the energy associated to the vibration is higher than the temperature scale, only a discrete unit or quantum of energy is exchanged. Consequently, as energy is transferred from one chromophore to the other, the collective vibration displays properties that have no classical counterpart.

At present, the challenge lies in discovering these vibrational modes in detail, i.e., how nature has managed to select them at the cost of others. In this respect, vibrionic coherences are likely to play contributing role during the energy transportation within the protein/pigment complexes i.e., when the vibrational energy is found to be iso-energetic with the excitonic energy gaps. Taken together, it appears that there exists a nontrivial quantum effect within the individual light harvesting and photosynthetic proteins. The presence of strongly coupled chromophores leads to the exciton states, having broader spectral coverage, and most probably coherent dynamics. The role of vibrionic coherence, in this picture, lies in exciton transfer through the protein's chromophore system, depositing the excitation in some form of final acceptor or trap state. But, in contrast to above scenario, there exists very little evidence for quantum effects being involved in energy transfer between individual proteins.

For the purpose of discussion related to the differences in between two types of coherence, i.e., vibrational and electronic coherence, we firstly refer to vibrational coherence, purely related to the difference in nuclear wave packet motion induced by photoexcitation and is characteristic of resonance in Raman active normal modes in the molecule on either the ground or excited electronic state. Electronic coherence here refers to the coherent excitation of two electronic states coupled by electronic interactions but uncoupled from any vibrational modes in the system. However, assignment of the nature of the modes is critical to a meaningful interpretation of the results. For example, in quantum systems, the type of coherence produced, is dependent on the basic set of states that harbor the coherence. These coherences can be over any state — electronic, vibrational or vibronic. Firstly, coherence in the site/chromophore basis refers to exciton states — the delocalised states that are stationary when evolved by the Hamiltonian. Excitons (site coherences) are ubiquitous in light harvesting systems and are generally produced when strong coupling between chromophores is present and generates splittings in the energy levels.

Secondly, coherence between excitons are states that are superpositions in the energy basis (basis of stationary states — excitons or vibronic states). Signature of such kind of coherences can be observed through quantum beats in spectroscopic experiments as they represent oscillations between two energy basis states (excitons) with different energies. A laser excites the system via its excitons and hence observed features of the system will also be between excitons. Lastly, it can be stated that undergoing coherent processes (when dephasing from the environment is smaller than the interaction energy of the said process) harbor process coherence itself. This is the case where the outcomes are quantum mechanical, producing quantum superpositions and can evolve superposition themselves at the same time. All of these types of coherence can be represented as off-diagonal elements of a density matrix in some basis. Furthermore, if we were to represent the density matrix of the system in the site basis, off diagonal elements would refer coherences between sites (excitons).

In most light harvesting systems, coherent quantum beats have been observed in the cases of coherence phenomena within light-harvesting complexes after short laser-pulse excitation. This outcome has inspired advances in our understanding of light-harvesting optimization where the coherences are found stable over a time scale, commensurating with the relevant energy transfer times. The advancement of ultrafast, femtosecond laser technology, have facilitated the development of spectroscopic techniques for probing the energy transfer pathways through light harvesting, photosynthetic systems as well as that of proteins. In fact, the presence of beats is a signature of underlying coherent phenomena, observed in most of biological light harvesting and photosynthetic systems/proteins with lifetime ranging from tens of femtoseconds through the picoseconds, highlighting the interplay of electronic excitations and vibrations. However, it remains unclear how these vibronic effects change or optimize the function of light-harvesting complexes — in other words, — what is the design principle we could learn?

Hence, the difficulty lies in determining the nature of these coherences and to establish the fact whether they play a substantial role in photosynthesis and/or light harvesting. The realization of how absorption-strength redistribution, induced by vibronic coupling, provides a multiplicative increase in the rate of energy funneling establishes a bio-inspired design principle for optimal light-harvesting systems. For example, Jacob C Dean *et al.* & Scholes's group[151] established from their works related to the study of vibronic enhancement while studying the Algae light harvesting

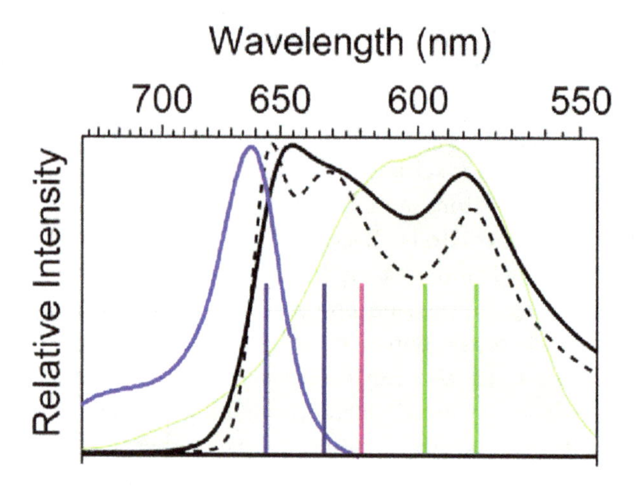

Figure 3.9. Steady-State Spectroscopy, Pigment Architecture, and Active Coherences in PC645 Absorption (black, solid) and fluorescence (red) spectra of PC645 at ambient temperature and absorption spectrum at 77 K (black, dashed) with estimated pigment absorptions shown as sticks. The laser pulse spectrum is also shown (blue, shaded).

that the occurrence of coherence between two remote pigments in a cryptophyte antenna is found to be linked through a vibrational resonance, transiently delocalizing the excitation and modifying energy transfer. This vibronic interaction unveils a molecular design principle probably utilized by cryptophytes to enhance transfer efficiency.

3.6.1 *Debate over the Interpretation of the 2DES Experiments*

There has been debate over the interpretation of the 2DES experiments. Particularly, it has been argued that oscillations owing to vibrational coherences can appear very similar to electronic coherences in 2DES data. In a typical photosynthetic protein, there are several vibrational modes accessible by stimulated Raman scattering. These modes lead to oscillations in pump probe and 2DES measurements at very similar frequencies to those reported to be exciton coherences. This being a significant challenge in the interpretation of the ultrafast measurements, for the purpose of discussion, the above mentioned group distinguished two types of coherence — vibrational and electronic coherence. They referred the vibrational coherence to pure nuclear wave packet motion induced by photo-excitation and as characteristic of resonance Raman active normal modes in the molecule, on either the ground or excited electronic state.

On the other hand, electronic coherence refers to the coherent excitation of two electronic states, coupled by electronic interactions but uncoupled from any vibrational modes in the system. But, in reality, the assignment of the nature of the modes is quite critical to a meaningful interpretation of the results. The enhancements and amplitude patterns, when studied carefully in detail, arrive at the key conclusion for the analysis of 2D coherence maps, stating that the vibrations do not build 'ladders' of energy levels independently on each chromophore's electronic absorption band. Instead, the electronic coupling between chromophores — even though only of the order of 40–80 cm^{-1}, changes the vibrational potentials, leading to distorted vibronic wave functions with amplitude spanning two or more molecules. In order to understand how it affects the rate of energy transfer, i.e., the mechanism and possible role of delocalization, the aforementioned group employed a simple model which accounts for the quantum effects of vibrations, addressing also how surface adiabaticity modifies Förster theory, thereby generalizing it in the context of transition-state theory.

Electron transfer theory of Jortner-Bixon[152] did put specific modification, for the application in electronic energy transfer. Also it provides relevant and valuable insight in this regard, i.e.,the states are not strongly mixed (modest electronic coupling), and the adiabatic correction to Fermi's golden rule (nonadiabatic) rate, k_{NA}, should be quite small, i.e., Förster theory is a reasonable to the extent of first level approximation. The Jortner-Bixon electron-transfer theory[152] is a modification of Marcus theory which treats high-frequency vibrations as additional channels for electron transfer. Each vibrational quantum on the acceptor (m with energy relative to that of the donor,

$$\Delta E_m = E_0 - mh\nu$$

yields an independent reaction pathway with rate k^m. The significance of each of such contributions to the overall rate depends on the Franck-Condon factors between the donor zero-point state and acceptor states, ensuring nuclear overlap at the state crossing. This leads to a set of parallel reaction channels, each of which still governed by the solvent reorganization (λ), driving force, and kT.

In this case, now the energy difference is specified by the vibrational state of the acceptor and the total electronic coupling dispersed among vibrational pathways, i.e.,:

$$k_{NA}^{0m} = \frac{2\pi J^2}{\hbar\sqrt{4\pi\lambda kT}} |\langle 0|m\rangle|^2 \exp\left[-\frac{(\Delta E_m - \lambda)^2}{4\lambda kT}\right] \tag{3.5}$$

Now, the next step is that the non adiabatic rates for channels $0 \to m$ should then be corrected by an associated adiabaticity parameter, and as a result, the electron-transfer rate turns out to be the sum over all high-frequency vibrational states. Scholles and his group applied this theory of electronic energy transfer, especially in the case where Förster theory breaks down, though not in a major scale, but slightly delocalized. Now, in order to sum explicitly over vibronic progressions in the donor de-excitation (m_D) and acceptor excitation (n_A), where $n(m)$ is the vibrational quantum number in the excited (ground) state, and A together with D designate energy acceptor and donor, respectively, we replace the Franck-Condon dressed electronic coupling, $J|\langle 0|m\rangle|$ in with the electronic coupling for energy transfer, dressed with a product of Franck-Condon overlaps for donor de-excitation and acceptor absorption, i.e., $J|\langle n_A|0\rangle\langle 0|m_D\rangle|$. The rate for energy transfer can then be written as

$$k_{EET} = \sum_n \sum_m \frac{k_{NA}^{nm}}{(1 + H_A^{nm})} \tag{3.6}$$

$$k_{NA}^{nm} = \frac{2\pi J^2}{\hbar\sqrt{4\pi\lambda kT}} |\langle n_A|O_A\rangle\langle O_D|m_D\rangle|^2 \exp\left[-\frac{(\Delta E_{nm} - \lambda)^2}{4\lambda kT}\right] \tag{3.7}$$

ΔE_{nm} depicting the energy difference between the specified donor fluorescence band and acceptor absorption band.

The adiabaticity parameter, H_A, here, gives the measure of the functional role for electronic delocalization at the cusp region of the two excited-state potentials in deciding the rate of energy transfer. This also gives insight into the physical meaning of coherence effects in energy transfer, which, basically is a coordinate-dependent delocalization, defined as follows:

$$H_A^{nm} = \frac{4\pi J^2 \tau_{vib}}{\hbar\lambda_{vib}} |\langle n_A|0_A\rangle\langle 0_D|m_D\rangle|^2 \tag{3.8}$$

Here, i.e., in the energy transfer case, the solvent relaxation time appearing in the adiabaticity parameter for electron transfer reaction has been replaced by the vibrational period, τ_{vib} because, here, in our case, the "reactive", coordinate is associated, specifically, with a vibrational mode. Likewise, λ denotes the reorganization energy for the vibrational mode under consideration (λ_{vib}). Next is the adiabaticity parameter which is analogous to the transmission coefficient k, meant for an single incidence, cross into the lower acceptor state from the donor state. It is valid for one period of vibration only, when considered in the context of transition-state theory. Scholles and his group calculated the vibronic coupling and surface characterization, supported by the analysis of the 2D coherence map which enabled them to hypothesize how vibronic mixing influences the rate of downhill energy transfer in PC645.

The effect turned out to be a robust manifestation of state mixing which is apparent in comparison with the Förster energy-transfer model. The operative term in the Förster rate expression is the spectral overlap between donor fluorescence and acceptor absorption spectra. In fact, spectral overlap ensures resonance, or energy conservation during the energy transfer process while inherently incorporating environmental fluctuations borne out in the spectral line shapes. The sensitivity of this effect depends on the electronic coupling and the vibronic enhancement will evidently be robust to line broadening. The above mentioned group evidently found the vibronic enhancement to be robust to line broadening, but, however, to be realized less by the system upon averaging over disorder. But, on the contrary, in the Förster limit of very weak coupling, a delta function at resonance is approached, and the vibronic modification is essentially never realized by the system.

In fact, for some light-harvesting complexes containing chlorophyll or bacteriochlorophyll pigments, this kind of disorder has been found, at ambient temperature, to wash out any significant vibronic improvement to the rate otherwise realized at lower temperatures.[152,153] In contrast, in the case of phycobiliproteins, for example, PC645, though small, a realizable rate enhancement acts as a means of directing energy flow across the large energy gaps separating chemically different chromophores within the protein which can be stated as an important facet of the energy-transfer mechanism in terminating the excitation at the protein periphery for subsequent transfer to membrane-bound acceptor complexes. For these proteins, such action enables a broad spectral cross-section for light capture while maintaining high energy transfer efficiencies. Indeed, Kolli *et al.*[183] have explored the energy propagation through an analogous phycobiliprotein, PE545, and have predicted a much larger initial spatial distribution of the excitation out of the central donor as a result of vibrationally activated transport to acceptor pigments, widely separated in energy. They argued that, in the presence of strong coupling between electronic excitations and quantized vibrations, a concrete and important advantage of quantum coherent dynamics is precisely to tune resonances that promote fast and effective energy distribution.

To address the vibronic effects further, Sholles and his collaborator compared the Förster limit to the energy transfer rate, calculated according to the modified Jortner-Bixon theory and considered only the $n_D = 0/m_D = 0/m_A = 0/n_A = 1$ spectral-overlap contribution to the rate, realising that all other terms are equivalent. However, two scenarios can be envisaged where the ratio has been calculated in order to obtain simple form for the vibronic "enhancement" factor compared to that from the Förster rate:

$$\frac{k_{vibronic}}{k_{Förster}} = \frac{J^2|\langle n'_A|O_A\rangle\langle O'_D|m_D\rangle|^2}{J^2|\langle n_A|0_A\rangle\langle 0_D|m_D\rangle|^2}\frac{1+H_A}{1+H'_A} \tag{3.9}$$

Primes, hereby denote the vibronic states. The calculated enhancement thus can be conveniently used to scale the Förster rate to avoid unnecessary approximations regarding the bath. Here, the rate of energy transfer, can in principle be enhanced by a factor of 3.5, solely, as a consequence of vibronic intensity redistribution. This, they concluded as a consequence of vibronic intensity redistribution caused by electronic coupling between the donor and acceptor. But, interestingly, this is over and above the energy-resonance enhancement already captured by the Förster spectral overlap and it is quite unanticipated because the total oscillator strength between states is conserved. The vibronic mixing redistributes intensity to the spectral bands, thus causing the intensity in being conserved. But, it should be noted that the multiplication of the renormalized bands in the spectral overlap evaluation produces the striking, unanticipated increase in the rate of energy transfer. The sensitivity of such an effect depends on the electronic coupling which, in the Förster limit, i.e., in the case of very weak coupling, a delta function at resonance is approached. As a result, practically, the vibronic modification is essentially never realized by the system.

Turner *et al.*[24] took a systematic approach to identify and differentiate signatures of different types of coherences present in their 2DES data. In their approach, one distinguishing feature has been pointed out which relied on the ability of 2DES in manipulating the time ordering of laser pulses to extract information from a sample. Though, this is like NMR experiments where different pulse sequences can be used to identify different contributions to the signal, however, a significant difference between NMR and optical two-dimensional spectroscopy exists. Here, in the optical excitation beams and signal beams in the optical case, all have well-defined directions. Here, they distinguished types of coherences in the three- versus four-level system. in which three-level system was modeled using a ground state, $\langle g|$ and two electronic excited states, α and β assumed to be excitonic states of the system (e.g. model dimer). A coherent superposition between these two electronic states, signaled by the off-diagonal density elements $\langle e_\alpha\rangle\langle e_\beta|$, was considered to give rise to signatures associated with purely electronic coherence between electronic eigenstates. The four-level system was assumed to be the representative of a single chromophore (two-level system) with an additional vibrational level, $|1\rangle$, in both the ground and excited states. In this case, vibrational coherence was possible as either a ground-state vibrational wave packet, $|g_0\rangle\langle g_1|$ or as an excited state vibrational wave packet, $|e_0\rangle\langle e_1|$.

According to Fassioli and his group's view, this comparison of the expected dynamics of a three-and a four-level system provides quite an effectively good test case for the simplest model of electronic or vibrational coherence. However, they emphasized that,[92] in reality, this procedure for discriminating between electronic and vibrational coherences is validated only in the strong electronic coupling limit. Thus, for a given spatial geometry, different pulse sequences can be chosen simply by alternating the time ordering of the pulses in the different beams. In one pulse sequence, a photon echo sequence, when achieved, referred to as the 'rephasing' spectrum, then by reversing the time ordering of the first and second pulses, the resulting spectrum is referred to as a 'non-rephasing' spectrum. The evolution of the cross peaks, appearing in the rephasing and non-rephasing spectra were predicted to behave differently for a three- and a four-level system. This happens due to the presence of an additional vibrational level in the excited state of the four-level system, opening several additional pathways, capable of creating a coherence on the ground vibrational level. Even though the interpretations have been still mired with controversies, when chromophore complexes form excitons, delocalised across the chromophore array, beats often arise from the superpositons of excitons states.

Importantly, in such case, the beat frequencies and decoherence times will be at per with that of vibrational and vibrionic states. Hence it does not seem likely that vibrational coherences are to play key roles in light harvesting. But, importantly, it appears that they might have a effective role in the function of reaction centers (RC) and bacteriorthodopsin. Another important point to be mentioned here is that not only the time scales of coherence and that of energy transfer do not appear separable but also not are the energy scales of couplings and environmental effects, in most of the light harvesting systems. Strictly speaking, energy transfer lies in an intermediate region which is neither strictly classical nor purely a quantum mechanical, thus still remaining with some degree of unresolved issues, i.e., whether it enhances or diminishes transfer. However, recent theoretical studies indicates the energy transfer occurring in this regime to be enhanced beyond that of classical hopping and that of purely quantum transport.[143, 144] Thus, **the present growing consensus is that most of observed quantum beat signals, relevant to energy transport, are of mixed vibrionic nature**. Strictly speaking, energy transfer lies in an intermediate region which is neither strictly classical nor purely a quantum mechanical and this still remains, with some degree, as an unresolved issues whether it enhances or diminishes transfer.

As stated earlier, the recent theoretical studies indicates the energy transfer occurring in this regime having enhancement beyond that of classical hopping and purely quantum transport,[143, 144] leading to the growing consensus that most of observed quantum beat signals, relevant to energy transport, are of mixed vibrionic nature. Not only that the long-lived vibrational coherences observed in photosynthetic light-harvesting systems can dramatically affect the efficiency by which reaction centers (RC) trap the excitations of the surrounding 200 chlorophylls. Next, the challenge lies in discovering these vibrational modes and how and why nature has managed to select them and suppress others. Here, the vibrionic coherences are likely to play contributing role during the energy transportation within the protein/pigment complexes i.e., when the vibrational energy is found to be iso-energetic with the excitonic energy gaps. Taken together, it appears that there exists a 'nontrivial quantum effect' within the individual light harvesting and photosynthetic proteins. Added to this fact, the presence of strongly coupled chromophores leads to the exciton states, having broader spectral coverage, and most probably coherent dynamics. "*The role of vibrionic coherence, in this picture, probably lies in exciton transfer through the protein's chromophore system, depositing the excitation in some form of final acceptor or trap state. But, in contrast to above scenario, there exists very little evidence for quantum effects being involved in energy transfer between individual proteins*".

The next crucial aspect to be mentioned is: it has been observed that the oscillations owing to ground-state vibrational wave packet motion has been observed to be insensitive to the choice of pulse sequence. However, for the three-level system, there are a limited number of excitation pathways, creating a coherence in the excited states and the observation of oscillations has been observed to be sensitive to the pulse sequence. Importantly, they will only occur in rephasing spectrum whenever the origin of the oscillations in the cross peak is from a three-level system. By contrast, oscillations owing to stimulated Raman scattering will be observed in both rephasing and non-rephasing spectra[118] [and references there in]. But, for discerning this signature, high-quality data is necessary as the non-rephasing signal is often significantly weaker than the rephasing signal.

For example, in the measured 2DES spectra for PC645, the oscillations of a cross peak observed below the diagonal were monitored. The oscillations were fit to a sum of eight damped cosine functions, identifying several oscillating components that contribute to the overall signal. As a next, the rephasing and non-rephasing contributions were separated and examined independently where components showed clear signs of oscillations Even with similarity in general, using PC645 further, and a related two-colour technique,[152] taking the Fourier transform of the separate components, indicated two modes, very close in frequency: 21 THz (700 cm^{-1}) and 26 THz (870 cm^{-1}), where, one of those frequency components was found clearly visible in the rephasing trace but absent in the non-rephasing trace in a related two-colour technique.[152]

Given the challenges of distinguishing coherences between electron oscillations, many experiments have been recently proposed[155–158] which compared the spectrally integrated pump-probe response of a vibrational coherence in a system with a single excited electronic state to that of a coupled electronic dimer in the limit of an infinite bandwidth pump — infinite bandwidth probe experiment. In ultrafast pump-probe spectroscopy, excitation of a two-level electronic system with accessible vibrational levels on the ground and excited states, creates a vibrational wave packet which oscillates on either the ground or excited state manifold. For a single probe-pulse wavelength, the wave packet oscillations increase or decrease the probability of absorption or emission leading to oscillations in the pump-probe spectrum.[156–158] Through a careful analysis of the two cases, i.e., vibrational and excitonic coherence, Fassioli-Scholles-group[92] demonstrated that as the bandwidth of both pump and probe pulses are increased in a spectrally integrated pump-probe experiment, the oscillations due to a vibrational coherence should decay to zero, linearly with the pump and probe pulse temporal durations, whereas the excitonic coherence should have residual oscillations. In principle, this technique can provide a relatively simple alternative approach i order to identify different modes, present in the ultrafast spectroscopic measurements.

3.6.2 *Quantum Coherence: Role of Reaction Center (RC)*

While different species of plants, algae and bacteria have evolved a variety of different mechanisms to harvest light energy, all of them share a common feature, known as a photosynthetic reaction center (RC). This play quite a vital and an essential role in the process of photosynthesis. The harvesting, i.e., the conversion of short-lived excitation energy, resulting from photon absorption, to a form that can be more leisurely used by a living cell, namely that of a charge gradient, takes place in this protein complex i.e., reaction center (RC), the function of which is to receive excitation energy, either by absorbing a solar photon directly, or by excitation transfer from the pigment molecules of nearby light harvesting complexes, then converting the excitation into a charge-separated state. In recent decades, the architecture of PSUs and their constituent proteins has been determined for several species, reflecting collectively the aforementioned structural motifs based on Förster theory. For example, in anoxygenic purple bacteria, the PSU contains the peripheral light harvesting complex II (LH2), excitonically coupled to a RC-light harvesting complex I (RCLH1) core complex, sometimes augmented by additional satellite complexes, expressed, depending on physiological conditions. Thus, during the early steps of photosynthesis, light-harvesting complexes absorb and transfer solar excitation energy to the reaction centre (RC) which, being stored in the form of an exciton, needs efficient harvesting of light energy this way, causing transfer of excitons to RCs well within the decay time of excitons (around a nanosecond), typically within tens of picoseconds. Next, in the RC, the excitation energy is converted into a trans-membrane electrochemical potential with near-unity quantum efficiency, causing almost every absorbed photon being converted into a charge-separated state. Despite the present knowledge about the pathways and timescales of charge separation, the precise mechanism, responsible for the high efficiency of this process, however, has still been under detail research.

In the 1990s, the remarkable direct visualization of coherent nuclear motion in the excited state of the primary electron donor in the bacterial RC[84] — a simpler version of the photosystem II reaction centre (PSII RC)), suggested a functional role for these motions, also, in the primary electron-transfer reaction, a view later supported by theoretical models.[159]

However, at that time, no coherence could be observed between the reactant and the product of that reaction. More recently, the observation of long-lived coherence in photosynthetic complexes (light-harvesting)[160–162] and oxidized bacterial RCs, unable to perform charge separation[160] has triggered an **intense debate** on the role of quantum coherence in promoting the efficiency of photosynthesis. As a result, no clear correlation between coherence and efficiency of energy or electron transfer has been presented so far. However, in this context, it could be mentioned that quantum coherence between the electronic states involved in energy or electron transfer has been found to introduce correlations between the wave functions of these states,[164] enabling the excitation to move rapidly and to coherently sample multiple pathways in space.[165]

3.6.3 *Quantum Coherence: Role of Pigments*

The need for rapid migration of energy among pigments, as governed by Förster's formalism, imposes constraints on inter-pigment separation as well as on the pigment energy levels. Additionally, nature appears to adapt actively utilizing the thermal disorder which facilitates efficient excitation transfer between pigments by maintaining a broad spectral resonance, a key factor, needed for the explanation of Förster's formulation.[160] The cooperating pigments display a hierarchical pattern of tight packing, exhibiting, as a result, a system of strong and weak electronic interactions, essential for efficient light harvesting. Within the most strongly interacting groups of pigments, electronic excitation is spread coherently following light absorption.[161] However, between weakly coupled pigment groups, electronic excitation is shared incoherently, namely through random excitation transfer.[160,166] This kind of coherent spread is known as exciton dynamics[167] and the incoherent spread as Förster resonant energy transfer (FRET).[161,167] Photosynthetic light harvesting interweaves both these behaviors. Additionally, some pigments fall into an intermediate coupling regime,[236] having a small amount of coherent spread of electronic excitation, though not well understood how much this influences the efficiency of light harvesting in purple bacteria. In purple bacteria, it has been examined how hundreds of pigment molecules, cooperate through quantum coherence to achieve such a remarkable light harvesting efficiency. This way, quantum coherent sharing of excitation modifies excited state energy levels, combining transition dipole moments, thus enabling rapid transfer of excitation over large distances. Here, co-operating pigments display a hierarchical pattern of tight packing and, as a result, exhibit both the systems of strong and weak electronic interactions, essential for efficient light harvesting. This way, within the most strongly interacting groups of pigments, electronic excitation is spread coherently, following light absorption. But, being weakly coupled between pigment groups, it has been observed to be shared incoherently, namely, through random excitation transfer.[161a,166]

At the present stage of developments, in all the phenomena connected to the functioning, working in the photosynthesis, hundreds of pigment molecules, present in purple bacteria cooperate through quantum coherence to achieve this remarkable efficiency in light harvesting. Quantum coherent sharing of excitation modifies excited state energy levels and combines transition dipole moments, enabling rapid transfer of excitation over large distances. Especially, in the case of purple bacteria, it exploits the resulting excitation transfer to engage many antenna proteins during light harvesting, thereby increasing the rate of photon absorption and corresponding energy conversion. Thus, the quantum coherence effect may render the process of energy and electron transfer, less sensitive to the intrinsic disorder of pigment-protein complexes, allowing these systems to successfully reach their final state, avoiding energy losses. With such understanding, the group of Romero and Van Grondelle,[135] followed by other investigations, addressed two key questions:

(1) Is electronic coherence present in the PSII RC?
(2) And, if it is so, in that case, does coherence promote the PSII RC charge-separation efficiency?

With the aim of gaining a plausible answer of these questions, they applied 2DES to the fully functional PSII RC. As we are aware by now, in photosynthetic light harvesting, quantum coherence can have different meanings depending on different circumstances. Quantum coherence can refer to quantum superpositions of localized molecular excitations occurring naturally because electronic couplings between molecular excitations lead to delocalized eigenstates (a.k.a. excitons). Sometimes, coherence in light harvesting, alludes to the coherent wave-like dynamics of energy transfer, which actually reflects the superposition of excitonic eigenstates. In the former case, the coherence is represented in the molecular site basis, whereas in the latter case, the coherence is represented in the delocalized exciton basis. But, both the site basis and the exciton basis are special because they are related to the spatial arrangement of chromophores and energy eigenstates of the Hamiltonian, respectively. But as both types of coherence effects play important roles in photosynthetic light harvesting, they must be discussed separately.

Thus, quantum coherence, manifested in the delocalized eigenstates of photoexcitation in photosynthetic complexes, plays a fundamental role in spectral properties, energy tuning, and energy transfer dynamics of photosynthetic light harvesting as well. However, the effects of excitonic coherence are more difficult to analyze and have been the subject of intensive research in the past few years. Spectroscopic measurements (termed 2DES) on the FMO complex at liquid-nitrogen[4] (77 K) and room temperatures[7,8] have shown time-dependent oscillations, presumed to be quantum beating, in the amplitudes of the spectral signals which matches the predictions of quantum theory. Inspired by these observations, it has been proposed that an intricate interplay of quantum coherent excitation transfer and environmental dephasing help in increasing the efficiency of the energy transport process.[27,36–39]

3.6.4 *Quantum Coherence: Role of Photosynthetic Excitons*

Biophysicists mapped the spectral signatures of the chlorophylls onto a structural model, working out the pathways and timescales of energy transfer among the chlorophyll, the crucial player in the process of photosynthesis. The outcome of such queries produced their spectroscopic signatures, often indicating coherently shared excitation over two or three chlorophylls, or in other words, shared delocalized excitation, known as an 'exciton'. Excitons can extend over multiple chromophores[92] having a profound impact on the electronic structure and optical properties of the system as well as on energy transfer dynamics. Clear examples of excitons have been discovered in ensemble-averaged linear and nonlinear spectroscopy experiments[92] including single-molecule spectroscopy of LH2 from purple photosynthetic bacteria.[92] Importantly, latter have provided strong evidence for delocalized excited states immediately after the excitation of a single LH2, whose implications on energy transfer can be twofold:- groups of strongly interacting chromophores can behave as effective donors and acceptors of excitation while transferring remains incoherent which aspect can be successfully explained by generalized F(ö)rster theory,[92] describing the incoherent transfer of energy between 'delocalized' electronic states.

Now, the vital role of exciton, related to the processes in photosynthesis, importantly, lead to further subtlety arising from quantum systems. This contains multiple, strongly coupled components due to which, the presence of the Hamiltonian and that of electronic orbital are both needed to be considered, as electron arising at an instant, is immediately spread over many states of position. The corresponding Hamiltonian, contains coupling terms relating in between position states (i.e., Hamiltonian has off diagonal elements). As a result, within a collection of chromophores, there exists an electronic coupling between these two position states, resulting in a new set of delocalised states, their probability lying on distinct, specific chromophore sites and not across the case of electronic orbit. In fact, the excitons are the natural states of the system in an energy basis as they do not evolve into a different state over time. Thus, being confluent with the energy of the state, the natural excitons, are able to posses this new set of states, i.e., to become the system's natural excitons, generated by excitation from a discrete packet of energy.

Again, importantly, if a system has uncoupled elements (i.e., chromophores), then the excitons for these elements will just be excitations of specific (uncoupled) chromophore sites, not being superpositions among them. Interestingly, in an energy basis, these excitons remain as the natural states of the system because they will not evolve into a different state over time. Broadly speaking, these excitations, referred to as excitons, are part of a broader class of objects called **'quasiparticle' but behave as if real particle**. This may sound like an odd and unnecessary distinction to make, but an important one to be noticed. Mathematically, it can be established that without a decohering environment, the Hamiltonian, considered for such kind of the description of the states, will act as mixer of the density matrix, i.e., mix them and then return exactly the same density matrix as an output (meaning that they are stationary in time and, hence not perturbed by the energy of interactions). Importantly, *all of these different possible exciton states give rise to the energy spectrum of a quantum system.*

Another quite important characteristics is to be noted here, i.e., the excitation is transferred between the excitons via Coulombic coupling whereby excitons hop between coherent clusters of chromophores (the excitons) in an incoherent fashion which is the essence of both supertransfer and generalized Förster theory. Hence, the quantum object moving between chromophores, created by the absorption of a photon is the excitation itself, transferred between chromophores. Initial absorption of sunlight in peripheral pigments, creates these excitons, i.e., electronic excitations, spread over several neighboring pigments, exhibiting strong quantum mechanical coherence. Thus, building on the work of Davydov,[169] the exciton model has been developed in order to describe the interaction of excitations in other molecular systems, particularly, in the interaction of chromophores in photosynthetic antennae.

Then, for both of the solid-state and molecular systems, one of the major differences in between the understanding of excitons is the extent of delocalization of the excitation. Photosynthetic excitons are fairly localized whereas the understanding of excitons is, in contrary, as far as the solid state is concerned, related to the typical delocalization length in the steady state, i.e., only being two to four chromophores (with notable exception in case of chlorosome antenna). Even then, in many cases, the excited states can be approximated to be fully localized to one molecule.[92] But, sometimes, caveats related to these facts, also could be observed, i.e., instead of considering the transfer and coherence between excited states of each chromophore, one should treat the transfer and coherence primarily between excitons (the number of exciton states in the system is equal to the number of chromophores that generate them). Thus, in a set of chromophore sites, possibility for the existence of an overlapping in between different excitons, is quite strong, also causing their population, transferred, i.e., funneled downhill with respect to energy. This fact implies that the efficiency of incoherent exciton dynamics is much more than chromophore to chromophore transference.

Importantly, Quantum coherence manifests itself, also in exciton states of BChl clusters that bunch up transition dipole moments of individual BChls together, *doing additional quantum coherence shifting of energy levels* and improving resonance (spectral overlap) between BChl clusters. In a system, quantum coherent sharing of excitation enables the modification of excited state energy levels combines transition dipole moments performing also the rapid transfer of excitation over large distances. However until recently, the answer to the queries related to survival of electronic coherence in these systems, firstly at physiological temperatures, and secondly, for long enough to influence the energy transfer process, had remained elusive. Remarkably, recent studies are providing evidence that indeed some of these fascinating molecular aggregates are 'typically designed' to sustain quantum coherent transfer for longer than expected time. Not only that, they can do so at temperatures as high as in biological conditions. A direct implication suggested by these studies is that, might be, quantum coherence affects light-harvesting processes in vivo,[119] but still, its exact role is yet to be understood.

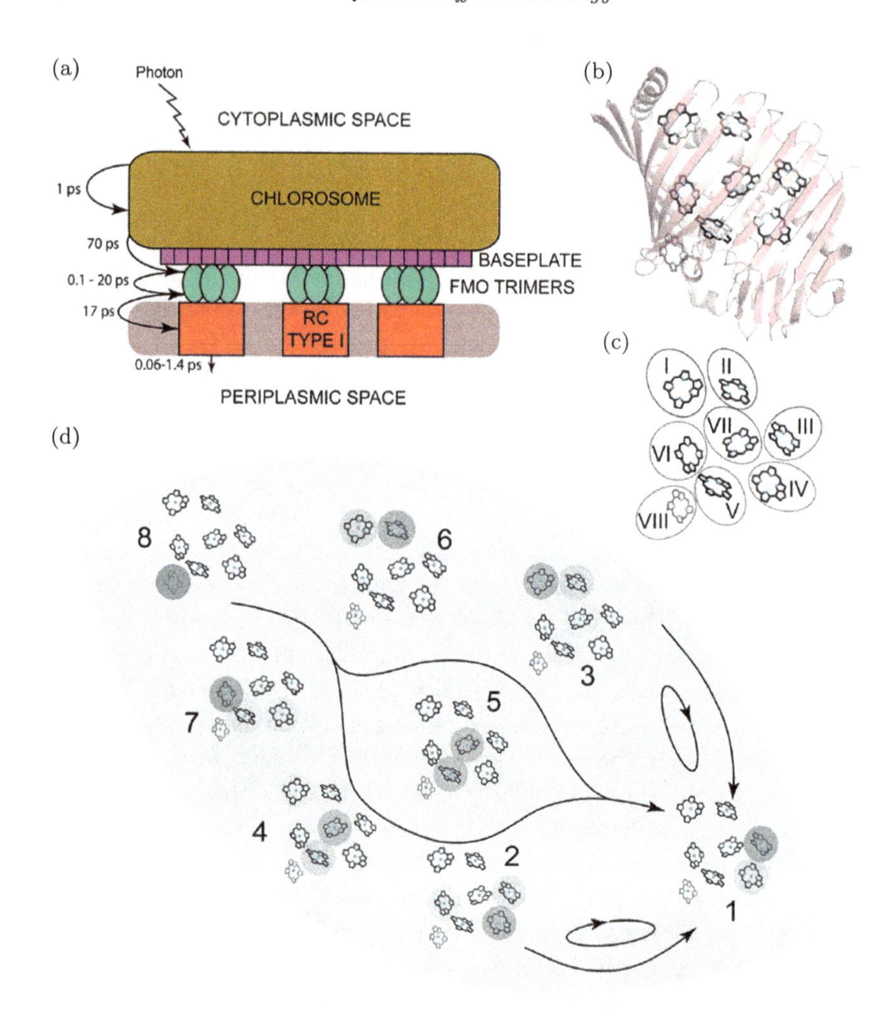

Figure 3.10. (a) Morphology of the green sulphur bacteria light harvesting apparatus. In green non-sulphur bacteria, the RC type I is replaced with RC type II and the FMO complex is missing. Transfer and charge separation times are given. (b) Structure of the FMO monomer (PDB 3eni). (c) Chlorophyll arrangement and chromophore labels for the FMO monomer. (d) Exciton states and energy flow downhill between them from state 8 to state 1. Acutely studied transfer rates and coherence times are labelled as arrows and ellipses respectively. Exciton states were calculated by finding eigenvectors of the FMO Hamiltonian (Adolphs and Renger 2006) inMathematica. Exciton states are calculated from the Hamiltonian given by. Chlorophyll tails are removed for clarity. Structures were rendered in PyMol.

However, the 'practical functionality' of quantum coherent phenomena may be subtler than efficiency maximization. For instance, the structure and arrangement of antenna complexes are not solely optimized for fast transfer but also for the incorporation of the ability of modulating their function under different environmental conditions.[170] Now, question may arise how electronic coherence with the help of possible coherent-evolution may help the requirement of having built-in mechanisms, with the purpose of modulating transfer efficiencies under different environmental parameters. Indeed, light-harvesting systems, most possibly, are able to take full advantage of unique quantum interference effects that include the possibility of efficiency regulation, according to initial state conditions[170,171] or through accumulative quantum phases in closed transfer pathways. An outstanding point of recently observed experimental results is that the observed electronic coherence expands several molecules across the whole antenna complex, including distant and weakly interacting pigments. This has motivated recent studies to investigate electronic energy transfer from the perspective of quantum entanglement. Importantly, *an intriguing question arises, i.e., whether correlations or interferences between distant pigments play a crucial role in the efficient functioning of a light-harvesting complex or they are just a consequence of exciton delocalization without having any kind of particular relation to transfer of the related efficiency.*

3.6.5 *Coherence: Role of Chromophores*

Light-harvesting complexes are comprised of light-absorbing molecules, chromophores, typically attached to a protein structure that holds them in place. The chromatophore, an exquisite quantum-biological device, rivals in its simplicity, efficiency and functionality of man-made solar energy devices. The present design, reached through biological evolution, packs a maximum number of BChls into the available membrane space, but, interestingly, leaving space in the membrane for key chemical reactions requiring diffusion of molecules, in particular, of quinones. Basically, photosynthesis is initiated by the absorption of light, only by the chromophores, which excites the molecules from the ground state to an electronic excited state. The highly efficient transfer mechanism takes place with the help of the electronic excitation through the LHC to the reaction center (RC) with almost unity quantum yield. This excitation is then passed to a reaction center (RC), being converted into useful chemical energy. Wide variety of light harvesting antenna structures can be found in nature, differing only in the arrangements, i.e., type of chromophores, (e.g. chlorophyll, bilins and carotenoids). Here, LHC complexes can be expressed as arrangements of pigments (most importantly chlorophylls) and protein molecules, functioning as a light gathering 'antenna' to absorb photons and become electronically excited.

Importantly, this protein system being extremely modular, following self-organized process, achieves its assembly having spherical shape and proper stoichiometry of RC, LH1, LH2. LHC complexes are arrangements of pigments (most importantly chlorophylls) and protein molecules functioning as a light gathering 'antenna', absorbing photons and become electronically excited. In chromatophore, the RC BChls acts as the ultimate recipients for the light energy absorbed which either absorb light directly or receive excitation energy from the LH1 BChls. The LH1 BChls, being most proximate to the RC BChls, furnish excitation energy to the RC BChls, leading to either direct absorption of sunlight or receive excitation energy from neighboring LH1 BChls or LH2 BChls. LH2 BChls, being the main entry route for light energy into the chromatophore, contribute most to the BChl pool, outweighing RC and LH1 BChls by a factor of about four, the exact value depending on light conditions in the growth medium. These are positioned within individual pigment-binding polypeptides, relatively constant compared to those calculated for the entire protein complex. Fluctuations in chromophore geometry arising from the dynamical motions of their polypeptide ligands, have been found to contribute relatively little to variation in the efficiency of energy transfer and calculated quantum yield, as determined with kinetic models. Compared to this, the PSII antenna system is robust and delivers virtually the same high efficiency despite comparatively large fluctuations of energy transfer rates from linker Chls to the core.

Nevertheless, all antennae complexes are able to convert the photogenerated excitations to charge separation with very high efficiency.[10] Quantum efficiency i.e., the probability of converting an absorbed photon into a charge separated state, is dependent on antenna size, light conditions as well as the organism, being documented to be in the range of 50–90%. For instance, the light harvesting to charge separation efficiency is in the range of 84–90% for photosystem II of higher plants. Besides that, the system is extremely stable, functionally against unavoidable radiation damage of its components and because of its modular form, repair or reuse of components being straightforward.

The excited state of a molecule such as chlorophyll is short lived when compared with usual biological processes. Experimentally, it has been observed that the process of relaxation to the ground state occurs, when, after about 6 ns, get dissolved in de-oxygenated diethyl ether at ambient temperature and about 4 ns, in vivo. The singlet excited state lifetime of chlorophyll is reduced compared with the radiative lifetime, largely caused by the inter - system crossing, yielding triplet excited states of chlorophyll. This process occurs with around a 10 ns time constant.[92] Before the molecule scan, relax, the electronic excitation must be 'harvested'. Specifically this means that, the excitation is transferred through space among the chromophores until it eventually reaches a RC, initiating charge separation. All of these aspects of the biological function of chromatophores need to have an in — depth study, mainly on the light absorption characteristics and mode of functionality involved, this being the very first as well as crucial step of light-harvesting.

3.6.6 *Exciton: What Happens in LH2?*

An important common feature shared by most of the photosynthetic organisms is that they capture photons in the form of excitons, typically delocalized over a few to tens of pigment molecules, embedded in protein environments of light harvesting complexes (LHCs). Spectroscopic signatures often indicate coherently shared excitation over two or three chlorophylls, thus "delocalized" excitation is known as an exciton. Many experiments and calculations have shown how excitons can substantially change the electronic structure of an antenna as well as that of energy transfer dynamics. Delocalized excitons created in such LHCs remain well protected despite being swayed by environmental fluctuations and delivered successfully to their destinations over hundred nanometer length scale distances, in about hundred picosecond time scales.

The LH2 complex, isolated from purple bacteria, is one of the most widely studied light-harvesting complexes providing clear evidence for delocalized excited states, containing only one type of chromophore, but comprising eight or nine subunits, circularly arranged, binding three bacteriochlorophylls, each giving rise to two units. Many experiments and calculations have shown how excitons can substantially change the electronic structure of an antenna as well as that of energy transfer dynamics. The strength of electronic couplings and the associated level of exciton delocalization within the B800 and B850 rings can clearly be seen in the single-molecule fluorescence-excitation spectra of LH2 complexes which reveals a fundamentally different structure in the absorption bands around 800 and 850 nm. This is consistent with absorption of the B800 ring occuring at 800nm, corresponding to the absorption of isolated bacteriochlorophyll-a molecules, while the B850 ring is red-shifted to 850 nm.

This shift is in part, a response to interactions between the chromophores and the protein; however, a significant contribution is owing to the strong coupling between molecules, which shifts the absorption resonance frequency to the new eigen energy of the exciton,[175] populating localized excited states of the different bacteriochlorophyll while absorption of B850 ring populates exciton states, delocalized over the entire ring.[172–174] However, this delocalization, in addition to changing the electronic landscape of the excitons in LH2, also takes the responsibility in changing the energy transfer dynamics by setting up a gradient, allowing excitation to flow downhill from the high-energy B800 ring to the lower energy B850 ring.

Importantly, in presence of all higher order plants, the PSI complex has been found to include sections of strongly coupled chlorophyll molecules.[155] Even if the majority of dynamics in PSI do not rely on strongly coupled chromophores, it has been speculated that the presence of excitons may affect some of the more subtler steps, i.e., the steps involving trapping of the excitation at RC. For example, strongly coupled chromophores are found in the light-harvesting machinery of some cryptophyte algae.[176] Due to the presence of high symmetry of the LH2 complex, the minimum number of independent parameters needed to fit its line-shape, is relatively small compared to the total number of pigment molecules involved. Indeed, the average values of excitation energies of α, β&γ - BChls, i.e., nearest neighbor electronic couplings, and the corresponding magnitudes of the disorder for each of the B800 and B850 units, seem sufficient for fairly accurate fitting of ensemble line shape. But, in fact, easiness of fitting the ensemble line-shape has been found to be a contributing factor for the 'lack of clear consensus', related to some details of the exciton Hamiltonian.

As has been found in FMO complex, real time coherent "beating signal" was observed in 2DES data of LH2 complex. Here, angle resolved coherent four wave mixing spectroscopy has produced the evidence for coherent dynamics that can be well separated from the relaxation signal due to energy transfer 2DES has also been used to determine parameters of the Hamiltonian for LH3 complex[16] under low light condition, quite similar to LH2. Clear peaks corresponding to exciton transfer from the B800 unit to the B850 unit were shown to emerge even at about 200 fs after excitation.[178]

Recently, newly developed 2DES spectroscopy have suggested new aspects detailing the possible role of dark states, present in carotenoids.[179] The process involved in exciton migrations, of LH2 complexes in aggregate, has important implication for understanding the design principle of efficient energy conversion, for which, detailed information on the arrangement of LH2 complexes is needed. AFM images have revealed various patterns of aggregates depending on light conditions.[180] Photon echo spectroscopy[6] of both PE545 and PC645 at 294K reported oscillatory cross peaks in the two dimensional representation, lasting more than 400 fs, initially interpreted as originating from electronic coherence. Further interrogation of the sources and corresponding implications of these signals have continued, and ensuing studies[24,181] clarified that most of them have vibrational origin, confirming the earlier suggestion by Doust *et al.* In fact, this did not appear surprising, considering the covalent nature of the pigment-protein bonding and rather strong vibronic coupling.

3.6.7 *Delocalized States of Exciton: Electronic Excitation of Chromophores & Vibrational Degrees of Freedom of the Environment*

Traditionally, it has been assumed that transfer of electronic excitation energy in light-harvesting systems happen in the Förster limit of RET where electronic excitations incoherently 'hop' between states. Because, in such condition, these states are localized on individual chromophores. Originally, the energy transfer phenomena had been considered in this way with the justification that in such kind of energy transfer, the electronic coherence between chromophores, i.e., quantum superpositions of excited electronic states of different chromophores, get rapidly destroyed by the environment of the antenna molecules;specifically, due to the presence of stochastic energy fluctuations, originating from the interaction between electronic excitations of the chromophores and vibrational degrees of freedom of the environment which are mainly caused from fluctuations in the protein and solvent surrounding the light-absorbing molecules. In such case, the physical fluctuations cause electrostatic fluctuations, eventually, becomes responsible for the modulation of the transition energy of the embedded chromophores.[45]

This, being the subsistence of the electronic coherence, is dependent on the strength of the electronic coupling between chromophores which can become significant in light harvesting antennae due to dense packs of chromophores, thereby enhancing the absorption cross section of the antenna. As a result, such an enhancement, i.e., stable 'excitons' become commonplace in light-harvesting antenna, raising the validity of Förster's theory in general, together with many of recent evidences indicating the possibility of survival of electronic coherence in the biological environment, even for weaker electronic coupling than previously thought. In order to get an appropriate microscopic descriptions for RET mechanisms going beyond Förster theory, applications of the RET mechanism in light harvesting[185] have been found necessary. Thus, out of convergence of interests on these topics, obtained from different fields of the influence of quantum optics research, these deviations from the Förster's incoherent picture of energy transfer, are now often referred to as *non-trivial quantum effects in photosynthetic light harvesting.*

Importantly, considered to be the key optimization in light-harvesting complex 'design', a light-harvesting antenna, can have the concentration of chromophores, chlorophyll, as high as 0.6M, pointing to the inter-chromophore separations, approximately of the order of 10 $\dot{A}$ centre-to-centre for the nearest neighbors. Hence, the electronic coupling scales quite steeply with distance (the inverse of chromophore separation cubed at long separations), pointing to the strong electronic interactions between chromophores in antenna complexes whereas instances of strong coupling is also possible, as observed in delocalized excited states or excitons. It should be noted here, that the electronic excited state is delocalized over two or more chromophores and the associated quantum-mechanical wave function possesses a typical pattern of amplitudes across these molecules. Formally, this should specify the typical case pointing to the fact that the excitation in a quantum superposition of the electronically excited states of different molecules, simply, are in a state, depicting the electronic coherence among chromophores.

Though this condition might not be always present, probably these are the most important delocalized excited states, possessing the most evident as well as an widely available signature of quantum coherence in photosynthetic light harvesting. In strongly delocalized systems, Scholes,[187] pointed to an important fact that similar to J- and H-aggregates, exceptional enhancements of energy transfer rates, sometimes called 'supertransfer', are possible. In addition to the normal quantum effect, additionally, a subtler effect may take place in such situations. He suggested that "*rather than excitation 'hopping' between, might be, delocalization of excitation transfer over donor and acceptor states occurs so that their relative phases evolve quantum coherently*". Also, the excitation transfer, appearing wave-like in a perfect periodic system, implies that different energy transfer pathways could also be realized. This might occur due to the simultaneous altering of transport properties through the active role, played by the process of "quantum interference".[92]

Recent results obtained from some breakthrough experiments, 2DES did put evidences of long-lived oscillatory features in the two-dimensional spectra of several light-harvesting complexes. First interpretation of this kind of oscillatory behavior had been interpreted as a signature of quantum coherent evolution of superpositions of electronic states. Not only that the vibrations also contribute to these oscillations, and hence, potentially play a role in modulating the 'electronic delocalization'. That is why this phenomena, in a more precise way, speak of vibronic coherences. As a resultant effect, all of these phenomenal non-trivial quantum effects could modify energy transfer dynamics to some extent. The observations suggesting long-lived quantum mechanical evolution of electronic coherences in photosynthetic antenna proteins have prompted a huge amount of theoretical research aiming at understanding how electronic coherence and in particular, quantum dynamics can be sustained in the complex biological environment. These theoretical studies explored whether coherences influence energy transfer dynamics, something not detected by the 2DES experiments. Several possible mechanisms supporting coherent dynamics have been identified. These include

(1) Weak electron–vibration coupling,
(2) Spatially correlated[101, 161] environmental fluctuations at different chromophores and a slowly relaxing vibrational environment.[134]

However, calculations so far indicate that environments at different chromophores are more or less independent.[44, 226]

3.7 Photosynthetic Exciton-Model and Electron–Phonon Interaction

Originally, the exciton model had been introduced for the description of electronic excitations in solid-state materials. Davydov[176, 195] adopted this model in the context of molecular system for treating excitations in molecular crystals. Building on the work of Davydov, the exciton model has been extended to describe the interaction of excitations also in other molecular systems[196] and, particularly, the interaction of chromophores in photosynthetic antennae. One of the major differences between the understanding of excitons in the context of solid-state systems and molecular systems is the extend of delocalization of the excitation. Photosynthetic excitons are fairly localized: the typical delocalization length in the steady state being only about two to four chromophores (with the notable exception of the chlorosome antenna), and in many cases the excited states can even be approximated fully localized to one molecule.[197,198]

It is an well known fact now that chromophores, when photoexcited, interact with each other by a Coulombic interaction between their transition densities[199,200] out of which, at large separations between the chromophores, a coupling exchange occurs in between molecules, caused by electronic excitation. As it is a dipole–dipole coupling between transition dipole moments and also, the exchange of excitation is slow compared with dephasing and relaxation processes accompanying the Stokes shift, the resultant excitation hops incoherently following Förster mechanism. When the electronic coupling is large compared to the homogeneous line broadening (or equivalently, the reorganization energy) and electronic energy differences between the chromophores, then, as an effect, the interaction in the stationary states of the system relative to the isolated chromophores are modified. The simple reason behind this is that now we cannot differentiate whether the excitation, i.e., the correct eigenstates become a linear combination of the possibilities.

The key assumption of the molecular (Frenkel) exciton model in the context of photosynthetic systems is that the Coulombic interaction between chromophores, is weak compared with the forces defining the electronic structure of the isolated molecules. As a result, considering that the electronic structure of individual chromophores is relatively unchanged by the interaction and the electronic states of single molecules remain a good basis, to describe the state of the system (under the Born–Oppenheimer approximation). Formally, this means that the interaction leads to eigenfunctions of the modified Hamiltonian that are given by a linear combination of the eigenfunctions of the isolated chromophores.

Here, the left-hand side of the figure shows the structure and absorption spectrum (in the spectral region of the third allowed singlet–singlet transition) of a naphthalene-type chromophore. On the right is a dimer held in fixed orientation by a 'norbornalog bridge'. The electronic coupling is found to be very strong, estimated to be about 890 cm^{-1}, causing the excited electronic states of the dimer to be delocalized over both the chromophores.[202] The two linear combinations of excitation being on

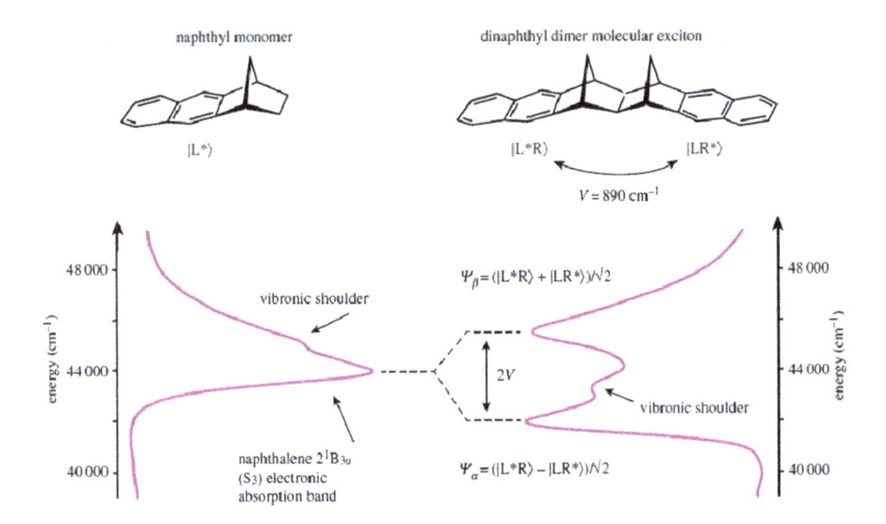

Figure 3.11. Exciton splitting in a naphthalene dimer. See text for details. Data are from the report published in J. Am. Chem. Soc. 115, 4345–4349.

the left or right chromophore define the phases of the excitation in each of the exciton eigenstates. These eigenstates are clearly observed as a splitting, only for one exciton state, because of the symmetries of the two states differing, and hence the way they couple to vibrations also becomes different.

3.7.1 *Excitons: Interaction with the Surrounding Environment*

In order to describe the behavior of excitons fully, i.e., how electronic states interact with each other, further description is needed about the ways of their interaction with the surrounding environment. Because, the interaction with the environment induces disorder in the system which might cause both heterogeneity and dynamical fluctuations in electronic parameters. As a result, this can influence substantially the extent of exciton delocalization and associated dynamics. Qualitatively, it is evident that exciton splitting, similar to that shown in Figure 3.12, will be obscured if the absorption bands are wider than the splitting. Of particular relevance is the interaction between electronic excitations and vibrational modes. Both the protein scaffold to which chromophores are attached and the solvent medium are usually modeled as a continuum of low frequency vibrational 'modes' interacting with the electronic excitations. Additionally, molecular vibrations of the chromophores, including high-frequency modes, are also typically identified in the vibrational spectra of chromophores. The vibrational degrees of freedom are usually modeled as a collection of harmonic oscillators. Each of these vibrations is quantized, in analogy to the concept of phonon, i.e., the elementary quanta of vibrational excitation in crystal lattices. Thus, borrowing then the language from solid-state materials science, *the interactions between the excitation and the vibrational degrees of freedom can often be described as electron–phonon interactions.*[204] However, it should be noted that phonons in the solid-state sense are not a sensible physical description of the fluctuations in solutions and complex environments, where the modes and their frequencies are continuously changing. Nevertheless, the concept of phonons in the phenomenological sense has been valuable as a basis of models for energy transfer.

3.7.2 *Electronic Structure of Excitons*

In light harvesting systems, with some simplification, each chromophore labelled i, can be modeled as a two-level system with single electronic transition energy, E_i, i.e., so-called site energy. The interaction between the transition dipole moments of two chromophores i and j is described by the electronic coupling parameter, V_{ij}. Within the single-excitation manifold, the basis for calculating the energies and wave functions for a system with N interacting chromophores is the Frenkel exciton Hamiltonian

$$H_{el} = \sum_{i=1}^{N} E_i |i\rangle\langle i| + \sum_{j\neq 1}^{N} V_{ij} |i\rangle\langle j| \tag{3.10}$$

denoting $\langle i|$ as the state with chromophore i excited and all others being in the ground state. The eigenstates of the electronic Hamiltonian, H_{el}, also commonly known as Frenkel excitons, are considered as delocalized among several molecules, owing to the electronic interaction between chromophores. This delocalization can have a profound effect in multichromophoric systems, creating new absorption bands and changing energy transfer dynamics. In order to analyze dynamics basically involving quantum coherence, it is useful to describe the probabilities of different states, being excited in terms of the density matrix. Basically, off diagonal terms contain information about superposition states (coherence in the quantum-mechanical sense). After photoexcitation, the general form of the density matrix $\rho(t)$ describing the state of N chromophores at time t, can be expressed as

$$\rho(t) = \rho_{00}(t)|0\rangle\langle 0| + \sum_{i=1}^{N}\sum_{j=1}^{N} \rho_{ij}(t)|i\rangle\langle j| \tag{3.11}$$

Here, $\rho_{00}(t)$ accounts for any losses caused by relaxing the system to the ground state $|0\rangle$, $\rho_{ij}(t)$ being the populations ($i = j$) and coherences ($i \neq j$) in the site basis. It should be noticed that coherences are basis dependent, i.e., they depend on whether we choose to write the density matrix in terms of localized molecular states or exciton states. The term 'electronic coherence' usually refers to coherence in the site-basis, while 'excitonic coherence' denotes coherence in the electronic eigen basis. It is important to note that for a 'pure' state, the differentiation of the site and excitonic basis is clear, i.e., the density matrix is an average over the ensemble. But, each member of the ensemble will be in a slightly different state, i.e., some excitons being more localized than others. However, the details of this distribution will be hidden once the average is taken. For example, it is not obvious, simply, just from the inspection of the density matrix of a three-molecule system whether excitation is coherently shared among all the three molecules or shared between pairs of molecules.

But, even then the final open question still remains, i.e., about the theory behind the viability of quantum coherent transfer as suggested by 2DES experiments, i.e., whether it might be present under natural sunlight illumination or not and also, its possible relevance for efficient transfer. According to Fassioli group, it is difficult to anticipate, at least at the present stage of developments whether the coherences are purely electronic act, promoting energy transfer in light-harvesting complexes. This fact arrives at the suggestions that in many of these systems, electronic transitions often couple strongly to selective 'vibrational modes', their frequencies matching those of electronic energy gaps, and therefore a significant amount of mixing between electronic and hence vibrational degrees of freedom can occur.[204] This is the main concept behind the famous Förster spectral overlap. Henceforth, frequency of oscillations in optical spectra, obtained from experiments, might posses mixed vibronic origin rather than the purely electronic or purely vibrational frequencies.[134] However, in this regard, certain limitations of the simple exciton model, in the presence of strong coupling to vibrational modes are well recognized.[205,206]

3.7.3 *Electron–Phonon Coupling*

The second and most important source of exciton localization is the unavoidable coupling between the electronic degrees of freedom of chromophores and the stochastic fluctuations in the surrounding environment. This means that the eigenstates predicted by the bare Hamiltonian H_{el}, can be delocalized over several molecules. As a result, the environment will tend to destroy phase relations between the excited states of different molecules, thus yielding an excited state, more localized than that of an isolated system. This interaction induces the so called dynamic disorder as it modulates electronic parameters on a timescale comparable to excitation dynamics. In such a case, the electron–phonon interaction is commonly assumed to be diagonal and given by

$$H_{el-ph} = \sum_{i=1}^{N} \sum_{k} g_k^i (b_k^\dagger + b_k)|i\rangle\langle i| \tag{3.12}$$

Here $b_k^\dagger$, (b_k) are phonon operators that create (annihilate) a phonon mode of frequency $\omega_{k'}$ and g_k^i denotes the coupling between the electronic transition and each of the mode existing thereby. Here, k is the coupling between the electronic transition and each mode. The effect of the environment in equation can be interpreted as a dynamical modulation of transition energies of the molecules. This kind of dynamical modulation happens during energy transfer, shifting the energy of the molecules constantly in and out of resonance. The distribution of 'phonon' frequencies in the environment and their coupling to the electronic transitions of the chromophores is characterized by the spectral density which is given by

$$J^i(\omega) = \sum_k |g_k^i|^2 \delta(\omega - \omega_k)$$

whereas, the energy associated with the equilibration of the environment after excitation is quantified by the reorganization energy

$$\lambda^i = \int_0^\infty J^i(\omega)/\omega d\omega$$

In the case of electronic interaction, taking place between the excitonic states, which is much stronger than the coupling to the environment, $V \gg l$, the relaxation comes into action in between the excitonic states, thus diagonalizing the electronic Hamiltonian H_el. At thermal equilibrium, the system is found to posses a statistical mixture of electronic eigen states,

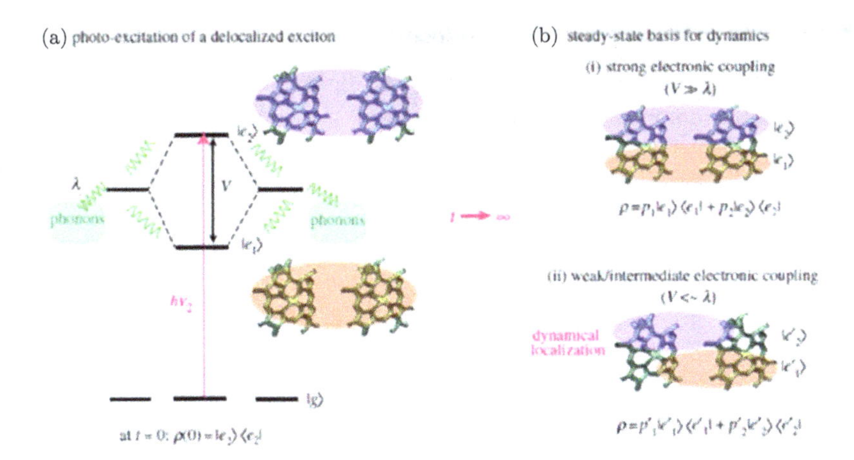

Figure 3.12. Dynamical localization in an electronic dimer. (a) At t =0 light excites a fully delocalized electronic state. (b) The interaction with the phonon modes induces relaxation and dephasing, and the steady state of the system corresponds to a statistical mixture of electronic states.

the excitations remain to some extent delocalized which is illustrated in Fig. 3.12.

For an electronic dimer consisting of two resonant molecules. At $t = 0$, light excites the highest exciton which is fully delocalized as the transition is assumed to obey the Franck–Condon principle of vertical transitions (phonon modes remain unaltered, immediately after the electronic transition). In the steady state, when the electronic system equilibrates with the phonon environment, the state of the electronic system is given as a classical mixture of excitonic states. In the very strong electronic coupling regime, these excitonic states correspond to the electronic eigenstates of H_{el}. Away from the limit of very strong coupling, the electron–phonon coupling effectively renormalizes the basis in which the electronic system relaxes and the excitonic states that diagonalize the density matrix in the steady state are more localized than the electronic eigenstates. This phenomenon, associated with polaron formation — an electronic state dressed by phonon modes — is referred to as dynamical localization.[207] Under very weak electronic coupling, electronic states are fully dressed by the environmental phonons and the excitation is completely localized. In thermal equilibrium, individual chromophores are populated according to a Boltzmann distribution, being under the regime of validity of Förster theory.

3.7.4 *Quantifying Exciton Delocalization*

Due to the importance of excitons in the light harvesting function, the considerable amount of research has been motivated to determine the problem related to the degree of delocalization or 'length' of excitons. The delocalization length of a pure state, i.e. a well-known state, can be expressed in a general form

$$|\psi\rangle = \sum_{i=1}^{N} a_i |i\rangle$$

This describes the excited state of a molecular aggregate of N chromophores which can be determined through the inverse participation (IPR), i.e.,

$$IPR = \sum_{i=1}^{N} |a_i|^4$$

ranging from 1 for a fully localized state to $1/N$ for a completely delocalized exciton. In practice, owing to interaction with the environment, the problem in determining the extent of delocalization of the excited electronic states is much more complex when dealing with a system, subjected to decoherence, occurence of loss in quantum coherence.

3.7.4.1 *Exciton-Length*

Next important point is the difficulty in quantifying "exciton length" under the electron–phonon coupling. Because, first of all, the environment induces mixedness in the system. Even though the initial photo-excited state is well known (pure, i.e. described by a wave function), generally, at later times, the state of the system is usually described in terms of an ensemble of pure states, adopting the formalism of density matrix, thus making the measures of exciton delocalization, say, the IPR, inapplicable. Furthermore, dynamical localization results in relaxation which has a basis, more localized than the bare electronic eigen-basis. In the quest for the measures in quantifying exciton delocalization under the electron–phonon coupling, recently, researchers have applied concepts from the field of quantum information, i.e., entanglement, to tackle the problem of multichromophoric energy transfer.[92,170] It has already been claimed that entanglement, itself being present in light-harvesting systems, might not be the best way to think about excitons in the context of light harvesting.[209] Because, measures of entanglement subjected to decoherence, are nevertheless useful to characterize the time evolution and spatial distribution of quantum correlations in molecular aggregates. This gives a detailed picture of how excitation is delocalized and how coherent sharing of excitation evolves with time. Added to it, such measures are also used to analyze the system's density matrix and capture the extent of delocalization of the basis in which the system relaxes, but, in effect, with the environmentally induced loss of purity. However, these measures are able to characterize quantum correlations between distant chromophores, especially while going beyond traditional measures, providing average delocalization lengths, but not corresponding to any physical length.

3.8 Theoretical Models of Coherence

In order to gain a better understanding of the various results together with the corresponding explanations and implications, described up till now, it becomes essential in considering a model commonly used for the simulation in some of these systems. One approach among them, is to simulate the quantum behavior of a single excitation in the FMO complex via an effective Frenkel exciton model.[10] For this, a photon is considered to create an excitation in an antenna molecule, eventually transferred to one of the sites in the FMO molecule, again to another along the chain, i.e., from chromophore to chromophore caused through the transition-dipole Coulomb interaction which process continues until reaching the reaction center (RC) where, it is employed for charge separation. The validity of such an assumption, i.e., of a single-excitation lies in vivo situations, i.e., when the bacteria [familiar with in such cases], lives in very low-light conditions. That is why it becomes exceedingly unlikely to have an experimental situations where the FMO complex, ever contains two excitations. Even though the single-excitation assumption has produced certain controversy,[209] this situation might very well arise i.e., when the rapid exciton-exciton annihilation processes make multiple-exciton states in a single complex extremely unstable one so that it tends to relax into a single-exciton state, even before energy transfer occurs.

However, in the present case, for the purposes of clarity, we restrict ourselves to the single-excitation model, though in principle, there may be straightforward method to construct a multiple-excitation model, as employed in[211], for example, for the description of an artificial photosynthetic system. Following this model, the system's Hamiltonian can be written as

$$H = \sum_{j=1}^{N} \epsilon_j |j\rangle\langle j| + \sum_{j<j'} J_{j,j'}(|j\rangle\langle j'| + |j'\rangle\langle j|) \quad (3.13)$$

where the states $|j\rangle$ represent the presence of an excited electron (exciton) at site, or BChl molecule j, where $j \in 1, \ldots, 7$ describing these sites in the FMO complex). The eighth BChl molecule has. It means specifically that, the whole FMO complex can be described with a single label which defines at which site the excitation is residing. Here, the parameter ϵ_j denotes the energy of that excitation for a particular site. This parameter, as has been pointed out by Lambert *et al.*[40] is dependent on the surrounding protein structure and varies quite a lot. $J_{j,j'}$ is the excitonic coupling between the j and j' sites. As has been opined by many stalwarts, working in this field successfully, determining and calculating related various energies and coupling strengths in a given LHC, is a quite complicated and difficult area of research, often involving both spectroscopy and *ab initio* modeling of the physical structure and related environment of the system.[28, 40, 44]

Firstly, accurate values of electronic couplings, found to be quite good in agreement with corresponding experiments, needs to be dependent on, again, in an *ab initio* basis, on related quantum chemistry methods, based mainly on the atomistic models of a LHC.[44, 211] Next, the site energies are far more difficult, faced in multiprong ways to determine, because, the modeling of the protein-pigment interactions is rather a non-trivial task, can be availed only through the successful and dependable accurate inputs from the experimental results. Recently, Renger and coworkers have[214] demonstrated the possibility for the calculation of site-energy parameters, applying structure-based theoretical approaches, carefully treating the electrostatic interactions between the electronic excitations and the surrounding protein environments for several photosynthetic complexes.[215]

However, many groups connected with this problem suggested after finding that as FMO LHC is one of the most studied examples, some understanding of the system's Hamiltonian already exists (even though often different values for the energies and excitonic coupling amplitudes are given in the literature.[40, 215] In this connection, Lambert in his paper,[40] presented the values of Hamiltonian, showing that the site-site electronic couplings $J_{j,j'}$, are of the same order as that of energy difference between sites, suggesting that site-site coherences could be strong. In addition, the magnitudes of site-site couplings are often found distributed broadly, with the possibility of the larger ones to be around the scale of $\approx$100 cm^{-1}, which, again, is comparable to the reorganization energy caused due to exciton-environment couplings.

3.8.1 *Approximations Used in Different Models for Coherence*

One of the questions, remaining to be answered for the system, described up till now, is the nature of its interaction with its environment. Here, we describe a simplest form of a model, trying to describe the above phenomena based on certain approximations which follows as:

(1) The first to assume is that each site is coupled to a bath of oscillators. The baths being coupled to each site, are independent of each other.
(2) The second assumption is that this coupling being the energy of each site means that the environment causes fluctuations of the energy in the site basis. In the exciton basis, the fluctuations have off diagonal matrix elements, capable of causing transitions between eigenstates.[37]
(3) The third is to assume the environment to be Markovian (i.e., without memory).
(4) Finally, the fourth assumption is to assume the exciton-bath coupling to be weak so that a second-order perturbation theory is applied for the description of the dynamics related to energy transfer.

The master equation, having minimal number of free parameters, as described above, gives the reduced density matrix of the FMO complex, simple to solve compared to others which have been more sophisticated by applying far more number of other criteria, crucial for the purpose to be served. On the contrary, this equation has a minimal number of free parameters having typically, a temperature-independent radiative relaxation rate for each site implying that if the excitation takes too long to traverse its way through the complex, eventually, it will be lost due to this relaxation. Fortunately, fluorescence decay of chlorophylls, occurs typically in the nanosecond time scale, a slow process compared to the other dynamics, thus, ultimately giving the FMO complex its ability to transport energy efficiently with a high success rate. This simple model has estimated some of the qualitative properties of experiments on FMO including the long coherence time and high transport efficiency.[38] The Markovian master equation approach, assuming a self-energy, can be cited as a specific example, i.e.,

$$\Sigma[\rho] = \Sigma_{ss}[\rho] + \Sigma_{deph}[\rho] + \Sigma_{diss}[\rho] \tag{3.14}$$

ρ depicts here the system density matrix. The first term is meant for irreversible excitation transport between site 3 and the sink. The first term of the above equation expresses the irreversible excitation transport between site 3 and the sink[37,215] [pointed out in Figures 3.5(a & b)–check], including RC in the FMO, in order to describe site 3 and sink respectively.

$$\Sigma_{ss}[\rho] = -\Gamma_D[s_D s_D^\dagger \rho - 2s^\dagger \rho s_D + \rho s_D s_D^\dagger] \tag{3.15}$$

where $s_D = |3 \times 0|$, and Γ_D is meant for sink tunneling rate. In the present situation, i.e., in FMO, site 3 is coupled to the reaction center as this consists of the lowest energy excitations, but in contrast, lies closest to the reaction center (RC). But, in the contrary, site 1 lies closest to the antenna. In this case, $|0\rangle$ represents an empty state, i.e., without any electronic excitation. The second term, $\sigma_{deph}[\rho]$ has been considered as the factor meant for temperature-dependent dephasing factor which has been dealt in detail by Rebentrost & Mohseni.[218]

The term, $\Sigma diss[\rho]$ describes the slow fluorescence relaxation process. A variety of insights can be obtained from this simplistic approach,[27,56] the essence of which is that the combination of coherent dynamics from the Hamiltonian H and level-broadening from $\Sigma deph[\rho]$ means that there remains a high probability of bringing an excitation from site 1 to site 3,(as shown in FMO complex), leading to the RC, before the excitation is lost due to the fluorescence relaxation $\Sigma diss[\rho]$. The efficiency of the transport can then be modeled considering the time-dependent sink or RC population, expressed as follows:

$$P_{RC}(t) = 2\Gamma_D \int_0^1 Q^{(3)}(t')dt' \tag{3.16}$$

In the above equation, $Q^{(3)} = Tr[|3\rangle\langle 3|\rho(t)]$ depicts the population of the site “3”, connected to the sink.

3.8.2 *Dephasing*

In most treatments of the FMO system, the coupling to the environment are considered for the modulation of the energy of each site which, following the language of quantum mechanics, is called as “pure-dephasing”. The radiative relaxation of each site is independent of temperature as the optical transition energy of each site is exceptionally high (>12,000 cm^{-1}) and consequently, the dephasing processes to be temperature dependent.

Added to this, each state mentioned in the above description, represents an exciton on a molecule transferring its energy to its neighbor. But in practice, no actual electron transport takes place, implying that the reorganization energy (the energy associated with a change in the surrounding protein environment due to the presence of an electron charge) is relatively low compared to processes where electron transport occurs. However, in such cases, the assumption that both of independent Markovian baths and of weak coupling to be valid, might not be true in reality. As stated earlier, ample evidences have already been found stating that the coupling to protein environments surrounding the FMO complex can be of the same order as that of electronic couplings[210,218](100 cm^{-1}).

Next, the bath (e.g., the protein scaffold surrounding the FMO complex) may have structure and associated dynamics, strongly correlated with, back-action, i.e., on the dynamics of the excitation in the FMO complex. To understand the effect of such intermediate coupling regime and complex environment, i.e., the system and bath, both have been treated with a range of non-Markovian and higher-order models, associating electron–phonon coupling.[5,134,192] In particular, non-perturbative approaches providing exact numerical results for certain models of excitation energy transfer have been successfully applied for the study of coherent quantum dynamics in photosynthetic light harvesting.[134,216,226] Though, constrained with practical difficulties related to these non-perturbative approaches, these methods appear to be computationally too expensive for applying the techniques involved to large photosynthetic complexes. Anyway, they provide valuable insight, shedding a different light on the coherence effects in light harvesting[134] in such cases, for example, the fact that bath memory effects conspire to enhance quantum coherence.

Going beyond the independent bath approximation, initial experiments indicated the presence of the surprising long-lasting quantum coherence between two electronic states which may arise because of coherence enhancing effects from coupling to common vibrational modes.[5,6,13] This idea has triggered the reexamination of the processes related to excitation transfer[218] and decoherence effects[219,220] in the presence of such an unusual environment. In principle, such a strongly coupled "common bath" should contain bath-induced fluctuations being spatially correlated.[221,222] However, as a counter-argument, some molecular dynamics simulations for the FMO complex[215] and reaction center[216] show only an weak correlations in between the movements (vibrations) of the chromophores, stating that, the uncorrelated bath approximations may be valid in such situation.

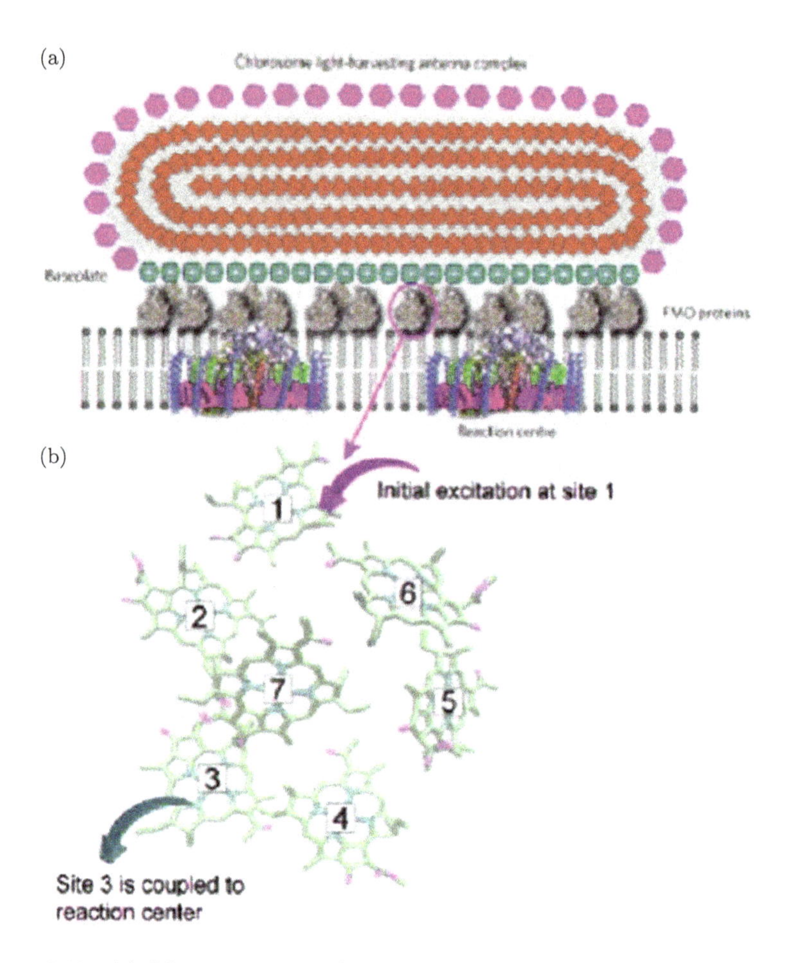

Figure 3.13. (a) Schematic diagram (from of roughly 200,000 BChl-c molecules (green rhombuses) encased in a proteinlipid structure form the chlorosome antenna. This is much larger than antennas seen in other species, possibly due to the low-light conditions in which these bacteria thrive. This antenna harvests sunlight, and then the excitation energy is transferred (via baseplate proteins, represented by the blue circles) to several FMO complexes. These act as a wire to transport the energy to reaction centers in the membrane. The reaction center shown here is from Chloroflexus aurantiacus. The red hexagons represent the chlorosome envelope. Below this, in (b), there is a rendering of the structure of one of the FMO pigment protein complexes. The function of FMO in green sulfur bacteria is to transport excitations from the sunlightharvesting LHC antenna (chlorosome) to the RC where it serves to initiate a charge-separation process. The FMO complex consists of eight bacteriochlorophyll a (BChl a) which are bound to a surrounding protein scaffolding. The excitation received from the antenna (here at molecule, or site, 1) is then passed from one BChl molecule to the next, until it reaches the molecule, or site, closest to the reaction center (here assumed to be site 3). It then irreversibly enters the reaction center, and ignites the charge separation process.

Recently, Sim,[225] comparing some of these master-equation models to a more complex many-body atomistic/molecular modeling scheme, found excellent agreement between the two approaches, suggesting that the assumption (based on relatively simple models), i.e., oscillations seen in experiment have a truly quantum origin, might be correct. However, whether the environment surrounding LHC systems has a structure or nature that preserves quantum effects in some way, is still a controversial and an open question. More experimental works might provide us with fruitful information required to answer this puzzle. Besides this, the quantum coherence manifests itself in characteristics, directly observable quantum beating signals among the excitons within the Chlorobium tepidum FMO complex at 77 K [give reference no. of Engel-quant-coher. nature05678.pdf/NewQBIO]. This wavelike characteristic of the energy transfer within the photosynthetic complex can explain its extreme efficiency, in that it allows the complexes to sample vast areas of phase space to find the most efficient path.

3.9 Important Insights About Some of Other Theoretical Modelings Related to Quantum Coherence

3.9.1 *Environment-Assisted Quantum Coherence & Photosynthetic Phenomena*

Apart from the direct observation in experiments of quantum coherent beating, some of the most intriguing insights that have appeared from the study of the FMO complex suggest that a combination of these quantum coherent oscillations of excitations between sites, and interactions with the environment, produce a transport efficiency higher than is possible with a Förster mode alone. If correct, this would fit the definition of functional quantum biology. Importantly, such a phenomenon was proposed and studied by Plenio *et al.*,[27] Mohseni *et al.*,[37] and by Lee *et al.*,[5] in detail. In each case, it is assumed that a single excitation is placed at a particular site in the FMO (usually site one), this excitation propagating then, through the FMO chain caused by the combination of coherent tunneling and environmental effects. As already stated, the goal of the FMO complex is to get this excitation efficiently to site three, which is coupled (incoherently) to a reaction center (RC). Thus the typical time, or efficiency, of reaching the RC is calculated as a function of environment temperature, site selective-couplings, and so on. In most cases it seems that the full quantum model will give a higher efficiency than a purely classical one. Below, we try to mention some of the physically intuitive explanations given for this phenomenon:

3.9.2 *Avoidance of Local Minima*

Plenio *et al.*[27] and the group of Mohesini,[37] applying Markovian quantum model, suggest both of the long-lived coherent oscillations between sites and that of an enhanced rate of excitation transfer reaching the "sink" site, over that predicted by the Föorster model. Specially, Plenio argued that essentially, the combination of coherent transfer and environmental "noise" causes level broadening which implies stating that the excitation can more easily escape local minima in the FMO network. (see Figs. 3.1 & 3.2 — i.e., the picture of FMO).

In other words, as mentioned also by Ishizaki and Flemming,[223] quantum delocalization can also help avoid and overcome local minima, or energy potential traps in the energy landscape of the FMO complex. According to their views, such a phenomena is even more important in higher plants due to the possibility of the presence of "up-hill" potential landscapes. These results can be related to earlier works by Gaab and Bardeen[225] who outlined how energy transfer can be optimized by carefully choosing the coherence time, together with the rate of transfer to the sink. As per their[225] works, firstly, there exists an optimum combination of trapping and coherence time, giving the most rapid population transfer to the trap. But, in general, these values are not either shortest trapping time or the longest coherence time, as would be expected based on the rate equation models and/or simple considerations from the analytical results which has already been derived with the Haken-Ströbl model for an infinite system. Secondly, in the coherent regime, where T_d, i.e., dephasing time is longer than the other relevant timescales, population trapping being automatically finite, is also finite through a network system and can be suppressed by quantum interference effects, it's magnitude being sensitive to the molecular geometry.

3.9.3 *Coherence Assisted Trapping*

Lee *et al.*,[5] after studying intensely, demonstrated that coherent quantum dynamics, together with rapid incoherent dissipation due to a trapping site can enhance the efficiency of the irreversible energy transfer between the initial energy donor and the sink, or reaction center (RC), over a purely classical model. Effectively, his model suggests that considering incoherent collapsing of exciton wave functions, it becomes possible in achieving a greatly enhanced energy-trapping efficiency, as it is an intricate combination of reversible quantum coherent evolution, the cause behind which is the effective use of the anti-Zeno effect to promote energy transfer. But, one should again notice that this "enhanced efficiency" is in comparison to that obtained with the Förster model. Anyway, several ambiguities remain in this–Firstly, is the Förster model the correct classical model to make a comparison to? Secondly, placing a single excitation at site "1" may match well certain experiments, but in normal light conditions, is this a correct approach?

Quite recently, Briggs and Eisfeld[243] showed that, for realistic coupling strengths, an alternative classical model of the FMO transport process can produce results identical to the quantum one. In addition, a recent analysis by Wu *et al.*[224,229] suggests that, as mentioned earlier, the efficiency enhancement gained by coherence in the FMO complex is only a 'few percent' compared even to that predicted by the Förster model. Finally, the true in-vivo conditions are not yet quite well understood, remaining a subject of controversy in current research. Perhaps, the only way to solve this issue is to show that in more complex components of some light harvesting complex, for example, in LH1 or LH2,[222,227] there exist significant energy traps which, without the assistance of quantum coherence, drastically impact/reduce the probability of an excitation through successfull navigation of its way to a reaction center (RC), before being lost to fluorescence relaxation.

3.10 The Heirarchy Model of Coherence

With the aim of gaining a deeper understanding of the interplay between the quantum coherent transport of energy in FMO and how it functions, interacting with its complex protein environment, serious efforts have been devoted a variety of non-Markovian and non-perturbative models being applied to these systems. One among them, i.e., the Hierarchy model, may be called as most successful, originally developed by Tanimura and Kubo.[231] This has been applied to both the FMO complex and other LHC components.[222] As this model has had a large impact on this field, we will try to place a brief description of its main components here, full derivations can be found in the literature.[224,228,229] Starting with the Hamiltonian that explicitly can describes the interaction term between site energies and phonon modes, i.e.,

$$H^{(e-p)} = \Sigma_g |j\rangle\langle j| q^{\hat{j}} \tag{3.17}$$

In the above expression, $q^{\hat{j}} = -\Sigma_k g_{j,k}\chi_k$ and $g_{j,k}$ depict the coupling constant of the jth site and kth mode in the bath. Here, the phonon modes are themselves considered as Harmonic oscillators, having the Hamiltonian $H^{(p)}$. Now, assuming $t = 0$, it is possible to separate the site (here, pigments) and the phonon modes, and obtain

$$\rho(0) = \rho_e(0) \otimes \rho_p(0)$$

so that the phonon modes are in a thermal equilibrium state which, finally can be expressed as,

$$\rho_p(0) = e^{-\beta H^{(p)}}/Tr[e^{-\beta H^p}],$$

where $\beta = 1/k_B T$. After taking average, the correlation function of the bath modes becomes, finally, as

$$C_j(t) = \langle q_j(t) q_j(0)\rangle = \frac{1}{\pi}\int_0^\infty d\omega J_n(\omega)\frac{e^{-i\omega t}}{1-e^{\beta\hbar\omega}} \tag{3.18}$$

The above equation offers sufficient description for the bath. Then, by applying Drude spectral density (specially meant for over-damped oscillators), we get

$$J_j(\omega) = \left(\frac{2\lambda_j\gamma_j}{\hbar}\right)\frac{\omega}{\omega^2+\gamma_j^2} \tag{3.19}$$

where, γ_j denotes the "Drude decay constant", indicating the memory time of the bath for the site j. In the present case, each site is assumed to possess its own independent bath, even though, generally, in case of Hierarchy method, it depicts one of it's power in having the ability to treat correlated baths.[222] Here, λ represents reorganization energy, related to the system-bath coupling strength. In this case, the implication of the Drude spectral density points to the fact that we can express an exponentially-decaying correlation function as

$$C_j = \sum_{m=0}^{\infty} c_{j,m} exp(-\mu_{j,m} t) \tag{3.20}$$

where

$$\mu_{j,0} = \gamma_j, \quad \mu_{j,m} \geq 1 = \frac{2\pi m}{\hbar\beta}$$

and the coefficients

$$c_{j,0} = \gamma_j \lambda_j \big(\cot(\beta\hbar\gamma_j/2) - i\big)/\hbar \tag{3.21}$$

and

$$c_{j,m\geq 1} = \frac{4\lambda_j \gamma_j}{\beta\hbar^2} \frac{\mu_{j,m}}{\mu_{j,m}^2 - \gamma_j^2} \tag{3.22}$$

Now, after having the information on the bath properties, i.e., the reorganisation energy, bath memory time together with that for the bath temperature, we can employ the Hierarchy equation with the purpose of describing the system-bath dynamics in the strong-coupling and non-Markovian regime, i.e.,

$$\begin{aligned} \dot{\rho}_{\boldsymbol{n}} = &- \left(iL + \sum_{j=1}^{N} \sum_{m=0}^{K} \boldsymbol{n}_{j,m} \mu_m \right) \rho_{\boldsymbol{n}} - i \sum_{j=1}^{N} \sum_{m=0}^{K} [Q_j, \rho_{\boldsymbol{n}^+_{j,m}}] \\ &- i \sum_{j=1}^{N} \sum_{m=0}^{K} \boldsymbol{n}_{j,m} \big(c_m Q_j \rho_{\boldsymbol{n}^-_{j,m}} - c^*_m \rho_{\boldsymbol{n}^-_{j,m}} Q_j \big) \end{aligned} \tag{3.23}$$

In order to explain the above equation, we can clarify the different mathematical terms and equations as follows: Here, $Q_j = |j\rangle\langle j|$ represents the projector on the site j, L being the Liouvillian for FMO, N=7. Added to this, there exists various parameters, defined above. In this picture, Hierarchy can be described by a large set of coupled equations, each being lebelled by $\boldsymbol{n}$. This denotes a set of non-negative integers, where each equation, specified are defined here as $\boldsymbol{n} = n_1, n_2, n_3, \ldots, n_N = n_{10}, n_{11}, \ldots, n_{1K}, \ldots, n_{N0}, n_{N1}, \ldots, n_{NK}$. It means that each site j, possesses an additional label m, from 0 to K. In turn, each of those labels can have from 0 to ∞ values. Specially, here, the system density matrix is expressed for the label $\boldsymbol{n} = 0 = 0, 0, 0, \ldots$ whose properties at any time t define those of the system. Also, in turn, being coupled to "auxiliary density matrices" this describes the complex bath fluctuations, expressed by the term in the equation where, $\boldsymbol{n}^{\pm}_{j,m}$. This implies the term in the index, defined by j, m, being increased or decreased by 1. For example, this can be explained as follows: if the terms in the term for m, are truncated for $K = 0$(say), then the system density matrix will be coupled to N other auxiliary density matrices, i.e, $\mathbf{0}^{+}_{1} = 1, 0, 0, \ldots, 0, \mathbf{0}^{+}_{2} = 0, 1, 0, \ldots, 0$, to be continued like this. As a result, each of such auxiliary equations gets coupled to the higher-tier equations in turn resulting into the formation of overall hierarchy of high dimensional simplex.

These hierarchy equations are infinite both in the sum of the m label, caused by the expansion of the correlation and also in the value of each label itself which needs to be truncated in some or other way possible. For example, the m label can be truncated for $\beta\hbar\gamma_m < 1$ and $2\pi(K+1)/\hbar\beta > \omega_S$ (here, ω_S is some characteristic system frequency). The overall truncation of the labels can be defined as the largest total number of terms in a label, or the tier $N_c = \sigma_{j,m} n_{j,m}$. Tan *et al.*[233] described this structure, and the meaning of each tier as the level of bath self-correlation, included in the simulation. But, the proper tier at which the truncation should be made is quite difficult to predict, though typically, truncation should be taken when $N_c\gamma \gg \omega_S$ (depending on the reorganization energy and temperature). Practically, truncation should be checked by looking for convergence in the system dynamics when changing N_c. A systematic and powerful algorithm to find convergence was introduced by Shi *et al.*[40] They proposed that a renormalization of the hierarchy equations allows a direct inspection of the elements of higher auxiliary equations to be made during the numerical solution, and thus the level of the tier can be truncated as necessary.

3.10.1 *Temperature Correction*

In addition, a temperature correction can be added to the equations of motion. This not only compensates for the truncation of the m label at level K, but also assists in overall convergence which is made by assuming that all higher-level auxiliary density operators for $m > K$ are uncoupled from each other so that they experience pure Markovian dynamics.[231] This results in an additional term in the equation of motion, sometimes called the "Ishizaki-Tanimura boundary condition",[232] which can be included in the Hierarchy equation of motion as,

$$L_{IT-BC} = \sum_{j=1}^{N} \sum_{m=K+1}^{\infty} \frac{c_{j,m}}{\mu_{j,m}} [Q_j, [Q_j, \rho_n]] \tag{3.24}$$

the double commutator in above equation represents a normal Lindblad form of equation where the summation can be written as, for $K = 0$,

$$\sum_{m=1}^{\infty} \frac{c_{j,m}}{\mu_{j,m}} = \frac{4\lambda_j}{\hbar^2 \gamma_j \beta}[1 - \gamma_j \hbar(cot(\gamma_j \hbar\beta/2))] \tag{3.25}$$

In other words, this term encapsulates additional environmental dephasing effects neglected by truncating K. Including this correction, even for $K = 0$, has been found to substantially assist the convergence of the Hierarchy results. As an example, in Fig. 16 [Lambert 5a] shows the results of an application of this model to the FMO system at 77 K and 300 K, for various γ_j and λ_j employed in the literature. The beauty of this method lies, to some degree, to the fact that all parameters can be extracted from experiments and the results correctly predict the long coherence time seen in experiments. Perhaps the most important point here is that long bath memory (implied by small γ_j) helps increase the coherence time of the coherent oscillations,[220] though this may not necessarily enhance the overall transport efficiency. In general, for each parameter, there is an optimal value which gives the largest transport efficiency from site 1 to site 3, and hence to the reaction center.[221, 226]

3.10.2 *Leggett-Garg Inequality and Entanglement Measures*

It becomes quite important for the complete and detailed verification about the resulting facts, i.e., whether the experimental observations of coherence truly come from quantum mechanics and not some alternative, classical, model.[243] Wilde *et al.*[235] proposed a method with the help of which it is possible to do so, i.e., the Leggett-Garg inequality can be applied, rather as a means to unequivocally ascertain whether the dynamics observed in these experiments are quantum or obey "macroscopic realism". However, as yet, such a test can not be performed using the measurements needed for such a test in the present stage of FMO experiments. However, their results show that if it were possible to perform such measurements, the currently accepted physical parameters for the FMO system would suggest a violation of the inequality at room temperatures. Not only the coherence, it is also interesting and necessary as well, to inquire about the appreciable amount of quantum entanglement present in the FMO complex during the exciton transfer. It is worth mentioning that Sarovar *et al.*[236] recently suggested that a small amount of long-range and multipartite entanglement also should exist even at physiological temperatures,[233] though, the role of entanglement in the energy transfer process has not yet been still clear.[27]

Wilde & McCracken *et al.*[234] framed a test for non-classicality in the FMO protein complex, the molecular complex responsible for the transfer of energetic excitations in a photosynthetic reaction. A tight-binding Hamiltonian for this simple system have been introduced by the quantum-chemists as phenomenological modifications to the standard Schrödinger equation for this purpose which allowed an open quantum system model associated with its dynamics.[27,37,38] Following the experimental results, obtained from the chlorosome antenna to the RC in green bacteria, several theoretical studies have computed the efficiency of energy transfer. The obtained result ultimately verified the fact that coherent quantum effects play a significant role. In order to asses this claim, Wilde *et al.* considered the application of the Leggett-Garg inequality to the FMO complex. Firstly, they discussed several observables that one might measure in a Leggett-Garg protocol, followed by the analytical results, i.e., when the dynamics involved are purely coherent in nature.

In these ultimate analytical results, the time intervals between measurements in a Leggett-Garg protocol were determined leading to the strongest violation of the inequality. They extended their studies by calculating the time intervals and applied it in a numerical simulation of the FMO dynamics that includes also the effects of noise. The results obtained stated that several observables exhibit a strong violation of the inequality for temperatures below room temperature, and the violation persists in some cases up to room temperature. The attached figure shows a theoretical violation of the Leggett-Garg inequality as a function of dephasing (and hence bath temperature) for measurements on each site, i.e., 1 to 7 in FMO. A value of <0 indicates a violation, and the results suggest that quantum phenomena persists up-to room temperature for measurements on sites 1, 2, 5, and 6. Still, there does not exist any way for the direct measurement for such kind of non-invasive observables in FMO. It is particularly difficult in this case, because unambiguous violations of the Leggett-Garg inequality, would require non-invasive measurements of excitonic populations. As most fluorescence and spectroscopic measurements are destructive, thus these are not appropriate for observing this inequality. However, for ruling out alternative equivalent explanations for different transport in FMO, ultimately, this kind of test would help.

After following through numerous numerical simulations, Wilde *et al.*[235] established that several choices of measurements lead to a violation of the Leggett-Garg inequality even in the presence of noise processes acting on an excitation in the FMO complex. Experimental confirmation could irrevocably exclude a class of macro-realistic theories so far as the description of the dynamics, related to the phenomena of the excitation is concerned. They framed tests that could be used to experimentally exclude such a class of macro-realistic theories, interestingly, including a classical incoherent hopping model, from describing the room-temperature dynamical behavior of an excitation in the FMO complex. For this purpose, they have introduced several examples of observables that one might apply in a test of macrorealism. According to their numerical simulation results, they predicted that these observables lead to a violation of the Leggett-Garg inequality and, furthermore, future work should be considered for assessing more realistic models of noise in the FMO complex, potentially including correlated noise[218] and non-Markovian effects.

3.10.3 *New Insights from Theoretical Modeling of Coherence*

Following the discussions for the Quantum coherence in photosynthetic light harvesting, we experience different meaning for different models, applied for the explanation of the coherence phenomena, meant for different circumstances. As has already been known by now, quantum coherence can refer to quantum superpositions of localized molecular excitations, occurring naturally because electronic couplings between molecular excitations lead to delocalized eigenstates (a.k.a. excitons). Sometimes, coherence in light harvesting alludes to the coherent wave-like dynamics of energy transfer, on the contrary, actually reflecting the superposition of excitonic eigenstates. In the former case, the coherence is represented in the molecular site basis, whereas in the latter case, the coherence is represented in the delocalized exciton basis. It is quite important to note here that the site basis and the exciton basis are special because they are related to the spatial arrangement of chromophores and energy eigenstates of the Hamiltonian, respectively.

Although both types of coherence effects play important roles in photosynthetic light harvesting, they must be discussed separately. For example,

Quantum coherence manifested in the delocalized eigenstates of photoexcitations in photosynthetic complexes, plays a fundamental role in spectral properties, energy tuning, and energy transfer dynamics of photosynthetic light harvesting. As already discussed, the effects of excitonic coherence are more difficult to analyze and have been the subject of intense research during past few years. Spectroscopic measurements (i.e., 2DES) on the FMO complex at liquid-nitrogen (77K) and room temperatures have shown time-dependent oscillations, presumed to be quantum beating, in the amplitudes of the spectral signals, which matches the predictions of quantum theory.

3.11 Discussion & Conclusion

The quantum coherence that we have discussed in the above perspective arises through "strong coupling between chlorophyll molecules making close contact with each other in the proteins of the photosynthetic apparatus", increasing the efficiency of the processes related. This kind of combination of experimental and theoretical evidence presented in the previous sections, provide a link between the presence of quantum coherence and its biological functional role, especially in the PSII RC. According to the disordered exciton-CT model which captures the 2D spectral shape and overall time evolution very well, in the specific realizations. In the averaged-over-disorder picture, analogous to the measured kinetics, the coherences are much weaker and decay in about 300 fs, owing to disorder-induced dephasing as a result of ensemble averaging. However, the experimental 2D traces show up to 1 ps long-lived oscillations. The most obvious reason for this discrepancy is the fact that model, developed in a pure exciton basis does not include the effects of coupled excitonic vibrational dynamics, generally containing a non-trivial interplay of electronic and vibrational coherences. This has been proposed as a possible origin for long-lived oscillations. The mechanism in which long-lived electronic coherences are sustained by non-equilibrium vibrational modes is strongly supported by two additional facts:

(1) The dominant 2DES frequencies observed, at both room temperature and 80K correspond to Chl, a vibrational modes that match the energy gaps between the exciton-CT states initiating charge separation,
(2) The 2D frequency maps show that the oscillating electronic states are the ones involved in each specific exciton-CT energy gap.

Therefore, it can be concluded that vibration assisted electronic (vibronic) coherence is present in the PSII RC and, as indicated by the relative quantum-beat amplitude, it also exists in a large fraction of PSII RCs. Furthermore, the correlation between electronic coherence and efficiency indicates that vibronic coherence is used by the system to drive ultrafast and efficient charge separation. Therefore, in line with recent theoretical works,[26, 38, 170, 236, 242] it is concluded that vibronic coherences survive the aggressive background noise, thus playing an essential role in charge separation dynamics. On the basis of the substantial evidence presented, Fassioli's group proposed that the PSII RC has evolved into a delicately tuned and robust environment, providing the required exciton-vibrational matching for efficient solar-energy conversion and, consequently, that the plant PSII RC operates as a quantum-designed light trap.[244]

3.11.1 *Limitations and Controversies*

In recent years, after going through quite serious, extensive, as well as committed studies regarding the experimental evidences, related different aspects of energy transfer have been considered for several antenna complexes. But, whether, at the present state of developments, it is plausible to provide a detailed as well as a constructive idea which could lead to the solid scientific processes, underlying presence or rather, involvement of quantum coherent evolution of electronic excitations, has raised many questions, raising some of basics, like:*what are the implications for light harvesting?* At the heart of all such kind of scientific discussions, the possibility that has been raised, undergoing up till now, is that corrections to hopping transport by coherence effects, even if they be of relatively subtle nature, may enhance light-harvesting function.

The numerous studies, both theoretical and experimental related to the involvement of quantum coherence which could claim figure of merit, address the connection between coherent excitation dynamics and consequently, its implications on energy transfer performance. Basically, this is related to the quantum yield, as well as to the probability, depicting whether an absorbed photon results in a charge separation at a RC. Within such framework, optimal transport is considered to follow an intermediate regime of coupling to the environment, allowing the transference of moderate strength, i.e.,

(1) not too weak, in case of purely coherent transference in order to have the migration in heterogeneous systems,

(2) not too strong, in case of disordered antennae, containing weakly coupled chromophore, or, finally,
(3) may be very strong in case of complete environmental localization of excitation.[27,37,38,105]

In principle, though it might be possible in having an regime of transport phenomena, in an optimal regime, scant evidence exists within the domain of quantum coherence as far as the energy transfer involving this kind of mechanism is concerned. Because, generally, this happens only when, it acts faster than transport, in the incoherent limits of Förster and generalized Förster theories. However, there could exist another possible scenario, i.e., enhancing coherent evolution of transfer to acceptor sites. This could happen in case of disordered antennae containing weakly coupled chromophores, under the presence of coupling to non-equilibrium discrete vibrational modes.[103] Some of authors also have argued that it is possible to construct a variety of alternative classical models that can, in principle, produce both classical beating[241] without reference to quantum coherence at all. Such alternative descriptions must be eliminated before one can unambiguously state that the FMO complex, or other LHCs, take advantage of quantum mechanics.

It appears, further studies are needed in order to clearly establish the role of coherent exciton dynamics in light harvesting and very importantly, whether and how far this exhaustive research efforts could be exploited in technological applications. New performance measures, in this direction, together with desired functionalities should be investigated, for example, in order to find any feasibility in achieving the control over energy transfer pathways.[92]

Lastly, it should be noted that, on a more fundamental level, the observation of coherence under 'laboratory conditions' of light harvesting—i.e., femtosecond laser excitation—has raised a few important, also quite a vital questions regarding the viability, i.e., in what extent such effects are relevant, or even present, in natural light harvesting.[22,238,239] Because, basically, energy transfer is not initiated by coherent laser pulses in nature but by continuous incoherent sunlight. Not only that another complexity comes from the environment where perturbations can cause decay and decoherence, potentially leading to reduced efficiency. All of such factors raises the need for robustness against these effects which many organisms seem to exhibit in their transfer processes.

Recently, the role that quantum coherence may play for efficient light harvesting has caught some notoriety (discussed also in a prior perspective,[240] particularly due to experiments results by Fleming, Engel and Scholes.[33,41,136] In time-resolved two-dimensional spectroscopy, it is possible to see oscillations of exciton state populations, special initial states prepared by carefully chosen laser pulses. The oscillations, lasting up to a few hundred femtoseconds, are attributed to quantum coherence emerging as a result of the initially prepared coherent quantum state which decay rapidly (compared to the typical lifetime of excitation in photosynthetic systems of one nanosecond). It is not presently known how much this phenomenon contributes to efficient light harvesting.[46] The quantum coherence that we discuss in this perspective arises through strong coupling between chlorophyll molecules making close contact with each other in the proteins of the photosynthetic apparatus. However, the focus of this review is to explain how this quantum coherence increases efficiency and architectural flexibility in the light harvesting system of a fundamental photosynthetic life form, namely purple bacteria.[4,172]

3.12 Concluding Remarks

The most important conclusion that we can draw from the vast amount of works over recent years on coherence and coherent dynamics in light harvesting is that both the suggestions by experiments and predictions by theory have been accumulating evidence that the mechanisms for energy transfer in light-harvesting involve quantum coherence. At this point, a key challenge in the theory of photosynthetic energy transfer is to formulate possible quantitative relations between electronic coherence and the light-harvesting function. Even if the origin of the experimental observations may not be clear, theoretical works that have not focused on the experiments but rather on microscopic theories for energy transfer, are also an important piece in the puzzle to understand whether energy transfer may be coherent or not. Indeed a vast number of theoretical studies have been developed and applied in developing theoretical framework to investigate energy transfer in light-harvesting complexes. Using realistic parameters for electronic structure and the environment, researchers have demonstrated that the regime of energies and coupling in pigment–protein complexes is of such a nature that coherent evolution can survive in the 0.1–1 ps timescale range (depending on temperature).

Therefore, even when precise elucidation of oscillatory dynamics in the two-dimensional spectra is still being debated, this does not mean that electronic coherence is not involved in light harvesting. Even if the origin of the experimental observations may not be clear, theoretical works that have not focused on the experiments but rather on microscopic theories for energy transfer are also an important piece in the puzzle to understand whether energy transfer may be coherent or not. Indeed vast numbers of theoretical studies have been developed and applied for varieties of theoretical framework in order to investigate energy transfer in light-harvesting complexes. Using realistic parameters for electronic structure and associated environment, it has been observed that the regime of energies and coupling in pigment–protein complexes is of such nature that, depending on temperature, the evolved coherence can survive in the 0.1–1 ps timescale range only. Therefore, *even when precise elucidation of oscillatory dynamics in the two-dimensional spectra is still being debated, this does not mean that electronic coherence is not involved in light harvesting.*

Again, it is interesting to investigate correlations between efficiency and the spatio-temporal distribution of coherence when pigment-protein complexes are exposed to different light intensities. Because, the ability to regulate transport properties under different environmental conditions is crucial for robust photosynthetic systems. Under high light-conditions, for instance, photosynthetic systems switch on photo-protection mechanisms, allowing themselves in controlling (reduce) how fast excitation energy is converted into chemical energy. The mechanism, i.e., how a convenient distribution of electronic coherence among pigments could allow control of the efficiency profile, may prove useful in adjusting transport properties under variations of light-intensity. Thus, finally, it can be stated that coherence is present at all times and is continuously being created, destroyed, and recreated again, out of the interaction of the electronic system with the surrounding nuclear degrees of freedom. Thus the picture referred, out of stationary eigenstates is most probably a result of coarse-graining via a convenient mean-field approximation.

Again, the wave function of an electronic excitation in a photosynthetic complex is never stationary, perpetually evolving under the influence of the fluctuations in its condensed-phase environment. As a result, these environmental fluctuations modulate the energy as well as couplings of the collective molecular system, leading to dynamical transitions of excitation energy transfer which gives physical picture, underpinning all quantum dynamical processes in the condensed phase. A potential as well as a crucial point, to be noted here is that for coherence to play a role it is still important for the relevant process to occur in a short time scale, because otherwise microscopic dephasing (decoherence) would be complete before population transfer has occurred. These considerations are directly relevant to recent discussions of the nature of photosynthetic energy transfer and other ultrafast optical biological processes such as vision under solar radiation.

In order to have a robust photosynthetic systems, it is crucial to posses the ability in regulating the transport properties under different environmental conditions. For instance, under high light-conditions, photosynthetic systems switch on photo-protection mechanisms, allowing them to control (reduce) the mechanisms, related to the conversion of fast excitation energy into chemical energy. For example, It will be interesting to investigate correlations between efficiency and the spatio-temporal distribution of coherence when pigment-protein complexes are exposed to different light intensities. In addition to what is known so far from two-dimensional spectroscopy (2DES), following theoretical works, new experimental techniques are desirable in resolving some of the current crucial issues because of the limitations faced by them, i.e., (2DES) which needs much more development following the demands of the research, related to this subject.

Next, whenever we mention two points related to this aspect:

(1) Firstly,- one of the big questions in this field is whether coherent dynamics, observed in experiment owes origin to the coherent and ultrafast nature of the excitation. Quite surprisingly, exciton dynamics appear to be dictated, at least partly, by quantum unitary evolution. Again, sometimes, the significance of coherent dynamics for light-harvesting function appear, either quite subtle or vague, hence demanding for clear understanding, to be clarified in a more convincing manner. But, nevertheless, these effects are likely to alter the microscopic picture of energy transfer and expected, as well, to provide a detailed picture about the functioning of the light-harvesting complexes. In this direction, the techniques of 2DES with incoherent light, as proposed by Turner *et al.*[24] may provide a decisive tool in clarifying, whether coherent dynamics are consequence, only of an 'artificial' excitation process or whether, instead, the characteristics of the pigment–protein complex allow such effects under solar-like light.

(2) Secondly, the conventional spectroscopic techniques, dealing with ensemble of different complexes, and therefore, taking the average over different complexes, may obscure certain dynamical features. In this regard, recent experiments with single molecule, developed by van Hulst and co-workers,[245] will help reveal how individual complexes harvest light and also about the significance of static disorder. Recently developed theoretical works have also suggested that three pulse photon echo techniques could be applied to single molecules for obtaining the dynamics hidden by the ensemble average.[246]

the question of whether or not quantum coherence is selected for, may not be the right one to ask. Nature is inherently quantum mechanical and, as such, any interaction must have some obligate quantumness that cannot be switched off. Teasing apart the query, i.e., whether efficiency in ordered chromophore assemblies, for example, is driven by coherence or by large interaction energies, still remains an unanswerable question. Analysis of similar light harvesting systems with different coherence time, temperature or chromophore separation (as in cryptophytes) may produce some fruitful results which will be beneficial in understanding the nature of quantum processes related to light harvesting. As photosynthetic systems are found to exhibit huge number of degrees of freedom and mixed strengths of interactions, a clear small parameter required by simple perturbative treatments, is often not attainable. So far, the amount of accumulated experimental data have shown that our conventional views of excitation energy transfer, based either on the Förster picture or the Redfield picture are inadequate for describing general coherent excitation energy transfer dynamics in photosynthetic complexes. In addition, issues such as the dynamics of coherence transfer, more general forms of system-bath couplings, how to treat high frequency vibrational modes, and dynamical localization effects are largely overlooked in present models of photosynthetic excitation energy transfer.

A accurate and quantitative theoretical and practical description of full coherent excitation energy transfer dynamics, especially in the intermediate coupling regime, is crucial for advancing our understanding of the true quantum effects in photosynthetic light harvesting. Added to this fact it still remains undecided as the accurate and quantitative theoretical descriptions of quantum processes in the condensed phase remain, still a formidable challenge in theoretical chemistry. In their new work, related to magnetoreception, Imamoglu and Whaley, while developing a general approach for looking at the interactions involved in the magnetic-field measurement, to work out whether the system is indeed a quantum meter, they concluded that the measurement process hinges on the long-lived quantum coherence of the radical pair. But, on the contrary, when they applied their analysis to photosynthesis, they came to a very different conclusion, appearing quite puzzling. In this case, the quantum meter is found to be a collection of chromophore molecules, transferring energy from absorbed sunlight to a RC centre, where, as we are aware, the energy is extracted only in the form of mobile electrons. Therefore, the quantum meter measures the intensity of the sunlight in terms of the rate at which electrons are produced.

The measurement process thus begins with sunlight, pumping the chomophores from their electronic ground state into an excited state, energy being transferred as a next, to the RC by excitons (electron-hole pairs) whereas the process, for continuance, 'must first find' their way through a labyrinth of chromophores. This involves the process of hopping from molecule to molecule, similar to a random walk, the transfer occurring more rapidly and more efficiently than expected. This findings has led some physicists to suggest that the excitons travel through the chromophores via a coherent quantum superposition of all possible pathways, which could allow the excitons to find the most efficient route to the RC with very few excitons being lost along the way. To decide whether coherent transference makes a difference, Imamoglu and Whaley looked at the relevant timescales. If the excitons remained coherent for relatively long periods of time, they should be more likely to reach their destination and therefore boost the performance of the quantum meter. What the researchers found, however, is that this enhancement is at best $5 - 10\%$, *and therefore photosynthesis could function without the need for quantum coherence. As commented by Gregory Scholes of Princeton University that the role of quantum coherence in photosynthesis is still a matter of scientific debate.*

Although a rigorous formulation of the system-bath model leads to a generalized quantum master equation for the reduced-system density matrix, in principle, it describes the exact time-evolution of the system through a memory kernel function.[247] But, in practice, calculating the exact memory kernel is unfeasible and perturbation theory must be employed to obtain approximate results, which leads to, for example, the Förster theory or the Redfield theory. Numerically, exact nonperturbative methods avoid this problem. However, they are often computationally expensive, limited only to a specific form of bath spectral density, thus hindering limitations in their ability to calculate dynamics in large pigment-protein complexes and to investigate system-bath correlations and environmental effects of photosynthetic light harvesting.

Applying, their analysis to photosynthesis, Imamoglu and Whaley, came to a very different conclusion, stating that, the quantum meter should be taken as a collection of chromophore molecules, transferring energy from absorbed sunlight to a RC. Therefore, the quantum meter measures the intensity of the sunlight in terms of the rate at which electrons are produced. Thus, the development of an accurate theory for photosynthetic excitation energy transfer that is numerically efficient and also applicable to a broad parameter regime, is crucial for the fundamental understanding of coherence quantum processes. Such new theoretical developments should then be benchmarked against nonperturbative calculations before being used for quantitative study. We note that the huge literature concerning exciton and charge transport in organic molecular crystals that were developed in the " '70s to the '80s" provide a valuable reference point.

Indeed, recently several research groups have applied the phenomenological Haken-Reineker-Ströbl model, which have originally been developed for organic molecular crystals to investigate geometry factors and the interplay of quantum coherent dynamics and dephasing in the efficiency of excitation energy trapping in photosynthetic complexes. Moreover, the small-polaron approach for coherent excitation energy transfer developed by Jang *et al.*[192] also has its root in the Grover-Silbey theory for exciton transport in organic molecular crystals. Thus, a potentially fruitful venue for developing an accurate theory for excitation energy transfer in the intermediate coupling regime is to follow the variational polaron method developed by Yarkony and Silbey[250], later generalized by Cheng and Silbey.[251] Note that these approaches can not be directly applied to photosynthetic excitation energy transfer, because in contrast to organic molecular crystals, photosynthetic complexes lack translational symmetry and also exhibit strong static disorder and a energetic landscape embedded in the site energies. All these additional complexities must be included in the theory in order to achieve an accurate description of light-harvesting excitation energy transfer.

References

[1] E. Schrödinger, (1992); What is Life?, Cambridge University Press.
[2] P.C.W. Davies; (2008); Quantum Aspects of Life; Imperial College Press, London.

[3] H.C. Longuet-Higgins; Quantum mechanics and biology; (1962) Biophys. J.; vol 2; 207.
[4] G.S. Engel, T.R. Calhoun, E.L. Read, T.-K. Ahn, T. Mancal, Y.-C. Cheng, R.E. Blankenship, G.R. Fleming; (2007); Evidence for wavelike energy transfer through quantum coherence in photosynthetic systems, Nature vol 446; 782.
[5] Lee H, Cheng YC, Fleming GR.; 2007; Coherence dynamics in photosynthesis: protein protection of excitonic coherence. Science; 316:1462–1465. [PubMed] [Google Scholar]
[6] E.Collini, C.Y. Wong, K. E. Wilk, P. M. G. Curmi, P. Brumer, G. D. Scholes, Coherently wired light-harvesting in photosynthetic marine algae at ambient temperature, Nature 453 (2010) 644.
[7] G. Panitchayangkoon, D. Hayes, K. A. Fransted, J. R. Caram, E. Harel, J. Wen, R. E. Blankenship, G. S. Engel; (2010); Long-lived quantum coherence in photosynthetic complexes at physiological temperature, PNAS, vol 107; 12766.
[8] G. Panitchayangkoon, D. V. Voronine, D. Abramavicius, J. R. Caram, N. H. C. Lewis, S. Mukamel, G. S. Engel; (2011); Direct evidence of quantum transport in photosynthetic light-harvesting complexes; PNAS; vol 108; 20908.
[9] T.-C. Yen, Y.-C. Cheng, Electronic coherence effects in photosynthetic light harvesting, Procedia Chemistry 3 (2011) 211.
[10] Blankenship, R. E. Molecular Mechanisms of Photosynthesis (Blackwell Science, Oxford/Malden, 2002).
[11] Vos MH, Rappaport F, Lambry J-C, Breton J, Martin J-L. Visualization of coherent nuclear motion in a membrane protein by femtosecond spectroscopy. Nature. 1993; 363:320–325. [Google Scholar]
[12] Parson WW, Warshel A. A density-matrix model of photosynthetic electron transfer with microscopically estimated vibrational relaxation times. Chem. Phys. 2004; 296:201–216. [Google Scholar]
[13] Novoderezhkin VI, Yakovlev AG, Van Grondelle R, Shuvalov VA. Coherent nuclear and electronic dynamics in primary charge separation in photosynthetic reaction centers: A Redfield theory approach. J. Phys. Chem. B. 2004; 108:7445–7457. [Google Scholar]
[14] Savikhin S, Buck DR, Struve WS. Oscillating anisotropies in a bacteriochlorophyll protein: Evidence for quantum beating between exciton levels. Chemical Physics. 1997; 223:303–312. [Google Scholar]
[15] Brixner T, *et al.*; 2005; Two-dimensional spectroscopy of electronic couplings in photosynthesis; Nature.; 434:625–628. [PubMed] [Google Scholar]

[16] Zigmantas D, *et al.*; 2006; Two-dimensional electronic spectroscopy of the B800-B820 light-harvesting complex. Proc. Natl. Acad. Sci. U. S. A.; 103:12672–12677. [PMC free article] [PubMed] [Google Scholar]

[17] Calhoun TR, *et al.*; 2009; Quantum coherence enabled determination of the energy landscape in light-harvesting complex II. J. Phys. Chem. B Lett.; 113:16291–16295. [PubMed] [Google Scholar]

[18] Collini E, *et al.*; 2010; Coherently wired light-harvesting in photosynthetic marine algae at ambient temperature. Nature.; 463:644–647. [PubMed] [Google Scholar]

[19] Schlau-Cohen GS, *et al.*; 2012; Elucidation of the timescales and origins of quantum electronic coherence in LHCII. Nat. Chem.; 4:389–395. [PubMed] [Google Scholar]

[20] Hildner R, Brinks D, Nieder JB, Cogdell RJ, van Hulst NF.; 2013; Quantum coherent energy transfer over varying pathways in single light-harvesting complexes; Science.; 340:1448–1451. [PubMed] [Google Scholar]

[21] Westenhoff S, Palecek D, Edlund P, Smith P, Zigmantas D.; 2012; Coherent picosecond exciton dynamics in a photosynthetic reaction center. J. Am. Chem. Soc.; 134:16484–16487. [PubMed] [Google Scholar)

[22] Mancal, T.; Bixner, O.; Christensson, N.; Hauer, J.; Milota, F.; Nemeth, A.; Sperling, J.; Kauffmann, H. F.; 2011; Dynamics of Quantum Wave Packets in Complex Molecules Traced by 2D Coher-entElectronic Correlation Spectroscopy. Procedia Chem., 3, 105–117.

[23] Christensson N, Milota F, Hauer J, Sperling J, Bixner O, Nemeth A, Kauffmann HF. 2011 High frequency vibrational modulations in two-dimensional electronic spectra and their resemblance to electronic coherence signatures. J. Phys. Chem. B, 115, 5383–5391. (doi:10.1021/jp109442b)

[24] Turner, D. B.; Wilk, K. E.; Curmi, P. M. G.; Scholes, G. D.; 2011; Comparison of Electronic and Vibrational Coherence Measured by Two-Dimensional Electronic Spectroscopy. J. Phys. Chem. Lett.; 2, 1904–1911.

[25] Scholes GD, Fleming GR, Olaya-Castro A, van Grondelle R.; 2011; Lessons from nature about solar light harvesting.; Nat. Chem.; 3:763–774. [PubMed] [Google Scholar]

[26] Cheng YC, Fleming GR.; 2009; Dynamics of light harvesting in photosynthesis; Annu. Rev. Phys. Chem.; 60:241–262.

[27] Plenio M.B. and S. F. Huelga S.F.; 2008; Quantum networks and Biomolecules; –arXiv-0807-4902v-30th July, 2008].

[28] Jonas, D. M.; 2003; Two-dimensional femtosecond spectroscopy; Annu. Rev. Phys. Chem.; 54, 425–463.
[29] Cowan, M. L., Ogilvie, J. P. & Miller, R. J. D.; 2004; Two-dimensional spectroscopy using diffractive optics based phased-locked photon echoes. Chem. Phys. Lett. 386, 184–189.
[30] Fenna, R. E. & Matthews, B. W.; 1975; Chlorophyll arrangement in a bacteriochlorophyll protein from Chlorobium limicola. Nature 258, 573–577.
[31] Camara-Artigas, A., Blankenship, R. E. & Allen, J. P.; 2003; The structure of the FMO protein from Chlorobium tepidum at 2.2 angstrom resolution. Photosynth. Res. 75, 49–55.
[32] Tronrud, D. E., Wen, J., Gay, L., and Blankenship, R. E.; 2009; The structural basis for the difference in absorbance spectra for the FMO protein from various green sulfur bacteria. Photosynthesis Research, 100, 79.;
[33] Cho, M. H. *et al.* Exciton analysis in 2D electronic spectroscopy; 2005; J. Phys. Chem. B109, 10542–10556.
[34] Perrin, F.: 1932; Thoérie quantique des transferts d'activation entre molécules de même espèce. Cas des solutions fluorescentes. Ann. Phys. (Paris) 17, 283–314.
[35] Knox, R. S. Electronic excitation transfer in the photosynthetic unit; 1996; Reflections on work of William Arnold; Photosynth. Res. 48, 35–39.
[36] Leegwater, J. A. Coherent versus incoherent energy transfer and trapping in photosynthetic antenna complexes. J. Phys. Chem. 100, 14403–14409 (1996).
[37] M. Mohseni, P. Robentrost, S. Lloyd, A. Aspuru-Guzik; 2008; Environment-assisted quantum walks in photosynthetic energy transfer, Jour. Chem.Phys. 129, 176106.
[38] F. Caruso, A. W. Chin, A. Datta, S. F. Huelga, M. B. Plenio; 2009; Highly efficient energy excitation transer in light-harvesting complexes: The fundamental role of noise-assisted transport, J. Chem. Phys. 131, 105106.
[39] H. Lee, Y.-C. Cheng, G. R. Fleming, Quantum Coherence Accelerating Photosynthetic Energy Transfer; 2009; Springer Series in Chemical Physics; 92, 60.
[40] Lambert, Neil; Yueh-Nan Chen, Yuan-Chung Cheng, Che-Ming Lid, Guang-Yin Chenb, Franco Nori; 2012; Functional quantum biology in photosynthesis and magnetoreception; Phys Reports; May 7; 2012.
[41] Fenna, R.E.; Mathews, B.W.; 1975; Chlorophyll arrangement in a bacteriochlorophyll protein from Chlorobium limicola, Nature 258; 573.

[42] Olson J M; 2004; Photosynth. Res.; 80, 181.
[43] Vulto, S. I. E., de Baat, M. A., Neerken, S., Nowak, F. R., van Amerongen, H., Amesz, J., and Aartsma, T. J.; 1999; Excited state dynamics in FMO antenna complexes from photosynthetic green sulfur bacteria: a kinetic model. Journal of Physical Chemistry B, 103(38), 8153.
[44] T. Renger, A. Klinger, F. Steinecker, M. Schmidt am Busch, J. Numata, F. Müh; 2012: Normal mode analysis of the spectral density of the Fenna-Matthews-Olson Light-Harvesting protein: How the protein dissipates the excess energy of excitons. J. Phys. Chem. B 116, 14565–14580.
[45] Renger, T.; 2009; Theory of excitation energy transfer: from structure to function, Photosynth. Res. 102; 471.
[46] Hayes, D., and Engel, G. S.; 2011; Extracting the excitonic Hamiltonian of the Fenna-Matthews-Olson complex using three- dimensional third-order electronic spectroscopy. Biophysical Journal, 100, 2043.
[47] A. W. Chin, S. F. Huelga, M. B. Plenio; 2012; Coherence and decoherence in biological systems: Principles of noise-assisted transport and the origin of long-lived coherences; 2012; Philos. Trans. A Math Phys. Eng. Sci. 370, 3638–3657.
[48] B. S. Rolczynski, H. Zheng, V. P. Singh, P. Navotnaya, A. R. Ginzburg, J. R. Caram, K. Ashraf, A. T. Gardiner, S.-H. Yeh, S. Kais, R. J. Cogdell, G. S. Engel; 2018; Correlated protein environments drive quantum coherence lifetimes in photosynthetic pigment-protein complexes. Chem 4, 138–149.
[49] A. Olaya-Castro, C.F. Lee, F. Fassioli Olsen and N.F. Johnson; 2008; Phys. Rev. B 78.
[50] P. Rebentrost, M. Mohseni, I. Kassal, S. Lloyd, A. Aspuru-Guzik; 2009; New J. Phys. 11, 033003.
[51] C. Olbrich, J. Strümpfer, K. Schulten, U. Kleinekathöfer; 2011; Theory and simulation of the environmental effects on FMO electronic transitions. J. Phys. Chem. Lett. 2, 1771–1776.
[52] E. Rivera, D. Montemayor, M. Masia, D. F. Coker: 2013 Influence ofsite-dependent pigment–Protein interactions on excitation energy transfer in photosynthetic light harvesting; 2013; J. Phys. Chem. B 117, 5510–5521.
[53] P. Brumer, M. Shapiro, 2012; Molecular response in one-photon absorption via natural thermal light vs. pulsed laser excitation. Proc. Natl. Acad. Sci. USA 109, 19575–19578. (doi:10.1073/pnas.1211209109)

[54] T. Mančal, L. Valkunas; 2010; Exciton dynamics in photosynthetic complexes: Excitation by coherent and incoherent light. New J. Phys. 12, 065044.
[55] J. Olšina, A. G. Dijkstra, C. Wang, J. Cao; 2014; Can natural sunlight induce coherent exciton dynamics? arXiv:1408.5385.
[56] H. C. H. Chan, O. E. Gamel, G. R. Fleming, K. B. Whaley; 2018; Single-photon absorption by single photosynthetic light-harvesting complexes. J. Phys. B At. Mol. Opt. Phys. 51, 05.
[57] G. Raszewski, T. Renger, Light harvesting in photosystem II core complexes is limited by the transfer to the trap: Can the core complex turn into a photoprotective mode?; 2008; J. Am. Chem. Soc. 130, 4431–4446.
[58] F. Müh, M. Plöckinger, T. Renger, Electrostatic asymmetry in the Reaction Center of Photosystem II; 2017; J. Phys. Chem. Lett. 8, 850–858.
[59] X. Hu, T. Ritz, A. Damjanovic, F. Autenrieth, K. Schulten, Photosynthetic apparatus of purple bacteria; 2002; Q. Rev. Biophys. 35; 1–62.]] for a review of the PSU in purple bacteria)]]
[60] J. S. Briggs, A. Eisfeld; 2011; Equivalence of quantum and classical coherence in electronic energy transfer, Phys. Rev. E 83, 051911.
[61] Johannes Kofler and Caslav Brukner; 2008; Conditions for quantum violation of macroscopic realism. Physical Review Letters, 101(9):090403.
[62] Kofler *et al.*; 2008; Quantum violation of macroscopic realism and the transition to classical physics. PhD thesis, University of Vienna, Vienna, Austria, June 2008.
[63] Francesca Fassioli, Ahsan Nazir, and Alexandra OlayaCastro; 2009; Multichromophoric energy transfer under the influence of a correlated environment. arXiv:0907.5183.
[64] Patrick Rebentrost, Rupak Chakraborty, and Alan Aspuru-Guzik; 2009; Non-Markovian quantum jumps in excitonic energytransfer. arXiv:0908.1961.
[65] Ahsan Nazir; 2009; Correlation-dependent coherent to incoherent transitions in resonant energy transfer dynamics.arXiv:0906.0592.
[66] R. Monshouwer, M. Abrahamsson, F. van Mourik, R. van Grondelle, Superradiance and exciton delocalization in bacterial photosynthetic light-harvesting systems, J. Phys. Chem. B 101 (1997) 7241.
[67] Y. Zhao, T. Meier, W. Zhang, V. Y. Chernyak, S. Mukamel, Superradiance coherence sizes in single-molecule spectroscopy of LH2 antenna complexes, J. Phys. Chem. B 103 (1999) 3954.

[68] D. Leupold, D. Stiel, K. Teuchner, F. Nowak, W. Sandner, B. Ucker, H. Scheer, Size enhancement of transition dipoles to one and twoexciton bands in a photosynthetic antenna, Phys. Rev. Lett. 77 (1996) 4675.

[69] R. van Grondelle, V. I. Novoderezhkin, Energy transfer in photosynthesis: experimental insights and quantitative models, Phys. Chem.Chem. Phys. 8 (2006) 793.]]

[70] J. Barber; 2009; Photosynthetic energy conversion: natural and artificial, Chem. Soc. Rev. 38, 185.

[71] R. E. Blankenship, D. M. Tiede, J. Barber, G. W. Brudvig, G. Fleming, M. Ghirardi, M. R. Gunner, W. Junge, D. M. Kramer, A. Melis, T. A. Moore, C. C. Moser, D. G. Nocera, A. J. Nozik, D. R. Ort, W. W. Parson, R. C. Prince, R. T. Sayre:2011; Comparing Photosynthetic and Photovoltaic Efficiencies and Recognizing the Potential for Improvement, Science 332, 805.

[72] J. Yuen-Zhou, A. Aspuru-Guzik, Quantum process tomography of excitonic dimers from two-dimensional electronic spectroscopy. I. Generaltheory and application to homodimers, J. Chem. Phys. 134 (2011) 134505.

[73] J. Yuen-Zhou, J. J. Krich, M. Mohseni, A. Aspuru-Guzik, Quantum process tomography of excitonic dimers from two-dimensional electronic spectroscopy. I. General theory and application to homodimers, Proc. Nat. Ac. Sci. 108 (2011) 17615.

[74] D. Abramavicius, B. Palmieri, D. V. Voronine, F. Sanda, S. Mukamel; 2009; Coherent multidimensional optical spectroscopy of excitons in molecular aggregates; quasiparticle versus supermolecule perspectives, Chem. Rev. 109, 2350.

[75] N. S. Ginsberg, Y.-C. Cheng, G. R. Fleming; 2009; Two-dimensional electronic spectroscopy of molecular aggregates, Acc. Chem. Res. 42, 1352.

[76] Cho, M. 2009. Two-Dimensional Optical Spectroscopy. Boca Raton: CRC Press.

[77] Pisliakov, A.V., Mancal, T., and Fleming, G.R.; 2006; Two-dimensional optical three-pulse photon echo spectroscopy II. Signatures of coherent electronic motion and exciton population transfer in dimer two-dimensional spectra. Journal of Chemical Physics, 124, 234505.

[78] Potts, D., and Kunis, S. 2007. Stability results for scattered data interpolation by trigonometric polynomials.

[79] Palmieri, B., Abramavicius, D., and Mukamel, S. 2009. Lindblad equations for strongly coupled populations and coherences in photosynthetic complexes. Journal of Chemical Physics, 130(20), 204512.

[80] Caram, J. R., and Engel, G. S. 2011. Extracting dynamics of excitonic coherences in congested spectra of photosynthetic light harvesting antenna complexes. Faraday Discussions, 153, 93.
[81] Caram, J. R., Lewis, N. H. C., Fidler, A. F., and Engel, G. S. 2012. Signatures of correlated excitonic dynamics in two dimensional spectroscopy of the Fenna-Matthews-Olson photosynthetic complex. Journal of Chemical Physics, 136, 104505.
[82] Heller EJ. 1981. The semi-classical way to molecular-spectroscopy. Accounts of Chemical Research 14: 368–375.
[83] Bitto H, Huber JR. 1992. Molecular quantum beats: High-resolution spectroscopy in the time domain. Accounts of Chemical Research 25: 65–71.
[84] Jonas DM, Fleming GR. 1995 Vibrationally abrupt pulses in pump-probe spectroscopy. In Ultrafast processes in chemistry and photobiology (Chemistry in the 21st century series IUPAC) (eds. MA El-Sayed, I Tanaka, Y Molin), pp. 225–256. Oxford, UK: Blackwell Science.
[85] Zewail AH. 2000. Femtochemistry: Atomic-scale dynamics of the chemical bond. Journal of Physical Chemistry A 104: 5660–5694.
[86] Butkus, V., Zigmantas, D., Valkunas, L., and Abramavicius, D.; 2012; 2012. Vibrational vs. electronic coherences in 2D spectrum of molecular systems. Chemical Physics Letters, 545, 40–43.
[87] Ogilvie JP, Kubarych KJ.; 2009; Multidimensional electronic and vibrational spectroscopy: An ultrafast probe of molecular relaxation and reaction dynamics. Pages 249–321 in Arimondo E, Berman PR, Lin CC, eds. Advances in Atomic, Molecular, and Optical Physics, vol. 57.
[88] Jessica M. Anna, Gregory D Scholles & Grondelle; 2013; Bioscience Advance Access, Dec 5.
[89] Myers JA, Lewis KLM, Fuller FD, Tekavec PF, Yocum CF, Ogilvie JP.; 2010; Two-dimensional electronic spectroscopy of the D1-D2-cyt b559 photosystem II reaction center complex; Journal of Physical Chemistry Letters 1: 2774–2780.
[90] Ostromov *et al.*; 2013; Ostroumov EE, Mulvaney RM, Cogdell RJ, Scholes GD. 2013. Broadband 2D electronic spectroscopy reveals a carotenoid dark state in purple bacteria. Science 340: 52–56.
[91] Tiwari V, Peters WK, Jonas DM. 2013 Electronic resonance with anticorrelated pigment vibrations drives photosynthetic energy transfer outside the adiabatic framework. Proc. Natl Acad. Sci. USA 110, 1203–1208. (doi:10.1073/pnas.1211157110)
[92] Francesca Fassioli, Rayomond Dinshaw, Paul C. Arpin and Gregory D. Scholes; 2013; Photosynthetic light harvesting: excitons and coherence; the Royal society Interface; 1–22. and references there in.

[93] Richards GH, Wilk KE, Curmi PMG, Quiney M, Davis JA. 2012 Coherent vibronic coupling in lightharvesting complexes from photosynthetic marine algae. J. Phys. Chem. Lett. 3, 272–277.
[94] Fragnito HL, Bigot J-Y, Becker PC, Shank CV. 1989 Evolution of the vibronic absorption spectrum in a molecule following impulsive excitation with a 6 fs optical pulse. Chem. Phys. Lett. 160, 101–104. (doi:10.1016/0009-2614(89)87564-5)
[95] Pollard WT, Fragnito HL, Bigot J-Y, Shank CV, Mathies RA. 1990 Quantum-mechanical theory for 6 fs dynamic absorption spectroscopy and its application to nile blue. Chem. Phys. Lett. 168, 239–245. (doi:10.1016/0009-2614(90)85603-A)
[96] A. J. Leggett, S. Chakravarty, A. T. Dorsey, Matthew P. A. Fisher, Anupam Garg, and W. Zwerger; 1987; Rev. Mod. Phys. 59, 1 – Published 1 January 1987; Erratum Rev. Mod. Phys. 67, 725 (1995); Dynamics of the dissipative two-state system.
[97] N.V. Prokofiev, P.C.E. Stamp, Phys. Rev. Lett. 80, 5794 (1998).
[98] Y. Cha, C. Murray, J. Klinm; Science, 243 (1989) 1325.
[99] A. Kohen, Prog. React. Kinet. Mec., 28 (2003) 119.
[100] Horodecki R, Horodecki P, Horodecki M and Horodecki K 2009 Rev. Mod. Phys. 81 865.
[101] Olaya Castro *et al.*; 2010; New J.Phys; 12; 08006.
[102] Fassioli Francesca & Alexandra Olaya Castro; 2010; arXiv: 1003.361Ov2[quant-phys]15July.
[103] Romero *et al.*; (2014); Published on line; NPHYS 3017, 13 July.
[104] R.P. Feynman and A.R. Hibbs, Quantum Mechanics and Path Integrals (McGee's Hill, New York, 1965).
[105] J. Itatani, J. Levesque, D. Zeidler, H. Niikura, H. Pepin, J.C. Kieffer, P.B. Corkum, and D.M. Villeneuve, Nature 432 (2004) 867; H.J.
[106] Worner, J.B. Bertrand, D.V. Kartashov, P.B. Corkum, D.M. Villeneuve, Nature 466 (2010) 604.
[107] T.J. Dunn, I.A. Walmsley, and S. Mukamel, Phys. Rev. Lett. 74 (1995) 884.
[108] F. Mintert, A.R.R. Carvalho, M. Kus, and A. Buchleitner, Phys. Rep. 415 (2005) 207.
[109] S. Lloyd, Science 261 (1993) 1569.
[110] V. Vedral, M.B. Plenio, M.A. Rippin, and P.L. Knight, Phys. Rev. Lett. 78 (1997) 2275.
[111] S. Mukamel, Principles of Nonlinear Optical Spectroscopy (Oxford University Press, New York, 1995).
[112] J. Rammer, Quantum Transport Theory (Westwiew Press, Colorado, 2008).

[113] Miller WH. 2012. Perspective: Quantum or classical coherence? Journal of Chemical Physics 136 (art. 210901).
[114] J. Cao, R.J. Silbey, J. Phys. Chem. A, 113 (2009) 13825.
[115] Prytkova T.R., I.V. Kurnikov, D.N. Beratan, Science, 315 (2007) 622.
[116] Nogly Przemyslaw; 2018; Retinal isomerization in bacteriorhodopsin captured by a femtosecond x-ray laser; Science; 361; 6398.
[117] Rebentrost, P, Mohseni, M, Kassal, I, Lloyd, S, & Aspuru-Guzik; A (2009); "Environment-assisted quantum transport." New J. Phys. 11, 033003.
[118] Horodecki R, Horodecki P, Horodecki M and Horodecki K 2009 Rev. Mod. Phys. 81 865.
[119] Vedral V 2008 Nature 453 1004]].–must be referred; HFSP Journal, Vol. 3, No. 6, December 2009, 386–400.
[120] van Grondelle R & Novoderezhkin; 2010; Nature, 463, 414.
[121] Sarovar M, Ishizaki A, Fleming G R and Whaley K B; 2010; Nature Physics; 6; 462.
[122] Hu X *et al.*; 1997.
[123] Philpott, M. R. Theory of Coupling of Electronic and Vibrational Excitations in Molecular Crystals and Helical Polymers. J. Chem. Phys. 1971, 55, 2039-&.
[124] Womick, J. M.; Moran, A. M. Vibronic Enhancement of Exciton Sizes and Energy Transport in Photosynthetic Complexes. J. Phys. Chem. B 2011, 115, 1347–1356.
[125] H.B. Gray, J.R. Winkler, Proc. Natl. Acad. Sci. USA, 102 (2005) 3534.
[126] J. Regan, J.N. Onuchic, Adv. Chem. Phys., 107 (1999) 497.
[127] D.N. Beratan, J.N. Onuchic, J.R. Winkler, H.B. Gray, Science, 258 (1992) 1740.
[128] J.N. Onuchic, D.N. Beratan, J.R. Winkler, H.B. Gray, Annu. Rev. Bioph. Biom., 21 (1992) 349.
[129] J. Regan, A. Dibilio, R. Langen, L. Skov, J.R. Winkler, H.B. Gray, J.N. Onuchic, Chem Biol, 2 (1995) 489.
[130] J. Regan, S. Risser, D.N. Beratan, J.N. Onuchic, J. Phys. Chem., 97 (1993) 13083.
[131] Butkus Vytautas, Hui Dong, Graham R. Fleming, Darius Abramavicius, and Leonas Valkunas; 2016; Disorder-Induced Quantum Beats in Two-Dimensional Spectra of Excitonically Coupled Molecules; J. Chem., Lett., 7; 277–282.
[132] Butkus, V.; Valkunas, L.; Abramavicius, D.; 2014; Vibronic Phenomenaand Exciton-Vibrational Interference in Two-dimensional Spectra of Molecular Aggregates. J. Chem. Phys. 2014, 140, 034306.

[133] Barford, W.; Trembath, D. Exciton Localization in Polymers with Static Disorder. Phys. Rev. B: Condens. Matter Mater. Phys. 2009, 80, 165418.

[134] van Amerongen, H.; Valkunas, L.; van Grondelle, R.; 2000; Photosynthetic Excitons; World Scientific: Singapore.

[135] Ishizaki, A.; Fleming, G. R.; 2009; Theoretical Examination of Quantum Coherence in a Photosynthetic System at Physiological Temperature. Proc. Natl. Acad. Sci. U. S. A., 106, 17255–17260.

[136] Fleming, G. R.; Scholes, G. D.; Cheng, Y.-C. Quantum Effects in Biology.; 2011; Procedia Chem. 2011, 3, 38–57. 22nd Solvay Conference on Chemistry.

[137] Savikhin S, van Noort PI, Blankenship RE, Struve WS. Femtosecond probe of structural analogies between chlorosomes and bacteriochlorophyll c aggregates. Biophys J. 1995; 69:1100–1104. [PMC free article] [PubMed] [Google Scholar]

[138] Savikhin S, van Noort PI, Zhu Y, *et al.* Ultrafast energy transfer in light-harvesting chlorosomes from the green sulfur bacterium Chlorobium tepidum. Chem Phys. 1995; 194:245–258. [PubMed] [Google Scholar]

[139] Savikhin S, Zhu Y, Lin S, *et al.* Femtosecond spectroscopy of chlorosome antennas from the green photosynthetic bacterium Chloroflexus aurantiacus. J Phys Chem. 1994; 98:10322–10334. [Google Scholar]

[140] Ma F, Yu L-J, Hendrikx R *et al.* (2017) Excitonic and vibrational coherence in the excitation relaxation process of two LH1 complexes as revealed by two-dimensional electronic spectroscopy. J Phys Chem Lett 8:2751–2756.

[141] Prokhorenko VI, Steensgaard DB, Holzwarth AR (2000) Exciton dynamics in the chlorosomal antennae of the green bacteria Chloroflexus aurantiacus and Chlorobium tepidum. Biophys J 79:2105–2120.

[142] Dostál J, Mančal T, Vácha F *et al.* (2014) Unraveling the nature of coherent beatings in chlorosomes.; 2014 J Chem Phys 140:1151.

[143] Rolczynski BS, Zheng H, Singh VP *et al.* (2018) Correlated protein environments drive quantum coherence lifetimes in photosynthetic pigment-protein complexes. Chem 4:138–149.

[144] Yin Y, Katsanos DE, Evangelou SN (2008); Quantum walks on a random environment; Phys.rev.A77; http://doi.org/10.1103/physreva.77.o22302.

[145] Rebentrost P., Mohseni M., kasal I *et al.* (2009) Environ-assisted quantum transport.new J. Phys. 11:033003 *et al.* (2009).

[146] Link G, Berthold T, Bechtold M *et al.*; 2001; Structure of the P700(+)A1(−) radical pair intermediate in photosystem I by

high time resolution multifrequency electron paramagnetic resonance: analysis of quantum beat oscillations. J Am Chem Soc 123:4211–4222.

[147] Thyrhaug E, Tempelaar R, Alcocer MJP, *et al.* (2018) Identification and characterization of diverse coherences in the Fenna–Matthews–Olson complex; Nat Chem https://doi.org/10.1038/s41557-018-0060-5.

[148] Thyrhaug E, Žídek K, Dostál J *et al.*; (2016); Exciton structure and energy transfer in the Fenna–Matthews–Olson complex; J Phys Chem Lett 7:1653–1660.

[149] Fuller FD, Pan J, Gelzinis A *et al.* (2014); Vibronic coherence in oxygenic photosynthesis. Nat Chem 6:706–711 Jacob C Dean, Tihana mirkovic, Zi S.D.Toa, Daniel G Oblinsky & Gregory D. Scholes–Chem 1, 858-872 Dec 8; 2016, Elsevier Inc.

[150] Huang, L. *et al.*; 2012; Cofactor-specific photochemical function resolved by ultrafast spectroscopy in photosynthetic reaction center crystals. Proceedings of the National Academy of Sciences.

[151] Jacob C Dean, Tihana mirkovic, Zi S.D.Toa, Daniel G Oblinsky & Gregory D. Scholes; 2016; Chem 1, 858-872 Dec 8; Elsevier Inc.

[152] Jortner, J. & Bixon, M (1988): Intramolecular vibrational excitations accompanying solvent-controlled electronic transfer reactions; J. Chem. Phys. 88, 167–170.

[153] Richards GH, Wilk KE, Curmi PMG, Quiney M, Davis JA. 2012 Coherent vibronic coupling in light harvesting complexes from photosynthetic marine algae. J. Phys. Chem. Lett. 3, 272–277. (doi:10.1021/jz201600f)

[154] Fujihashi, Y., Fleming, G.R., & Ishizaki, A (2015); Impact of environmentally induced fluctuation on quaantum mechanically mixed electron and vibrational pigment states in photosynthesis energy transfer and 2D electronic spectra; J. Chem. Phys. 142; 212403.

[155] Monahan D.M., Whaley-Mayda, L.Ishizaki, A.and Fleming, G.R.(2015); Influence of weak vibrational-electronic spectra and inter-site coherences in weakly couplings on 2D electronicspectra and inter-site coherence in weakly coupled photosynthetic complexes.; J. Chem. Phys., 143; 065101.

[156] Yuen-Zhou J, Krich JJ, Aspuru-Guzik A. 2012 A witness for coherent electronic vs vibronic-only oscillations in ultrafast spectroscopy. J. Chem. Phys. 136, 234501. (doi:10.1063/1.4725498)

[157] Fragnito HL, Bigot J-Y, Becker PC, Shank CV. 1989 Evolution of the vibronic absorption spectrum in a molecule following impulsive excitation with a 6 fs optical pulse. Chem. Phys. Lett. 160, 101–104. (doi:10.1016/0009-2614(89)87564-5)

[158] Pollard WT, Fragnito HL, Bigot J-Y, Shank CV, Mathies RA. 1990 Quantum-mechanical theory for 6 fs dynamic absorption spectroscopy and its application to nile blue. Chem. Phys. Lett. 168, 239–245. (doi:10.1016/0009-2614(90)85603-A)

[159] Vos, M. H., Rappaport, F., Lambry, J-C., Breton, J. & Martin, J-L.; 1993Visualization of coherent nuclear motion in a membrane protein by femtosecond spectroscopy; Nature 363, 320–325 (1993).

[160] Deisenhofer J. *et al.*, Deisenhofer J, Epp O, Mikki K, Huber R, Michel H. Structure of the protein subunits in the photosynthetic reaction centre of Rhodopseudomonas viridis at 3Å resolution. Nature. 1985; 318:618–624. [PubMed: 22439175]

[161] Sener MK, Olsen JD, Hunter CN, Schulten K. Atomic level structural and functional model of a bacterial photosynthetic membrane vesicle. Proc. Natl. Acad. Sci. USA. 2007; 104:15723–15728.[PubMed: 17895378]

[162] Şener, MK.; Schulten, K. The Purple Phototrophic Bacteria. Hunter, CN.; Daldal, F.; Thurnauer, MC.; Beatty, JT., editors. Vol. 28. Springer: Advances in Photosynthesis and Respiration; 2008. p. 275–294.

[163] Strüumpfer, J. Deisenhofer J, Epp O, Mikki K, Huber R, Michel H. Structure of the protein subunits in the photosynthetic reaction centre of Rhodopseudomonas viridis at 3Å resolution. Nature. 1985; 318:618–624. [PubMed: 22439175]

[164] Timpmann K, Trinkunas G, Qian P, Hunter CN, Freiberg A.; 2005; Excitons in core LH1 antenna complexes of photosynthetic bacteria: Evidence for strong resonant coupling and off-diagonal disorder. Chem. Phys. Lett. 2005; 414:359–363.

[165] Herman P, Kleinekathöfer U, Barvík I, Schreiber M. Exciton scattering in light-harvesting systems of purple bacteria. J. Luminesc. 2001; 94–95:4.

[166] Sumi H. Bacterial photosynthesis begins with quantum-mechanical coherence. Chem. Rec. 2001; 1:480–493. [PubMed: 11933253]

[167] Caycedo-Soler F, Rodríguez FJ, Quiroga L, Johnson NF. Interplay between excitation kinetics and reaction-center dynamics in purple bacteria. (2010) New J. Phys. 12 095008.

[168] Ishizaki A, Fleming GR. Unified treatment of quantum coherent and incoherent hopping dynamics in electronic energy transfer: Reduced hierarchy equation approach. J. Chem. Phys. 2009; 130:234111–2341110. [PubMed: 19548715]

[169] Elisabet Romero, Zavier Prior *et al.*; 2017; Quantum — coherent dynamics in photosynthetic charge separation revealed by wavelet analysis; Scientific Reprts 7; Article 2890.

[170] Davydov, A.S.; 1982; Biology and Quantum Mechanics, Pergamon Press, New York, 1982.
[171] Fassioli F.; Olaya Castro A., Scheuring *et al.*; 2009; Biophys. J.; 97; 2464.
[172] Ahn T.K., Avenson T J *et al.*; 2009; Plant Cell; v21(6); 1798–1812.
[173] Cogdell RJ, Gall A, Koöhler J. 2006 The architecture and function of the light-harvesting apparatus of purple bacteria: from single molecules to in vivo membranes. Q. Rev. Biophys. 39, 227–324. (doi:10.1017/S0033583506004434)
[174] A M Van Oijen AM, Ketelaars M, Kohler J, Aartsma TJ, Schmidt J. 1999 Unraveling the electronic structure of individual photosynthetic pigment –protein complexes. Science 285, 400–402. (doi:10.1126/science.285.5426.400)
[175] Read EL, Schlau-Cohen GS, Engel GS, Georgiou T, Papiz MZ, Fleming GR. 2009 Pigment organization and energy level structure in light-harvesting complex 4: insights from two-dimensional electronic spectroscopy. J. Phys. Chem. B 113, 6495–6504. (doi:10.1021/jp809713q)
[176] Sundstrom V, Pullerits T, Van Grondelle R. 1999; Photosynthetic light-harvesting: reconciling dynamics and structure of purple bacterial LH2reveals function of phyotosynthetic unit. J. Phys.Chem. B 103, 2327–2346. (doi:10.1021/jp983722+)
[177] Jang, S.J. *et al.*; (2018).
[178] Mercer, Ian P.; C. El-Taha, Yasin, Kajumba, Nathaniel; Marangos Jonathan P., Tisch W.G. John, Gabrielsen, Mads; Cogdell J. Richard, Emma Springate, and Edmund Turcu; 2009; PhysInstantaneous Mapping of Coherently Coupled Electronic Transitions and Energy Transfers in a Photosynthetic Complex Using Angle-Resolved Coherent Optical Wave-Mixing.; Rev. Lett. 102, 057402 – Published 6 February, 2009.
[179] Harel Elad and Gregory S. Engel; Quantum coherence spectroscopy reveals complex dynamics in bacterial light-harvesting complex 2 (LH2); PNAS January 17, 2012 109 (3) 706–711.
[180] Ostromov *et al.* (2013); Phytotoxicity of a surfactant-containing product towards macrophytes; [Article in English]; https://www.researchgate.net/publication/263456086; Russian Journal of General Chemistry 83(13):2614–2617.
[181] Simon Scheuring & Sturgis James N.; 2005; Chromatic adaptation of photosynthetic membranes; Science; 2005; July 15; 309(5733); 484-7.
[182] Mc Clure *et al.* (2014).
[183] Kolli A, Nazir A, A. Olaya Castro, Octo. 21, 135(15) 154112.

[184] Kolli A, O'Reilly EJ, Scholes GD, Olaya-Castro A. The fundamental role of quantized vibrations in coherent light harvesting by cryptophyte algae, Journal of Chemical Physics, 2012, vol. 137 (pg. 174109–174115).

[185] O'Reilly EJ, Alexandra Olaya Castro; 2014; Non-classicality of the molecular vibrations assisting exciton energy transfer at room temperature; Nature Communications 5(1):3012.

[186] A.Ishizaki & G. R Fleming,; 2012; "Quantum coherence in photosynthetic light harvesting, " Annu. Rev. Condens. Matter Phys. 3, 333–361.

[187] Gregory D. Scholes; 2020; Limits of exciton delocalization in molecular aggregates; Faraday Discuss; V 221, 265–280.

[188] Hemelrijk PW, Kwa SLS, van Grondelle R, Dekker JP. 1992 Spectroscopic properties of LHC-II, the main light-harvesting chlorophyll a/b protein complex from chloroplast membranes. Biochim. Biophys. Acta Bioenerg. 1098, 159–166.

[189] Szalay L, Tombaćz E, Singhal GS.; 1974; Effect of solvent on the absorption spectra and Stokes' shift of absorption and fluorescence of chlorophylls. Acta Phys. Acad. Sci. Hungaricae 35, 29–36. (doi:10.1007/BF03159738)

[190] Ishizaki A, Calhoun TR, Schlau-Cohen GS, Fleming GR. 2010 Quantum coherence and its interplay with protein environments in photosynthetic electronic energy transfer. Phys. Chem. Chem. Phys. 12, 7319–7337. (doi:10.1039/C003389H)

[191] Schroöder M, Kleinekathoöfer U, Schreiber M.; 2006; Calculation of absorption spectra for light-harvesting systems using non-Markovian approaches as well as modified Redfield theory. J. Chem. Phys. 124, 084903. (doi:10.1063/1.2171188)

[192] Chin AW, Prior J, Rosenbach R, Caycedo-Soler F, Huelga SF, Plenio MB. 2013 The role of nonequilibrium vibrational structures in electronic coherence and recoherence in pigment –protein complexes. Nat. Phys. 9, 113–118. (doi:10.1038/ nphys2515)

[193] Jang S, Newton M, Silbey R. 2004; Multichromophoric Förster resonance energy transfer. Phys. Rev. Lett. 92, 218301. (doi:10.1103/ PhysRevLett.92.218301)

[194] Hossein-Nejad H, Curutchet C, Kubica A, Scholes GD. 2011; Delocalization-enhanced long-range energy transfer between cryptophyte algae PE545 antenna proteins. J. Phys. Chem. A 115, 5243–5253. (doi:10.1021/jp108397a)

[195] Kassal I, Yuen-Zhou J, Rahimi-Keshari S. 2013 Does coherence enhance transport in photosynthesis? J. Phys. Chem. Lett. 4, 362–367. (doi:10.1021/jz301872b)

[196] Davydov AS.; 1962; Theory of molecular excitons. [translated by M. Kasha and M. Oppenheimer, Jr.]. New York, NY: McGraw-Hill Book Company.

[197] Kasha M, Rawls HR, Ashraf El-Bayoumi M. 1965; The exciton model in molecular spectroscopy.; Pure Appl. Chem. 11, 371–392. (doi:10.1351/pac196511030371)

[198] Novoderezhkin VI, Doust AB, Curutchet C, Scholes GD, van Grondelle R. 2010 Excitation dynamics in phycoerythrin 545: modeling of steady-state spectra and transient absorption with modified Redfield theory. Biophys. J. 99, 344–352. (doi:10.1016/j.bpj.2010.04.039)

[199] Pullerits T, Chachisvilis M, Sundstrom V. 1996; Exciton delocalization length in the B850 antenna of Rhodobacter sphaeroides. J. Phys. Chem. 100, 10 787–10 792. (doi:10.1021/jp953639b)

[200] Andrews DL. 1989 A unified theory of radiative and radiationless molecular energy transfer. Chem. Phys. 135, 195–201. (doi:10.1016/0301-0104(89) 87019-3)

[201] Shipman LL, Housman DL. 1979 Forster transfer rates for chlorophyll a. Photochem. Photobiol. 29, 1163–1167. (doi:10.1111/j.1751-1097.1979.tb07835.x)

[202] Krueger BP, Scholes GD, Fleming GR. 1998; Calculation of couplings and energy-transfer pathways between the pigments of LH2 by the abinitio transition density cube method. J. Phys. Chem.B 102, 5378–5386. (doi:10.1021/jp9811171) 81.

[203] Scholes GD, Ghiggino KP, Oliver AM, Paddon-Row MN. 1993 Through-space and through-bond effects on exciton interactions in rigidly linked dinaphthyl molecules. J. Am. Chem. Soc. 115, 4345–4349. (doi:10.1021/ja00063a061)

[204] May V, Kuhn O. 2005 Charge and energy transfer dynamics in molecular systems, 2nd edn. Weinheim, Germany: Wiley-VCH

[205] Fulton RL, Gouterman M. 1961 Vibronic coupling. I. Mathematical treatment for two electronic states. J. Chem. Phys. 35, 1059. (doi:10.1063/1.1701181)

[206] Heid CG, Ottiger P, Leist R, Leutwyler S. 2011 TheS1/S2 exciton interaction in 2-pyridone.6-methyl-2-pyridone: Davydov splitting, vibronic coupling, andvibronic quenching. J. Chem. Phys. 135, 154311.(doi:10.1063/1.3652759)

[207] Ottiger P, Leutwyler S, Koöppel H. 2012 Vibrational quenching of excitonic splittings in H-bonded molecular dimers: the electronic Davydov splittings cannot match experiment. J. Chem. Phys. 136, 174308. (doi:10.1063/1.4705119)

[208] Silbey, R.; 2011; Description of quantum effects in the condensed phase. Procedia Chem. 3, 188–197. (doi:10.1016/j.proche.2011.08.026)

[209] Novoderezhkin VI, Romero E, Prior J, van Grondelle R.; 2017; Phys Chem Chem Phys.; 19(7):5195-5208. doi: 10.1039/c6cp07308e.PMID: 28149991

[210] Dekker, J. P. & van Grondelle, R.; 2011; Multiple charge separation pathways in photosystem II: Modeling of transient absorption kinetics. Chem. Phys. Chem. 12, 681–688.

[211] Tiersch M, Popescu S, Briegel HJ.; 2012; A critical view on transport and entanglement in models of photosynthesis. Phil. Trans. R. Soc. A 370, 3771–3786. (doi:10.1098/rsta.2011.0202)

[212] P. K. Ghosh, A. Y. Smirnov, F. Nori; 2011; Quantum effects in energy and charge transfer in an artificial photosynthetic complex, J. Chem. Phys.134; 244103.

[213] B. Mennucci, C. Curutchet; 2011; The role of the environment in electronic energy transfer: a molecular modeling perspective, Phys. Chem. Chem. Phys. 13; 11538.

[214] J. Adolphs, F. Möuh, M. E.-A. Madjet, T. Renger; 2008; Calculation of pigment transition energies in the fmo protein: From simplicity to complexity and back, Photosynth. Res. 95, 197.

[215] F. Mueh, M. E.-A. Madjet, T. Renger; 2010; Structure-based identification of energy sinks in plant light-harvesting complex II, J. Phys. Chem. B114; 13517.

[216] J. Adolphs, F. Mueh, M. E.-A. Madjet, M. S. A. Busch, T. Renger; 2010; Structure-based calculations of optical spectra of photosystem i suggest an asymmetric light-harvesting process; J. Am. Chem. Soc.; 132; 3331.

[217] Olbrich, T. L. C. Jansen, J. Liebers, M. Aghtar, J. Strumpfer, K. Schulten, K. Knoester, U. Kleinekathofer; 2011; From Atomistic Modeling to Excitation Transfer and Two-Dimensional Spectra of the FMO Light-Harvesting Complex, J. Phys. Chem B 115; 8509.

[218] P. Rebentrost, M. Mohseni, I. Kassal, S. Lloyd, A. Aspuru-Guzik, Environment-assisted quantum transport; 2009; New. J. Phys. 11; 033003.

[219] J. Jin, X. Zheng, Y. Yan; 2008; Exact dynamics of dissipative electronic systems and quantum transport: Hierarchical equations of motion approach, J. Chem. Phys. 128; 234703.

[220] P. Nalbach, D. Braun, M. Thorwart; 2011; Exciton transfer dynamics and quantumness of energy transfer in the fenna-matthews-olson complex, Phys. Rev. E 84; 041926.

[221] Nazir, A.; 2009; Correlation-Dependent Coherent to Incoherent Transitions in Resonant Energy Transfer Dynamics, Phys. Rev. Lett. 103; 146404.

[222] H. Hossein-Nejad, G. D. Scholes; 2010; Energy transfer, entanglement and decoherence in a molecular dimer interacting with a phonon bath, New J. Phys. 12; 065045.
[223] A. Ishizaki, G. R. Fleming; 2011; On the Interpretation of Quantum Coherent Beats Observed in Two-Dimensional Electronic Spectra of Photosynthetic Light Harvesting Complexes, J. Phys. Chem. B 115; 62.
[224] J. Wu, F. Liu, Y. Shen, J. Cao, R. J. Silbey; 2010; Efficient energy transfer in light-harvesting systems, I: optimal temperature, reorganization energy, and spatial-temporal correlations, New Journal of Physics 12; 105012.
[225] J. Strümpfer, K. Schulten, The effect of correlated bath fluctuations on exciton transfer; 2011; J. Chem. Phys. 134; 095102.
[226] C. Olbrich, J. Strumpfer, K. Schulten, U. Kleinekathofer, Quest for Spatially Correlated Fluctuations in the FMO Light-Harvesting Complex, J. Phys. Chem. B 115 (2011) 758.
[227] S. Shim, P. Rebentrost, S. Valleau, A. Aspuru-Guzik; 2011 Microscopic origin of the long-lived quantum coherences in the Fenna-Matthew-Olson complex, arXiv:1104.2943.
[228] K. M. Gaab, C.J. Bardeen; 2004; The effects of connectivity, coherence, and trapping on energy transfer in simple light-harvesting systems studied using the Haken-Strobl model with diagonal disorder, J. Chem. Phys. 121 (2004) 7813.
[229] J. Wu, F. Liu, J. Ma, R. J. Silbey, J. Cao, Efficient Energy Transfer in Light-Harvesting Systems, II: Quantum-Classical Comparison, Flux Network, and Robustness Analysis, http://arxiv.org/abs/1109.5769.
[230] J. Ströumpfer, K. Schulten; 2009; Light havesting complex II B850 excitation dynamics, J. Chem. Phys. 131 (2009) 225101.
[231] Y. Tanimura, R. Kubo; 1989; Time Evolution of a Quantum System in Contact with a Nearly Gaussian-Markoffian Noise Bath, J. Phys. Soc. Jpn. 58, 101.
[232] A. Ishizaki, Y. Tanimura; 2005; Quantum Dynamics of System Strongly Coupled to Low-Temperature Colored Noise Bath: Reduced Hierarchy Equations Approach, J. Phys. Soc. Jpn. 74, 3131.
[233] Y. L. Y. A. Tan, F. Yang, J. S. Shao; 2004; Hierarchical approach based on stochastic decoupling to dissipative systems, Chem. Phys. Lett. 395.
[234] Q. Shi, L. Chen, G. Nan, R.-X. Xu, Y. Yan; 2009; Efficient hierarchical Liouville space propagator to quantum dissipative dynamics, J. Chem.Phys. 130, 084105.

[235] M. M. Wilde, J. M. McCracken, A. Mizel; 2010; Could light harvesting complexes exhibit non-classical effects at room temperature?, Proc. R.Soc. A 446, 1347216.

[236] M. Sarovar, A. Ishizaki, G. R. Fleming, K. B. Whaley; 2010; Quantum entanglement in photosynthetic light-harvesting complexes, Nature Phys. 6; 462.

[237] Wilde M.Mark & McCrecken, James M.; 2009; Could light harvesting complexes exhibit non-classical effects at room temperature? arXiv:0911.1097v1[quant-ph], 5Nov; 2018.

[238] R. W. Schoenlein, L. Peteanu, R. A. Mathies, C. Shank; 1991; The first step in vision: femtosecond isomerization of rhodopsin, Science 254, 412.

[239] M. A. van der Horst, K. J. Hellingwerf, Photoreceptor proteins, "star actors of modern times": A review of the functional dynamics in the structure of representative members of six different photoreceptor families, Acc. Chem. Res. 37 (2004) 13.

[240] A. Vaziri, M. B. Plenio; Quantum coherence in ion channels: resonances, transport and verification, New J. Phys. 12 (2010) 085001.

[241] Fassioli F, Olaya-Castro A, Scholes GD.: 2012; Coherent energy transfer under incoherent light conditions. J. Phys. Chem. Lett. 3, 3136–3142. (doi:10.1021/jz3010317)

[242] Scholes G. Quantum-coherent electronic energy transfer: Did Nature think of it first?; 2010 J. Phys.Chem. Lett.; 1:2–8.

[243] J. S. Briggs, A. Eisfeld; 2011; Equivalence of quantum and classical coherence in electronic energy transfer, Phys. Rev. E 83 051911.

[244] Beljonne D, Curutchet C, Scholes GD, Silbey RJ. 2009 Beyond Foörster resonance energy transfer in biological and nanoscale systems. J. Phys. Chem. B 113, 6583–6599. (doi:10.1021/jp900708f)

[245] Borrego C M, Gerola P D, Miller M and Cox R P (1999) Photosynth. Res. 59, 159.

[246] Ruban A V, Berera R, Ilioaia C, van Stokkum I H M, Kennis J T M, Pascal A A, van Amerongen H, Robert B, Horton P and van Grondelle R 2007 Nature 450, 575.

[247] Dawson C M, Hines A P, McKenzie R H and Milburn G J 2005 Phys. Rev. A 71 052321.

[248] Hildner R, Brinks D, Nieder JB, Cogdell RJ, van Hulst NF. 2013 Quantum coherent energy transfer over varying pathways in single light-harvesting complexes. Science 340, 1448–1451. (doi:10.1126/science.1235820)

[249] Dong H, Fleming GR. 2013 Three-pulse photon echo of finite numbers of molecules: single-molecule traces. J. Phys. Chem. B 117, 11 318–11 325. (doi:10.1021/jp402768c)

[250] Zwanzig, R Nonequilibrium Statistical Mechanics, Oxford University Press, Oxford, 2001.

[251] Chen, D., Ye, J., Zhang, H., & Zhao, Y. (2011). On the Munn-Silbey Approach to Polaron Transport with Off-Diagonal Coupling and Temperature-Dependent Canonical Transformations, The Journal of Physical Chemistry B, 115(18), 5312–5321.

Chapter 4

Theory of Olfaction

"And what in fluctuating appearance hovers, ye shall fix by lasting thoughts" —Gothe

4.1 Introduction

Sensing the smell, i.e., Olfaction process, is accepted as one of the most ancient and again at the same time, one of the most intriguing characteristics of living organisms, maintaining a typical contrast nature when compared with the associated environment. It is the oldest and most fundamental aspect of chemical sensing which are being applied by almost all kind of lifeforms in interpreting their surroundings. The process of smelling is caused by certain kind of small molecules, neutral, volatile in nature, known as odorant. For example, in human beings, the odorant molecules, binding to specific sites on olfactory receptors in nasal cavity, causes finally the olfaction process. This process has certain typically interesting and fascinating characteristics, thus attracting the science community greatly, deliveing a great number of unique theories each of which tried to define and explain henceforth the mechanism behind such process.

Even though there have been considerable knowledge of structure of ORs (Olfactory Receptors), detailed knowledge about the molecular mechanisms needed in discriminating different odorants has not yet been fully understood[1] except the already accepted as most obvious characteristic of an odorant molecule, is its shape. Amoore[2–5] was the first who did put a conjecture that the response to scent is caused through the initiation of a mutual structural fitting, happening in between the receptor and the odorant (Lock and Key model). An idea, i.e., a more flexible modification of the existing one, proposes that the whole system adopts a certain kind of distortion in inducing a more appropriate mutual fit (hand and glove model). A further modification of such a shape-based theory requires a particular receptor, responding to only one structural feature, such as, a functional group, but opposed to the main body of the odorant (odotope model).[6,6,7]

Richard Axel and Linda B. Buck reported in one of their most pioneering and the landmark paper on the discovery including also the cloning of the genetic code for several mammalian olfactory receptors (ORs), belonging to a larger gene family. Even though this kind of sensing process might appear as an immediate and intimate but the mechanism proposed, still remained not well understood. For both fundamental science and industry,[9–13] this critical point makes this subject crucially important problem in its own right. The olfactory system in human beings has been found to be triggered by binding the small, neutral, and volatile molecules, known as odorants related to specific sites on olfactory receptors (ORs) within the nasal cavity.

Olfaction, i.e., the phenomena related to sense of smell, contains two parts; (i) the reception and detection of a chemical, and (ii) how that detection is sent to and processed by the brain. Importantly, there exists basic differences in the Olfaction processes within vertebrates and insects, so far the types of receptors used. In human, olfaction is performed via G-protein coupled receptors (GPCRs), whereas, in the case of olfaction in insects, primarily insect olfactory receptors (insect ORs) are used. At present, the most accepted proposal related to the associated theory is that the operation of olfaction process is based on the electrostatics and van der Waals surface of the odorant which permits binding to the receptor. After such a stage, the receptor undergoes a conformation change from its inactive state to its active state. The interplay between active and inactive conformations was validated as a likely description of activation in central nervous system (CNS) GPCRs through the analysis of the dynamics of the histidine and adenosine receptors. However, this process of detecting an odorant has been still under question. Among them, one theory named the "shape theory of olfaction"[14] suggests that certain olfactory receptors are triggered by certain shapes of chemicals and those receptors send a specific message to the brain. Despite considerable knowledge of structure of ORs, the detailed molecular mechanisms for discrimination between different odorants are not yet fully understood.

In 1963, Amoore[4] conjectured that such molecular mechanism is primarily related to the shape of the odorant and accordingly it is initiated

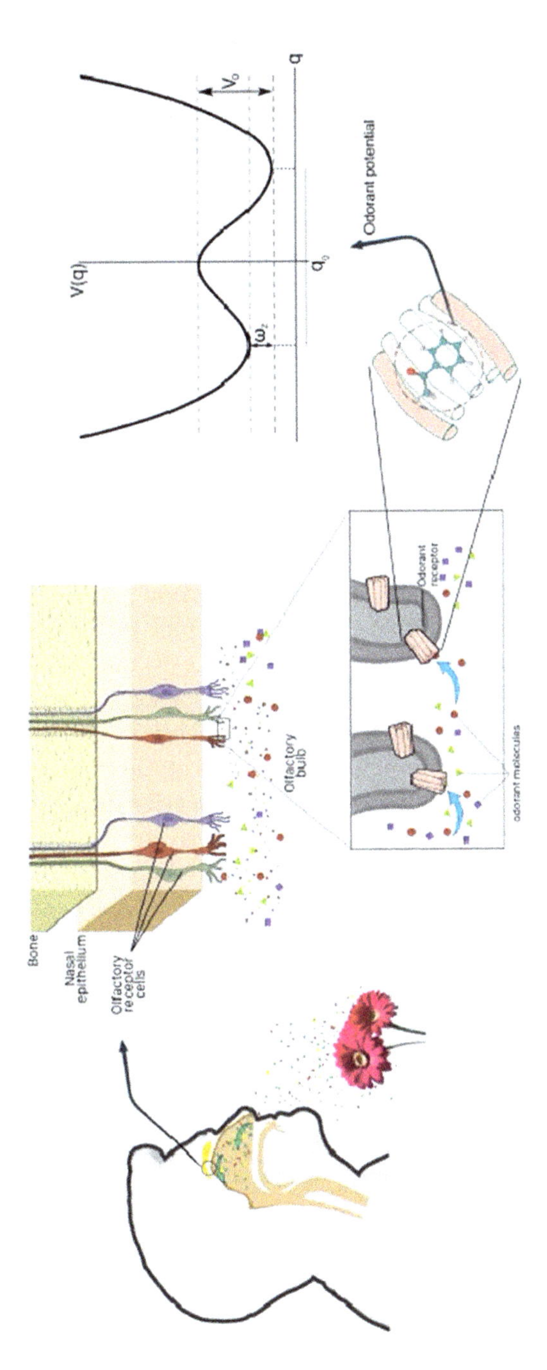

Figure 4.1. A scheme for the sense of smell in which odorants are absorbed by odorant receptors (ORs) in the olfactory receptor cells in the nasal cavity. In the quantum model, each odorant can be simulated as an asymmetric double-well potential for odorant recognition. The signal transduction relies on the success of an electron tunneling from a donor site of an OR to an acceptor site of the same or another OR, facilitated by a vibrational transition in the odorant according to the energy difference between the donor and the acceptor sites.

by a mutual structural fit between the odorant and ORs (i.e. lock and key model). In fact, the idea was motivated from the molecular mechanism of the enzyme behaviour. The model can be modified by introducing a distortion of the whole system to induce a more appropriate mutual fit (i.e. hand and glove model). A more refined demonstration of the idea requires ORs responding to only one structural feature, such as a functional group, instead the main body of the odorant (i.e. odotope model).[7] There is plenty of evidence for cases where the structure does seem important to an odorant's detection.[14,43] Despite the predictive power of these structure-based models, there are some evidence against them: odorants that smell similarly whilst being structurally different (e.g. benzaldehyde versus hydrogen cyanide), and odorants that smell differently whilst being structurally the same (e.g. ferrocene versus nickelocene).[16,17,19] All such shape-based models are primarily based on mechanical mechanisms.

The quantum model of olfaction, firstly proposed by Dyson[19] and refined by Wright,[20] is based on the idea that the signature of scent is caused due to the odorant's unique vibrational spectrum not its structure. An unique scent is attributed to its unique spectrum in the same way a colour is associated to its unique frequency of light. At present, it is an well known fact that the Quantum effects in biology, a continually growing field of interest, includes, for example, coherent energy transfer within photosynthetic bacteria proteins, mechanism of the avian magnetic compass and the possibility of inelastic electronic transfer (IET) occurring in olfactory receptors,[67] to name a few. Considering the possible importance of quantum mechanics in biology, each suspected case of nontrivial quantum effects in biology should be examined skeptically.[23] Our intention in compiling this review is 'trying to present certain glimpse of the vital findings and models used during the validation and examination of a few of the contemporary theory of olfaction', especially, Vibrational Theory of Olfaction (VTO) theory (based on quantum phenomena), suggesting that the olfactory receptors detect the vibration of the molecules reaching them and the associated "smell", is caused by the different vibrational frequencies, aptly called the "vibration theory of olfaction", and, thereby, address an important question, suggested by Barwich[24]: What, if anything, is so special about smell?

Substantial controversies and intense debate have still been continuing regarding the mechanism by which the chemical identity of odorants is established by olfactory receptors. For a substantial period of time, quite a number of theories relating odorant quality to molecular structure have already been proposed. Presently, our honest conjecture will be to try to give a short review of the recently proposed and heavily discussed two most prominent theories and add another theoretical approach together with the proposition of the direct participation of certain neurotransmitters or their hydrolysates in assisting the docking of odorant molecules with the olfactory receptor protein.

4.2 Chemical Olfactory Stimulation & Different Theories

There have been many a number of theories over a large period which proposed a molecular structure relating odorant quality. A concise review have been placed, following the characteristics, related to the two most prominent theories together with another one which involve the direct participation of certain neurotransmitters or their hydrolysates. These factors supposed to be responsible in assisting the docking of odorant molecules with the olfactory receptor protein.

4.2.1 *The Steric Theory of Odor*

Linus Pauling,[24] indicated in 1946, that a specific odor quality is the crucial reason for causing the presence of the typical molecular shape and size of the chemical. Similarly, John Amoore in their book, "Molecular Basis of Odor", extended the idea of a "Steric Theory of Odor" which has originally been proposed by R.W. Moncrieff in 1949[25,26] which, importantly put a novel idea, stating that air borne chemical molecules are smelled only when they *fit* into certain complimentary receptor sites, placed on the olfactory nervous system. In fact, this "lock and key" approach can be termed as an extension from enzyme kinetics. Amoore proposed certain primary odors (ethereal, camphoraceous, musky, floral, minty, pungent and putrid). Also other characteristics, for example, the molecular volume and similarity in shape of various odor chemicals were also compared (even though, interestingly, only by making hand prepared molecular models, physically measuring volume and creating ultimate silhouette patterns — it should be remembered that there had been no computer molecular modeling programs developed yet in that era). It should be mentioned that the Steric theory has been found well suited to the idea which states that the odorant receptor proteins accept only those odorants, present at a specific receptor sites. These, then, after being activated (by conformation deformation?), form couples to the G-protein which finally starts the signal transduction cascade.

4.2.2 *Neurological Origin of Olfaction: Olfactory Receptor Structure: GPCR*

In neurobiological studies, 'a key question', i.e., one of so called many mechanisms of cell signaling, — has always been found revolved around a question, i.e., how cells recognize specific signals. In this context, over the past 25 years, Olfaction has become an important experimental system as the olfactory receptors play the role of mediator in this aspect because of its primary interaction of the brain with the external world. Any odor stimulus, initially, is represented as activation of one to many different olfactory receptors. Vice versa, anything that binds and activates an olfactory receptor is, — as per definition, — an odor, or odorant (as the single compounds are often called). Among the receptors (OR), Olfactories constitute one of the largest families of G-protein-coupled receptors. Throughout the genome, one can notice the presence of many such large and small clusters of olfactory receptors. Not only this, these receptors have been observed, expressed in a highly specific manner in which monogenic expression is the general rule, i.e., depicting one neuron–one receptor only.

(ORs) constitutes one of the largest family of the G-protein-coupled receptor. These play a crucial and vital role as well, in recognizing thousands of odorant molecules in the olfactory sensory system. These have been found to comprise a multigene family in various species, from fish to mammals. The odorant-binding site in ORs has been elucidated, showing that a binding pocket constructed by transmembrane helices provides the molecular basis for odorant sensitivity and specificity. The one-neuron/one-OR rule and OR-instructed axonal convergence result in neural circuits that ensure precise transmission, from the olfactory epithelium to the olfactory bulb, of the combinatorial receptor codes for odorants.

During the journey of research on olfaction, the strict shape matching can no longer be accepted as individual receptors because these are known to accommodate multiple already accepted as odorants.[27–29] A common metric for odorant description, usually known as a common metric, is the **shape**, most probably defined via a van der Waals space-filling model. This may be designed for particular odorants for fitting binding pockets of the receptor. Besides this, any other odor stimulus is initially represented as activation of one to many different olfactory receptors. however, there are certainly some correlations between the shape of a molecule and odorant receptor (OR) response, but, on the contrary, many cases have been found to have very different shapes, producing the same pattern of activationmeant for the odorant receptor (OR) repertoire. This may happen as many ORs are broadly tuned. Imprantly, studies on the sense of smell have already advanced to the point with the aim of elucidating mechanisms which underlies integration of OR-mediated signals for odor perception. Finally, in turn, this can again be related in turn for the survival and continuation of chemical communication.

The OR family, among many, especially possess two characteristics, making them an excellent model system to understand GPCRs,:especially, its size and associated structural diversity among its members. Research on the OR binding site investigates what amino acid sequences determine the receptor-binding capacity. This promises a better understanding of how the basic genetic makeup of GPCRs relates to their diversification in ligand-binding capacities. The structure of an olfactory receptor will necessarily be strongly connected to the mechanism it employs to establish odorant chemical identity. The vertebrate olfactory receptors, for example, i.e., there are GPCRs,[7] constitute the largest subclass within GPCRs. However, we lack good atomic level structural information for olfactory GPCRs yet, though X-ray structures have been found for other GPCRs, for example, bovine rhodopsin. However, the sequence of amino acids is known for many receptor proteins,[37–39] and general structural features, and the general manner in which they operate, are already quite well established.[39]

In reality, the secondary structure of any olfactory receptors (OR) have not yet been sufficiently known quite clearly, only we do know that vertebrate olfactory receptors are GPCRs.[7] With the help of already gained experimental results, the structures of some other GPCRs have been come into the grasp of our knowledge, to work out a number of aspects related to their operational activities.[84] Out of such results, it has naturally been assumed that olfactory receptors (OR) also share their characteristics with these other receptors. However, Olfaction within vertebrates and insects differs in the types of receptors used. For example, in humans, olfaction is performed via G-protein coupled receptors (GPCRs), whereas olfaction in insects primarily uses insect olfactory receptors (insect ORs).

Recognizing these stimulli, the discovery of the basic olfactory receptors (ORs) in 1991,[6] established the olfactory pathway in a rather specific way, i.e., suggesting this mechanism as a greater group of signaling mechanisms. This has been found to be mediated by G-protein-coupled receptors (GPCRs), involved in sensing neurotransmitter.[31] Also, there has been a rising tide of consensus about the presence of molecular recognition. Interestingly, this idea made it increasingly unlikely that olfactory receptors

(OR) detected vibrations when all other GPCR receptors did not. Also, quite clearly, it could be opined that the typical ligands of odorant receptors are quite different from the typical ligands of other GPCRs which occur, both with respect of affinity and specificity, the diagrams of which have been beautifully and successfully, shown by Poseview. In fact, the dynamics of a GPCR is very important to know i.e., how it operates. The crucial role, played by the odorant, could be to perturb the dynamics through the action of interactions with the residues of the receptor, thus causing it to explore configurations with probabilities, importantly, differing from those of the empty receptor. Hence, strict shape matching is no longer to be held onto, as individual receptors are known to accommodate multiple found odorants.[27–29]

Even though sufficient clarity about the knowledge related to the secondary structure of any olfactory receptors has not yet been achieved, we do know clearly that vertebrate olfactory receptors are GPCRs. Very recently, with the help of a combination of experiment and computer simulations, structures meant for some GPCRs, has recently been possible to work out a number of aspects related to their operational activities.[84] Naturally, it could naturally be assumed that olfactory receptors also have the similar characteristics with these receptors, but, depending on the specific features of the odorant. The act of recognition, then could be thought of as happening as and when the signals of configurations related, supporting the binding and activation of an associated G-protein so that it could enter sufficiently often in order to activate with high probability.[31] By now, however it is clear that the this kind of typical ligands of odorant receptors are quite different from those of other GPCRs, so far as the affinity and specificity is concerned, as has been shown beautifully through the diagrams by Poseview. In fact, it is crucially important to know the the dynamics of a GPCR, i.e., how it operates.[32] Thus, The role of the odorant may be to perturb the dynamic through its interactions with the residues of the receptor, exploring configurations with probabilities, differing from those of the empty receptor, also depending on the features of the odorant.

Once activated, there is thought about exchange of GDP for GTP by the G-protein, and at some later stage, the release of its α subunit feedng into the signal amplification. In general, GPCRs' transmembrane signaling proteins are comprised of seven helical domains crossing from the extra — to the intra-cellular environments. Each receptor possesses a ligand binding pocket near the extracellular surface and an intracellular heterotrimeric G-protein complex, which gets dissociated after the activation of GPCR.

Members of the family mediate extracellular chemical signals, thus sharing homologous structure.[33]

Olfactory system, in fact, responds to an array of stimuli which are structurally different by nature. Recognition is then thought to occur only through the configurations by supporting the binding and the activation of a G-protein which are entered sufficiently often. This activation has been found to occur with high probability.[33] Once activated, the idea of exchange of GDP for GTP by the G-protein has been proposed which at some later stage, causes the release of its α subunit, this way feeding into the signal amplification. The most accepted theory for the operation of olfaction is based on the electrostatics and van der Waals surface of the odorant which permits binding to the receptor so that it undergoes a conformation change from its inactive state to its active state. The interplay between active and inactive conformations was validated as a likely description of activation in central nervous system (CNS) GPCRs through the analysis of the dynamics of the histidine and adenosine receptors.[6, 32, 40–43]

As already stated, receptors of vertebrate olfactorysystem are g-protein coupled receptors, or GPCRs, involved in numerous fundamental physiological processes. Within GPCRs, Olfactory receptors constitute the largest subclass. Even then, we are still short of more satisfactory structural information in atomic level itself, for olfactory GPCRs. However, by now, informations for the sequence of amino acids have already been known for many receptor proteins,[40] and also for the general structural features. Not only that the manner in which they operate, are well established[41] by now.

GPCRs work by transducing the signal received on the outside of the membrane into an intracellular signal,[41] being of enormous biological and medical importance. Excellent reviews of their structure, evolutionary origin and pharmacology have been written, to which the reader is referred for background.[42] It is generally assumed that a conformational change in the receptor, releases an intracellular assembly of proteins called g-proteins, even though the details are still in a an active research mode. These in turn activate other enzymes, for example, kinases, which then phosphorylate ion channels to activate them. Importantly, a kind of multi-step system achieves gain at the expense of speed. Structurally, GPCRs appear to be remarkably conserved in the transmembrane part and highly diverse in the cytoplasmic loops, especially, the extracellular ones where in many cases, the ligand binding takes place.[40] The most conspicuous structural motif of GPCRs is a sequence of seven transmembrane helices spanning back and forth across the membrane, abbreviated to 7-TM. Depending on the type of GPCR, ligands can either bind close to the extracellular face of the receptor, or in other locations on the extracellular loops.

It coild be concluded that there might be something more subtle underlying the use of electrons in olfactory receptors, for example, the mechanism of activation of non-olfactory GPCRs might indeed be the same, even though it has not been elucidated yet. Specifically, the related problem is not yet clearly settled, i.e.,how binding causes the crucial receptor activation step. However, one proposition is stating that, might be, electron transfer is an ancestral general mechanism in GPCRs, and the additional trick of making electron transfer contingent on vibrations may be a later evolutionary advance.[43]

4.2.3 *Odorant Recognition*

By now, we have been successful in gaining the knowledge quite satisfactorily about many of the biological mechanisms related to the molecular structure of most odorant molecules. As a result, it has been possible to quantify a smell response which becomes measurable especially in two ways: (1) gaining knowledge from the receptor level about the depolarization of the cell (triggered by receptors) or (2) by fluorescent magnetic resonance imaging (fmri) of the brain. Even with such possibility, it is measurable only through an individual's perception. However, there remains much less possibility of doing so. Again, even after the presence of such puzzling parts of the problems understood, it appears extremely in managing the number of degrees of freedom related to the associated problem, appears extremely frustratingly vast:for example, the number of possible odorants might be in excess of 100,000 whereas type preset in the number of functional human receptor currently found to be around as 390. Many programs, for example E-DRAGON, have been developed in calculating molecular descriptors, based on the submitted e-molecular structure of the odorants.

But, practically, the crucial problem arises when the cross-correlation between molecular descriptors and response patterns fail to reveal any kind of particular metric. Usually, in such cases, number of carbons or the presence of a particular functional group, acts as truly faithful response patterns. Usually, a common metric considered for odorant description is the shape, expected to be defined via a van der Waals space-filling model. This may be designed for particular odorants but only in fitting within binding pockets of the receptor. But even there is possibility for having certain kind of correlations in between the shape of a molecule and odorant receptor response. In many cases, i.e., very different shapes can also produce the same pattern of activation of the odorant receptor repertoire caused by the presence of many broadly tuned ORs. This kind of observation[45] suggested the Infrared (IR) vibrational spectra to be better predictors of smell than shape. But in practice, programs like E-DRAGON, together with those used for pharmaceutical design, do not directly implement vibrational spectra as an odorant metric. Not only that, they explore the minimum energy (usually in vacuo) *geometry of the odorant* and do not account for effects at the binding site of the receptor or environment.

However, in case of IR spectra, the problem that whether the correlation is any better than for shape for all known cases have not yet been settled. Recent work[78] probes whether drosophila melanogaster identify chemical species on the basis of shape or not. The experiment considered four odorants, along with their deuterated counterparts in which the flies were trained to avoid one or other of the isotopic versions. Their observation stated that the flies can generalize their response to the other molecules based on which isotope of hydrogen was used'. The results point to the fact that the flies are capable of responding to the presence or absence of deuterium, rather than molecular shape. This important evidence suggest that drosophila melanogaster can distinguish odorants by their molecular vibrations.[78] However, the drosophila olfactory receptors are of a different type to human receptors. In such case, the only commonality existing here could support the swipe card model which applied hydrophobic receptor environment and an acceptable energy tuned gap (to the odorant vibrations).

In any quest for the development, meant for the theory of odorant recognition, the shape of olfactant molecule, must play a crucial role if only to let the scent molecule access key parts of the receptor. Broadly speaking, theories of the initial actuation event can be divided into two categories. Among them, one class relies on the shape of the olfactant molecule, covering many structural activities related descriptions. Even though some level of fit could be clearly seen as necessary, no such fit could be claimed as sufficient (see Figure below).

When considered the aspects of affinity and specificity, the typical ligands of odourant receptors posses quite distinct and at the same time different characteristics than those of other GPCRs. When considered from left towards right, characteristically, it possess three ligands, in decreasing order of affinity and specificity. The left being called norbiotin (pdb 1LDO), a ligand with a near-covalent binding energy to the protein avidin, consists

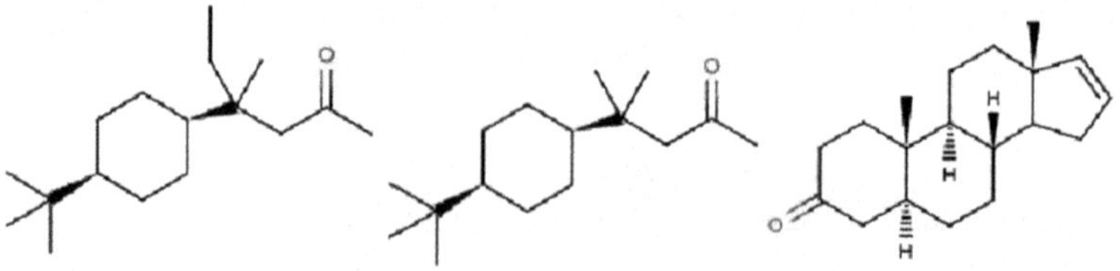

Figure 4.2. Contrast these three odorants: according to shape theory, which would you predict smell the same? From left to right, cis-ketone (4-(4-tert-butyeyelcyclohezxy)-4-methylhexan-2-one), cis-nor-ketone(4-(4-tert-butyeyelcyclohezxy)-4-methylhexan-2-one) and 5 alpha-androst-16-en-3-one, Cis-ketone and 5 alpha-androst-16-en-0ne have the same "penetrating urine odour" and cis-nor-ketone is practically/totally odourless.[27]

of seven hydrogens bonded to protein, each contributing between 2 and 5 kcal/mole. At the center, lies an emblematic GPCR ligand, noradrenalin which is depicted for its interactions with the β-adrenergic receptor, having five hydrogen bonds. At right, the diagram depicts an insect pheromone, tetra-decadien-1-ol, attached to its odorant-binding protein. safely, it could be assumed that this protein has evolved for high affinity, as the insect can perceive the odorant only at low concentrations. Interestingly, even the pattern of interactions has been observed to be quite different, for example, the odorant interacts mostly, by dispersion forces, and only the terminal hydroxyl forms a single hydrogen bond with a nearby carboxylate. In fact, *the ability to form hydrogen bonds influences bulk physicochemical properties as well as binding to receptors.*

At this point, most of odorants — by definition, have been considered as volatile and hence, expected to have few features available for molecular recognition. Thus, even with the absence of availability of a picture of an odorant in a receptor, known knowledge of human odorant-binding proteins serve us as second best. The dynamics of a GPCR is very important regarding it's operations.[35] The role of the odorant may be to perturb the dynamics through its interactions with the residues of the receptor. This makes possible the exploration of it's configurations with probabilities, differing from those of the empty receptor which, however, depends on the features of the odourant. As a result, the recognition is thought to occur whenever configurations, supporting the binding and activation of a G-protein are entered sufficiently often so that the activation occurs with high probability.[84] Once activation starts, there is thought to be an exchange of GDP for GTP by the G-protein, and at some later stage the release of its α subunit, which feeds into the signal amplification, starts.

4.3 Mechanism of Molecular Recognition & Different Theories of Olfaction

As an answer to the question related to the working procedure of the olfactory receptors, two proposed mechanisms can be mentioned as the initial molecular recognition steps: (1) First one: based purely on shape and weak bonding interactions between the odourant and the receptor (onwards referred to as the Docking theory); (2) that the receptors are also capable of identifying vibrational frequencies (which we shall refer to as the Vibrational theory). Quite interesting point here is that both of these theories acknowledge the necessity that in recognizing an event, some crucial level of matching is essential in between the odorant and the receptor. Importantly, in doing so, involvement of both, i.e., some shape compatibility (with the aim of avoiding high energy repulsive interactions), and alignment of atomic groups to support weak bonding interactions (in order to ensure so that the time the odorant resides in the receptor is sufficient to trigger recognition) even though there could be difference in the way how this contributes to the final recognition. From the basic perspective, the docking theory proposes that a change in configuration of the receptor follows the arrival of a molecule that binds sufficiently strongly, so that an initial signal starts the binding and release of a trimeric G protein in mammals happens. On the other hand, the vibrational theory proposes that the signal can **only** occur once the receptor recognizes a part of the vibrational spectrum of the odorant,[40] *as a staring step.*

4.3.1 *Docking Theory of Olfaction*

R.W. Moncrieff, in 1949,[25] proposed a shape-based theory of odor,[25] by publishing an article in American Perfumer called "What is odor: a new theory". There, he applied the notion of Linus Pauling's idea related to the shape-based molecular interactions. Eventually, this idea superseded the older theory of vibration, relating the theory of olfaction, renaming it as the docking theory of olfaction. In this approach, in addition to shape, more precisely, a range of non-covalent interactions have been reflected. However, this remained also as the mainstream theory, in both commercial fragrance chemistry as well as in academic molecular biology. Following this idea, John Amoore[2] speculated further the number of over ten thousand smells as distinguishable by the human olfaction system of human kind. He argued that this fact is the resulting effect, as a result of the combination of basic seven primary odors, having correlation to odor receptors for each.

Interestingly, most of these are found as the spectrum of perceived colors in visible light, generated by the activation of three primary color receptors.[2] Not only that, John Amoore speculated further that over ten thousand smells, distinguishable by the human olfaction system, can be obtained also from the combination of seven basic primary odors. He also emphasized that these correlate to odor receptors for each, much as the spectrum of perceived colors in visible light which can be generated by the activation of three primary color receptors. It is important to mention that his most convincing work included the camphoraceous odor, in which experiment, he posited a hemispherical socket in which spherical molecules, such as camphor, cycloctane, and naphthalene, could bind.

4.3.2 *Smell Alters with Increasing Concentration*

We are already aware with a important fact that to study the difference between the affinity of an olfactant in case of a specific receptor with its efficacy being quite crucial. This can be determined by the signals, initiated to the brain, can be quite crucial. One underlying question is whether the receptors acts only as a binary, having only on and off states. Even though Rhodopsin receptors are known to be binary, i.e., on or off states only, in the case of the β_2-adrenergic receptor (one of the better characterized GPCRs), dopamine (a weak partial agonist) is just as efficacious as isoproterenol (full agonist) in disrupting. This fact appears to be the molecular switch, but this can not be taken as enough to induce the full activation of the receptor, for example, the case of isoproterenol.[98] Also, using fluorescence resonance energy transfer (FRET) for the bimane-tryptophan quenching system that different types of agonists induce different types of conformational states, this way contradicting the binary proposition: the ligands do not simply modulate the equilibrium between an active on and inactive off state, but there could be many degrees in between.

Thus the conclusion arrives at the point that receptors might not be always binary in nature. But, then, the question might arise: are olfactory receptors binary? In a swipe card model, quite similar to the inelastic tunneling model, there remains still there are some vital issues, i.e., related features are yet to be resolved. We have not yet decided about the problem i.e., when an odorant binds to a receptor, more than one electron can be present in electron tunnel, but not sure what happens when limited only by electron supply to D or removal from A? Or does the odorant requires leaving and be replaced before the next electron can contribute to the signal?

An extra degree of complexity is caused due to the concentration dependence of odour, introduced in the related process. The next crucial point arises in case of binary olfactory receptors where the potency of an odorant's signal could directly be attributed to the number of receptors, occupied by odorant molecules. This makes the potency linearly varying with concentration, occurring at least at low concentrations. But, in olfaction, this is notoriously absent. On the contrary, in many cases, the odorant is found more likely to change its character[6,7,99] with increasing concentration. This act implies that at saturation, certain "wrong" odorants are likely to find their way into an olfactory receptor and, whilst they may fit and bind inefficiently, they still activate the olfactory receptor to a certain degree. Smell change with increasing concentration suggests that, as absolute receptor saturation is approached, some odorants can activate non-parent receptors. Receptors that are unimportant at low concentrations become significant when some other receptors are saturated (Figure 4.3).

Proposition of docking theory of olfaction states that an odorant molecule is caused as a result of a range of weak non-covalent interactions caused between the odorant [a ligand] and its protein odorant receptor (found in the nasal epithelium), for example, electrostatic and Van der Waals interactions, H-bonding, dipole attraction, pi-stacking, metal ion, Cation–pi interaction, and above all, the hydrophobic effects. Not only that these also include odorant conformation.[45,46] Typically different previous type of recognition has been termed as the shape theory of olfaction.[39]This primarily considers molecular shape and size but the latter model have been oversimplified because of the fact that as two scent molecules may have similar shapes and sizes but different sets of weak intermolecular forces activate different combinations of odorant receptors. Thus, earlier "lock and key" and "hand in glove" models of protein, i.e., a more nuanced picture of ligand binding has been replaced by applying the concept of the distortion of flexible molecules, thus capable of forming the optimal interactions with binding partners which is proposed in the case of molecular docking of non-olfactory G-protein coupled receptors.

It is worth to be mentioned that, in 1991, Linda Buck and Richard Axel,[9] published in their Nobel Prize winning research r about the olfactory receptors.There they identified in mice 1,000 G-protein-coupled receptors

Figure 4.3. Hexanal, smell changes with increasing concentration.

n mice, which has been used for olfaction. Since, currently recognized all types of G-protein receptors are considered to be activated through binding (docking) of molecules with highly specific conformations (shapes) and non-covalent interactions. Olfactory receptors are also assumed to be operated in a similar fashion. Around 347 olfactory receptors have been identified on further research on human olfaction systems. Importantly, a recent version of the previously named 'shape theory', also known as odotype theory or Weak Shape Theory proposes a combination of activated receptors to be responsible for any one smell, in contrary to the older model which consists of one receptor, one shape, one smell only. Receptors in the odotype model recognize only small structural features on each molecule, and in this model, brain is considered as responsible for processing the combined signal into an interpreted smell. Much of recent works on the docking theory focuses on 'neural processing', rather than the specific interaction between odorant and receptor, considered responsible in generating the original signal.

Extensive studies have been conducted to explain and elucidate the complex relationship in explaining the the docking of an odorous molecule by characterising its character related to perceived smell. Based on this model, fragrance chemists have proposed structure models for the smells of amber, sandalwood, and camphor and many others. Studying in detail, Leslie B. Vosshall and Andreas Keller, (published in Nature Neuroscience in 2004), tested several key predictions of the competing vibration theory but failed in finding substantial experimental support for it.[46,49] Even though the data were described by Vosshall as "consistent with the shape theory", however, she opined that "they don't prove the shape theory". Other studies, i.e., when experiments performed with Drosophila,[50] also showed that in the case of olfactory receptors, molecular volume of odorants are capable of determining the upper limits of neural responses. Quite importantly, it is to be noted that — at 2015, in "Chemical & Engineering News",the article on the debate related to the "shape" versus "vibration", points to the fact that in the "acrimonious, nearly two-decade-long controversy, — on the one side are a majority of sensory scientists who argue that our odorant receptors detect specific scent molecules on the basis of their shapes and chemical properties. On the other side, are a handful of scientists who posit that an odorant receptor detects an odor molecule's vibrational frequencies[50] only.

In that new study, the article, led by Block *et al.*,[47] takes aim at the vibrational theory of olfaction, even not finding any kind of evidence in

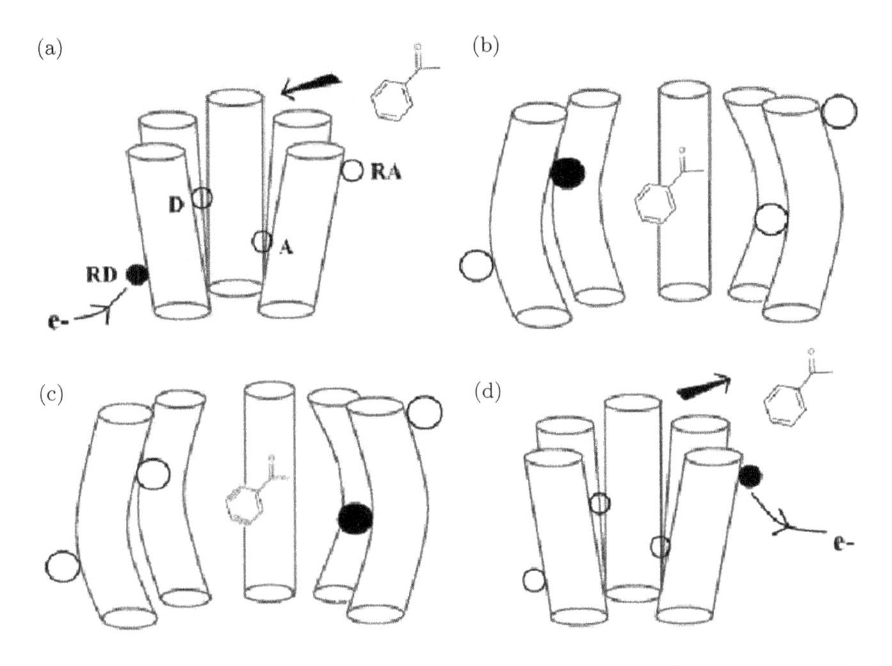

Figure 4.4. A scheme for the proposal of electron transfer in the olfactory receptor with intra-protein electron transfer.

olfactory receptors distinguishing vibrational states of molecules. Specifically, according to the report of Block *et al.*, the human musk-recognizing receptor, i.e., OR5AN1, has been identified in using a olfactory receptor expression system which is robustly heterologous, responding in vitro, to cyclopentadecanone and muscone but fails to distinguish isotopomers of these compounds. Furthermore, the mouse (methylthio) methanethiol recognizing receptor, MOR244-3, together with other selected human and mouse olfactory receptors, responded similarly to normal, deuterated, and carbon-13 isotopomers of their respective ligands. These results were found paralleling with the musk receptor OR5AN1. After such findings, the related authors negated the possible validity or applicability of the proposed vibration theory to the human musk receptor OR5AN1, mouse thiol receptor MOR244-3, or other olfactory receptors examined. Importantly, after detailed theoretical analysis, these authors showed that the proposed electron transfer mechanism of the vibrational frequencies of odorants could easily be suppressed by quantum effects of nonodorant molecular vibrational modes. Authors concluded also: "*These and other concerns*

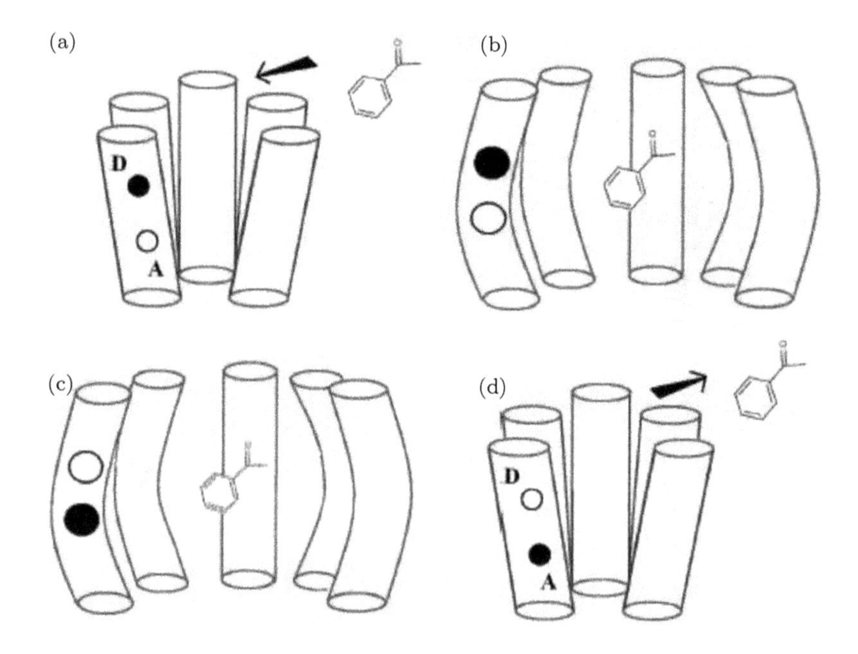

Figure 4.5. A scheme for the proposal of electron transfer in the olfactory receptor with intra-protein electron transfer.

about electron transfer at olfactory receptors, together with our extensive experimental data, argue against the plausibility of the vibration theory."

While criticizing the vibration theory, proposed by Turin, Vosshall commented on this, in PNAS: "Block *et al*.... shift the "shape vs. vibration" debate from olfactory psychophysics to the biophysics of the ORs themselves. The authors mount a sophisticated multidisciplinary attack on the central tenets of the vibration theory using synthetic organic chemistry, heterologous expression of olfactory receptors, and theoretical considerations to find no evidence to support the vibration theory of smell".[38] In this regard, Turin, the main architect of the vibrational theory comments that Block used "cells in a dish rather than within whole organisms" and that "expressing an olfactory receptor in human embryonic kidney cells, doesn't adequately reconstitute the complex nature of olfaction...". Vosshall responds: "Embryonic kidney cells are not identical to the cells in the nose ... but if you are looking at receptors, it's the best system in the world.[50]

Anyway, even after numerous studies being carried out, essentially, the docking theory has faced an indefensible argument in order to discover and establish the structure — odor relations; a necessity, must for having substantially strong footed predictive power.[51] By now, we are already aware of the already established knowledge that different smell can exist for even similarly shaped molecules with different molecular vibrations, (metallocene experiment and deuterium replacement of molecular hydrogen).[45] It is important to note that Turin, during his experiment with metallocene, observed the sandwich structure in ferrocene and nickelocene which possess almost the same molecular structure as the reason for having such distinct odors. Thus he opined that "because of the change in size and mass, different metal atoms give different frequencies for those vibrations, involved in the metal atoms"[40] which observation is *compatible* with the vibration theory. However, in contrast to ferrocene, nickelocene has been observed to decompose rapidly in air and the cycloalkene odor observed for nickelocene, but not for ferrocene which could reflect simply the decomposition of nickelocene giving trace amounts of hydrocarbons, cyclopentadiene, for example.

The challenge regarding smell of molecules with similar structures is contrary to the results obtained with silicon analogues of bourgeonal and lilial. Because, despite their differences in molecular vibrations, they have similar smells and also activate the most responsive human receptor also in a similar way, i.e., hOR 17-4.[85] Extensive studies showed the human musk receptor OR5AN1 responding identically to deuterated and non-deuterated musks but with single-neuron comparison of the olfactory receptor responses both to deuterated and non-deuterated odorants.[53] We are well aware that differently shaped molecules with similar molecular vibrations posses similar smells (replacement of carbon double bonds by sulfur atoms and the disparate shaped amber odorants). But, interestingly, hiding functional groups do not hide the group's characteristic odor. In such cases, very small molecules of similar shape, appearing most likely to be confused by a shape-based system, have extremely distinctive odors, hydrogen sulfide, for example.

Again, metals have been suggested, for example, in olfaction, Cu(I) may be associated with a metallo-receptor site with the purpose of having strong-smelling volatiles', also being good metal-coordinating ligands, forexample, thiols.[54] This hypothesis has been confirmed for some of the specific cases of thiol-responsive mouse and human olfactory receptors.[55,56] However, finally, this concept has claimed limited success for the predictive power of docking theories. Not only that, after the new understanding about the highly dynamic nature of the recognition process, it is quite difficult to accept the possibility of having a strong link between the structure of the odorant, that of the receptor, and the corresponding recognition. Odor descriptions in the olfaction literature has been found to have correlate more strongly with their vibrational frequencies than with their molecular shape.[57] However, we are still at an early stage in the research into the consequences of this dynamic behavior, and more of new insights may be forthcoming.

4.3.3 *Theory of Olfaction by Brookes's et al.: Swipe Card Model of Odor Recognition*

Theory by Brookes *et al.*[17,18] involved their model of odor recognition with a core issue of physics, i.e., in our noses — what kind of microscopic interactions enable the receptors, rather small protein switches — to distinguish different scent molecules, related to discrimination of different odors, surveying the different physical processes when smelling something. Not only that, finally, it is essential to highlight the difficulties in developing a full understanding of the mechanics of odorant recognition, together with clearly defined "swipe card" mechanism, This idea has been discussed here by the above group, i.e., Turin's theory of inelastic electron tunneling is used to discern olfactant vibration frequencies. This theory is explicitly quantal, since it requires the molecular vibrations to take in or give out energy only in discrete quanta. Obviously, these ideas lead to many of experimental tests and challenges, but finally, even after emerging as capable of explaining many observations, remain hard to reconcile in other ways. And yet, it should be admitted that there remained still some important gaps in a comprehensive physics-based description of the central steps, especially in odorant recognition.[6,7]

In general, in order to respond to an odorant molecule, the olfactory receptor G (a transmembrane protein), releases a subunit of a neighboring G protein, thus, starting a chain reaction of Ca ion influx into the cell (ion channels for this process may also have quantum features). The model, by Brookes *et al.* proposed that an electron source X arrives and exchanges charge with the receptor protein. This charge, as a next, travels to a donor D in one part (helix) of the transmembrane receptor protein, also having the liberty of hopping, or tunneling to the acceptor A which could be placed in another helix of the protein which, importantly, assisted inelastically by an odorant phonon emission. Then, after arriving at D, the charge triggers the release of the neighboring G protein subunit. But, many of precise features of this model are still to be developed, i.e., yet to be clarified. For example,

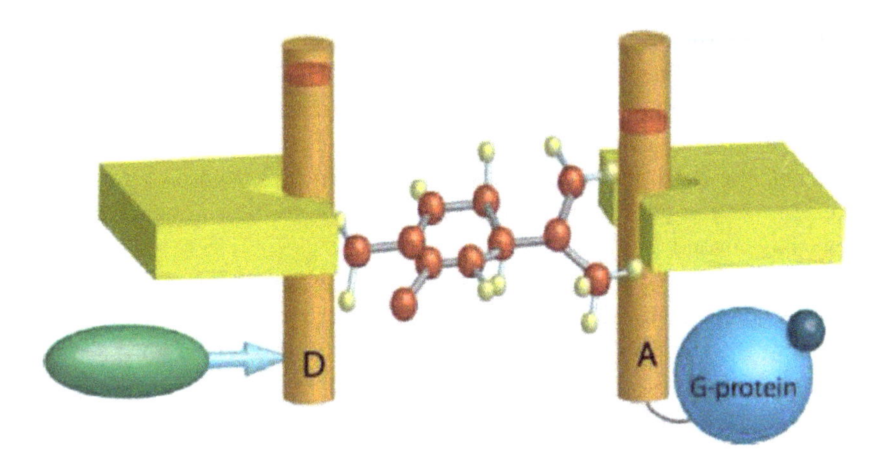

Figure 4.6. The schematic shows outlines the model proposed by Brookes *et al.* to describe charge-mediated odorant discrimination.

the origin of the electron source X is unknown, but is conjectured to be an oxidizing agent, present in the cell fluid, which diffuses throughout the cell and randomly arrives at the transmembrane to initiate the charge-transport process. Anyway, the goal of this model is to have the release of the neighboring G protein subunit which depend on vibrational properties of the odorant molecule, and thus capable of discriminating different molecules beyond features, like size and shape.

The scheme of the model, as shown [Fig. 4.6] above, has been proposed by Brookes *et al.* where the process of charge-transport-mediated odorant discrimination has been described. In general, a transmembrane protein, called the olfactory receptor G, responds to an odorant molecule through the release of a subunit of a neighboring G protein. This way, a chain reaction of Ca ion influx starts into the cell (the related ion channels show that this process may also have quantum features). In their model, Brookes *et al.* demonstrated the process as follows: after arrival, an electron source X, exchanges charge with the protein as the receptor. This is followed by the next step when the charge travels to a donor D in one part (helix) of the transmembrane receptor protein. The charge can either hop, or tunnel, to the acceptor A, placed in another helix of the protein which is inelastically assisted by an odorant phonon emission. As a next, the e-charge, after arriving at D, triggers the release of the neighboring G protein subunit. However, many precise features of this model are not well understood yet.

For example, the origin of the electron source X is unknown, but, however, is conjectured to be an oxidizing agent in the cell fluid so that it diffuses throughout the cell and randomly arrives at the transmembrane in order to initiate the charge-transport process. The goal of this model is to have the release of the neighboring G protein subunit which depend on vibrational properties of the odorant molecule, and thus able to discriminate different molecules beyond features like size and shape.

In science, for humans in most senses, at least a reliable idea or some novel mechanisms involved. Even with that, we, most of the time, yet do not completely understand the basic determinants (metrics), for example, how smell works at the odorant recognition level. Broadly speaking, we try to describe this process — where small molecules meet large receptor proteins (factors of 1000's larger in size!) and like — depending on the combination of David and Goliath, — there is (might be not) a triggering of a signaling cascade, resulting into a smell — perceived by the brain. *But still remains the basic question — unsolved: How do particular molecules cause (or inhibit) this process?.* Of course, it is not just the problem with olfaction phenomena that the effect of one specific small molecule can cause a cascade of important processes. Many of similar examples can also be cited, say, the triggering of cells by hormones or the signal transmission in nerves by acetylcholine.[55] This kind of combination of sensitivity (one molecule can initiate a complex chain of events) and selectivity (different molecules generate distinct perceived odours) is quite remarkable[27] Thus arises the crucial and vital querry: In principle, how this works?

In order to achieve the goal, firstly, one must understand the physics of the mechanisms, controlling especially the initial activation step, i.e., when an odorous molecule meets one olfactory receptor. Though the crystal structure of soluble proteins can easily be determined, the detailed structure of olfactory receptors still remains quite unclear because GPCRs are membrane proteins. Despite substantial progress[58, 59] in producing large quantities of olfactory receptors (ORs), the ambitious aim of crystallizing these elusive proteins has not yet been fully achieved as atomic structures of ORs not yet settled in detail. Thus, whilst full structural information will surely be highly illuminating, a static picture of structure, — alone might not tell us how odorant recognition is achieved.

4.3.4 *Problems Related with Odorant Recognition*

It has been observed that humans are gifted with capacity in perceiving thousands of molecules through smell, all of them being small enough in size for being volatile, each of having the capacity to activate a few olfactory receptors (it has already been observed that it is almost impossible to find one odorant, activating only one receptor). Not only that, humans practically have been observed to have the capabilities of detecting odorants even at very small concentrations in air, for example, even 1 parts per trillion. Again these olfactory receptors have been noticed in possessing special capabilities in selectivity, even if some of the odorants may agonize or antagonize a receptor.[60] Next, arises a crucially important question related to odorant recognition: by now, the molecular structure of most odorant molecules and odorant receptor response are known quite precisely; yet, in most of the cases, many broadly tuned ORs are found to be present. Not only that very different shapes has been found to produce the same pattern of activation of the odorant receptor repertoire leading to a crucial suggestion that might be, Infrared (IR) vibrational spectra are better predictors of smell than shape.[61] By now, a common metric for odorant description, taking usually the shape, possibly defined via a van der Waals space-filling model. This may be designed with the aim of a particular odorants so that it fits within binding pockets again, inside the receptor, this points to a significantly important point, i.e., there might be some correlations between the shape of a molecule and odorant receptor response.

But, many cases have been noticed, i.e, the real controversy surfaced when, for the odorant receptor repertoire, different shapes have been observed which, surprisingly, produce same pattern of activation, i.e., broad tuning have been found in many of ORs. But, so far as the underlying physics concerned, related to the odorant recognition at the receptor level, this means how an odorous molecule (which we shall call M) initiates a measurable signal. Similar casees having potential analogies with vision can be mentioned where a photon causes an initial molecular transformation.[62] Almost certainly, important processes are there controlling the overall perception[63] whenever the brain builds a scent perception out of a number of receptors. At this stage, from a range of receptors, it becomes possible for any one molecule of them, to initiate a signal to the brain. However, the crucial point still remains, i.e., the brain must receive distinctive information to work with. The group of Brookes *et al.*[64,98,99] tried to elaborate, exactly, what molecular information determines whether a given receptor responsible for initiating a signal after it being activated in the brain.

Another important factor in any theory of odorant recognition is the shape of olfactant molecular playing a crucial role. It has been observed that it is needed to let the scent molecule access key parts of the receptor.

Theory related to the initial actuation event falls into two broad categories. One among them relies on olfactant molecular shape alone, a class covering many structure activities related descriptions. For this, even though some level of fitting is clearly necessary, but not sufficient enough as the odorant must some how activate the receptor. Then, who takes the responsibility for turning the key in the lock? One of natural assumptions points that, might be, odorant causes a *mechanical deformation of the receptor.* To illustrate this problem, the case of ferrocene and nickelocene could be mentioned where one could find molecules having different odours, and yet similar shapes.

Among other possible models, the proposed picture in swipe card model states the importance also of other factors, for example, the shape as important enough, capable of providing other information, for example, characterizing the odorant which plays substantially crucial roles. For example,, in Lock-and-key models, a key of the right shape provides all the information necessary for the opening of the lock. On the contrary, in a swipe card (or key card) model, even though the shape is needed to be good enough to fit the machine, additional information is to be conveyed in a different manner. In case of the specific swipe card model of odorant recognition, a 'molecular vibration frequency' is used as the additional information in the shape-based lock and key ideas. Even though the role of molecular vibrations has been around for many years, only in 1996, a specific mechanism for signal transduction proposed. Ayway, most of our present discussion of the swipe card model has been concentrated on Turin's specific proposal which states that molecular vibrations provide the necessary information for activation — and the inelastic tunneling for recognizing relevant vibration frequency.

Turin's mechanism, specifically in this context, can be categorized as an approach, involving quantum idea, as partly because of tunneling, but primarily because of a 'quantum oscillator' which has been considered to be responsible for the purpose of only receiving or giving energy as *quanta* of specific energy. This way, inelastic electron tunneling has to cover a long journey in making itself acceptable in the physical sciences,[65,66,68,69] especially, related to biological context in reactions.[69] However, this idea, being basically a new conception so far as the application in the field of biological signaling concerned, has led to a misunderstandings[66] many a times. But in the positive side, Turin's basic idea, leads directly to possible experimental and theoretical tests. This proposal together with simple shape constraints, as Brookes *et al.* opined,[18,66] goes a long way towards understanding how odorants activate olfactory receptors. Nonetheless, Brookes *et al.*, established the vibration frequency as the crucial part, capable of dominating smell, and the swipe card description appears to be a more useful paradigm than lock and key.

Again, shape-based theory has been found not been satisfactory enough providing a full description of signal signatures, whereas the swipe card paradigm can. It is an well known fact that, for instance, conformational mobility correlates with different odours for enantiomers. Brookes *et al.*'s observation, related to protein studies, has concluded that the receptor itself will be undergoing larger length-scale motions. Following this it has been suspected that other dynamical aspects, for example, starting from promoting modes to stochastic resonance, may have certain conclusive roles. Nonetheless, in this regard, some form of inelastic electronic process seems a fully viable and crucially important part of our sense of smell.

4.3.5 *Quantum Theory of Olfaction*

Dyson[19] proposed "The quantum model of olfaction", firstly in 1938, then refined in a more detailed way by Wright.[20] The model has been based on the idea which states that the signature of scent is caused basically, due to the unique vibrational spectrum of the odorant — not its structure. Importantly, a specific scent is attributed to its unique spectrum which, surprisingly in the same way, as a colour, is associated to its unique frequency of light. Influenced by such phenomena, i.e., the inelastic electron tunneling (ET) in metals,[65,70] Turin proposed the mechanism of olfactory detection by stating it as basically an odorant-mediated biological inelastic ET.[45] The underlyig process, i.e., in such process, the signal transduction depends on the success of an ET from a donor (D) site of an OR to an acceptor (A) site of the same or another OR. Added to this, the main criteria behind this mechanism is facilitated by a vibrational transition in the odorant, corresponding to the energy difference between D and A sites. In the long run, basic support for the related evidence for the vibration-based mechanism was obtained from sophisticated quantum chemistry calculations.[71,72] This way, Turin's theory, basically requires a source of electrons or holes in order to allow charge-flow to take place. The precise biological origin has not yet been known, but, intuitively could be suggested that it may well consist of reducing (oxidizing) species (X) in the cell fluid.[74]

Importantly, Chěcińska and co-workers[74] examined this aspect exclusively, i.e., the dissipative role of the environment could be dynamically involved in this kind of vibration-based model. Specifically, it is this group who proposed the presence of strong coupling to be responsible for enabling the environment, capable of enhancing the frequency resolution of the olfactory system.[74] However, the main evidence against Turin's theory is pointrd to the presence of the differentiable smells of chiral odorants: they have identical spectra in a chiral solution, but with different smells.[17] Importantly, very recently, using the master equation approach, this problem has also been addressed by another group, led by Salary[73a] (2017). They pointed out about the possibility of the chiral recognition in olfaction to be dependent on the detection of the chiral interactions between the chiral odorant and ORs.[76] Anyway, In vibration based theory, one of the major prediction is the isotope effect, i.e., isotopes should smell differently. However, even though the recent behavioral experiments have revealed that fruit flies,[72,78] honeybees[77] and humans[79,80] can distinguish isotopes, interestingly, experimental evidence against isotopic discrimination keeps yet the debate still open.[46,80]

In fact, major difficulties have being faced by scientists, dealing with the problems related to smell, because, few pieces of the crucial puzzles appearing, not only unclear to them but also quite confusing about the way they fit into the big picture. Only knowledge possessed by them is that odorant molecules in our noses, actuate in the air several types of receptors with that in our noses, eventually triggering nerve cells in the brain for analyzing. But quite a serious kind of problem appeared to scientists facing the fact that *the shape and size of molecules* can make odors smell differently but on the contrary, some molecules with nearly identical shapes, smell nothing alike. The reason behind this apparent conundrum is caused due to the 'lack of understanding' of what happens before and during the interaction of the odorant molecules with the nasal receptors. Thus, some form of initial atomic-scale processes must include needed selection criteria which might explain why receptors react differently to molecules of the same (or different) shapes.

Physicists Jennifer Brookes, Filio Hartoutsiou, Andrew Horsfield and Marshall Stoneham. possibly, have looked into a specific criteria, i.e., the electrons in the receptors are found to be able to be triggered, i.e., **tunnel** between energy states, provided, the odorant molecule's vibration frequency matches the energy difference of these states. The LCN group tested the physical viability of such a mechanism in a rigorous way and suggested in 1996 for the first time by a scientist named Luca Turin which established that a general model of such an electron tunneling is consistent with physics laws together with already known features of 'smell'. The field of the science of smell seems tantalizingly close in demonstrating the reality of **quantum biology**. However, exactly how our noses are capable of distinguishing and identifying a myriad of differently shaped molecules, remain as a big challenge for conventional theories of olfaction.

In physics, quantum tunnelling, barrier penetration, or simply tunnelling can be defined as a quantum mechanical phenomenon in which an object, an electron or atom, for example, passes through a potential energy barrier, in which case, according to classical mechanics, the object does not possess sufficient energy to enter or surmount that barrier. Thus, tunneling can be depicted as a consequence of the wave nature of matter, where the quantum wave function describes the state of a particle or other physical system, and wave equations, for example, the way with the help of which Schrödinger equation describe their behavior. The term "Tunneling" in Quantum mechanics, defines a typical process which, however, in several ways, often found to be exploited in technology. But, this kind of typical problems appears, especially, whenever a particle is considered to cross the barrier, even after being considered forbidden by classical physics. For example, this could happen for small-scale objects, say, electrons — due to their, rather extraordinary property, i.e., wave-like properties. The probability of transmitting a wave packet through a barrier decreases exponentially with the height, width of barrier and, particularly, the mass of the tunneling particle. That is why tunneling is notable most prominently in low-mass particles, for example, electrons or protons, tunneling through microscopically narrow barriers.

Now, if an odorant molecule's vibrations (or phonons) cause electrons in a nasal receptor to tunnel between energy states, eventually they send nerve signals to the brain. As a result of this, finally, different vibrational frequencies are detected by receptors of diverse nature. At this point, different frequencies cause odorants, also having difference in smelling. Even though in the past, the basic chemical image of smell was a "lock and key" model, due to the presence of the diverse shaped molecules, fitting in different receptors, the above team succeeded in explaining how the electron tunneling mechanism turns to be more like a "swipe card" model. Similarly, just like a credit card, an odorant molecule would be "read" by receptors, picking up its vibration spectrum, matching along with its shape.

But, it still remains an well known problem that no one can entirely claim about the exact happenings next, i.e., whenever a smelly molecule wafts into one of our nostrils. Somehow, the molecule is supposed to interact with a sensor, i.e., — a molecular receptor — embedded delicately with inner skin of our nose. However, It has already been an established fact that thousands of different smells are distinguishable to a well-trained human nose. Thus, in a ultimate sense, it still remains quite a puzzling question, i.e., how this information is carried in the shape of the smelly molecule. Quite interestingly, many molecules, even though have almost identical shape, posses very different smells for swapping around an atom or two. A few of interesting examples can be mentioned, i.e., Vanillin smells of vanilla, but eugenol, quite similar in shape, smells of cloves. Some molecules even when could be called as a mirror image of each other — just like one, s right and left hand — also have different smells. On the contrary, equally interesting and intriguing fact is that some very differently shaped molecules can smell almost exactly the same.

In The history of olfactory science, even though different theories but of similar nature have been there for several years, recently, above theory, proposed and developed by Luca Turin, has come to prominence. He suggested that the process of inelastic electron tunneling is the method through which vibrations are detected by the olfactory receptors within the hose. In 1938, Dyson suggested the idea of possible association with infrared resonance (IR) when measuring molecular vibration. R.H. Wright popularized this idea in the mid 1950's. However, by that time, infrared spectro-photometers became generally available for such kind of spectral measurements. Taking the advantage of this, Wright was able to correlate with certain odorants.[67] However, during the 60's and early 70's, vigorous debate raged as to the validity of each such theory, i.e., for classifying chemical odorants.

Luca Turin argued while proposing his theory that the molecule's shape alone is not enough to determine its smell, instead, it's the *quantum properties of the chemical bonds present in the molecule*, which provides the crucial information. According to Turin's quantum theory of olfaction, when a smelly molecule, after entering the nose, binds to a receptor, immediately a process called quantum tunneling starts within the receptor. Typically, at this stage, when a smelly molecule enters the nose and binds to a receptor, it allows quantum tunnelling to happen. In the process of quantum tunneling, an electron can pass through a material by jumping from point A to point B in a way that appears to bypass the intervening space. As with the bird's quantum compass, in this process too, the resonance plays the crucial role. As per Turin's proposal, a particular bond, present in the smelly molecule, can resonate with an electron having right energy, needed to help an electron

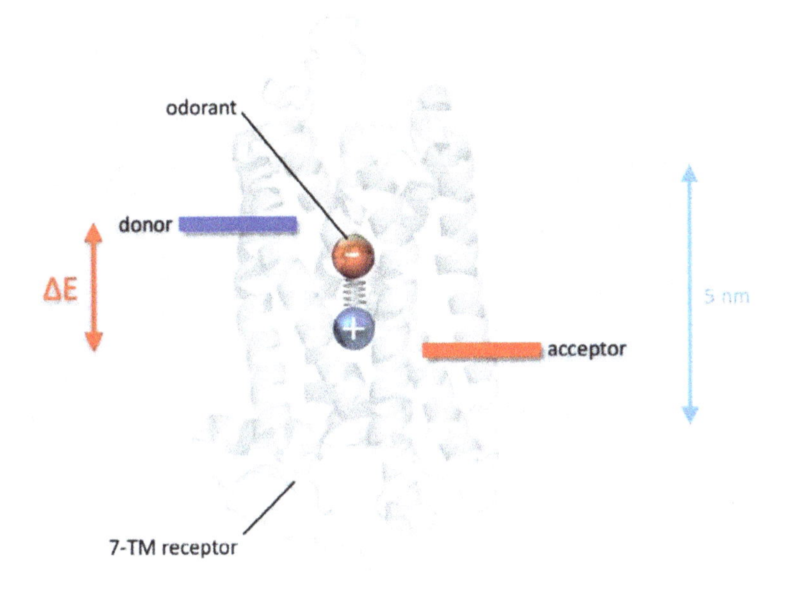

Figure 4.7. This cartoon illustrates the mechanism of the Turin theory of olfaction.

on one side of the receptor molecule, for taking the leap to the other side. Importantly, the electron can make this leap, only through the so-called quantum tunnel, if the bond is vibrating with just the right energy. When the electron leaps to the other site on the receptor, it could trigger a chain reaction that ends up sending signals to the brain so that the receptor comes into contact with that particular molecule, being an essential part in the process, giving a molecule its smell, thus defining the process fundamentally to be a '**quantum process**'.

As depicted in figure, the odourant molecule resides in a receptor (helices embedded in a membrane). By hopping from one electron level (donor D) to another (acceptor A), a signal is generated (caused by the activation of a G protein). At the initial state, the electron resides on D, having an energy ΔE, higher than the final electronic state of the electron, being in A. Initially, the odourant molecule is in its lowest vibrational state, but in the final state, the electron on A, i.e., the odourant vibrational mode has a higher energy, coming from the electron. To maintain the conserve energy conserved overall, the difference in energy between the two vibrational states must also be ΔE. Now, if the splitting between the electronic states be increased by 50%, then it would be impossible to maintain the conservation of energy, as a result of which, the electron jumps between D and A, with the hopping becomes forbidden for the mere necessity of maintaining the conservation of energy. As a result, the receptor becomes unable in signaling the presence of the odourant. Importantly, the presence of such a signal indicates the presence of a molecule which has a vibrational mode of frequency, defined by the receptor (the energy splitting between D and A), as very much needed for the validity of the vibrational theory.

Basically, Olfaction requires certain mechanism in order to involve somehow the actual chemical composition of the molecule. In quantum tunneling, this particular factor plays a crucial role for natural explanation. *The strongest evidence for the theory in Turin's discovery lies in the fact that two molecules with extremely different shapes can smell the same if they contain bonds with similar energies.* Interestingly, he also predicted that boranes — relatively rare compounds and hard to come by — smelled very like sulphur, or rotten eggs, even if he never smelt a borane before, so the prediction turned out to be quite a gamble.

4.3.6 *Vibration Theory of Olfaction: Involvement of Quantum Mechanics*

It was Luca Turin[43] who has reinvigorated the vibration theory in 1996. He proposed the mechanism for the sense of smell as caused by the presence of G-protein receptors, broadly responsible for the detection of molecular vibrations. Also, he stated the process as the inelastic electron tunneling where the electron loses energy across molecules, then filling a binding site with a G-protein receptor, as a next. Here, interestingly, the chemical to the receptor would be acting as a bridge, allowing for the electron to be transferred through the protein. But, usually, as this kind of electron transfers would be a barrier for the electrons, it, s energy would be lost, caused by the vibration of the molecule. In this way, recently bound to the receptor, this electron, ultimately gains the ability to smell the molecule.[45,64]

But, even though the vibration theory possess some important and crucial experimental proofs of the concept, multiple controversial results still have been in many other experiments. For example, animals are shown to be capable in distinguishing smells between molecules having different frequencies but same structure,[45,46] while on the contrary, due to distinct molecular frequencies,[46] some other experiments showing people unaware of distinguishing smells. Thus it has not been dis-proven, and has even been shown to be an valid effect in olfaction of animals other than humans such as flies, bees, and fish. This way, the present vibrational theory can be stated just as an iteration in a historical movement within olfactory science, to propose a relationship between molecular vibration and osmic properties. The proposition of vibrational modes controlling the osmic mechanism originates from the works of Malcolm Dyson. Coming off the contemporaneous development of Raman spectroscopy, Dyson believed that probing molecular vibrations of a molecule would elucidate correlations between these vibrations and the osmic properties of the odorant. The proposition was that the thermalized vibrations inherent to a molecule at a given (physiological) temperature, would activate the receptor protein.

4.3.7 *Turin's Vibrational Theory of Olfaction*

There exists a long history related to the developments of the theory of olfaction probing the molecular vibrations of odorants. Soon after the discovery of Raman spectroscopy, rather many of vague proposals,[81] inspired by Nature philosophie with a hope-for unity of 'spectral' senses (hearing, vision and smell), the first clear proposal for a vibrational mechanism in olfaction is due to Dyson. This theory had been revived in the late 1950s by Robert Wright,[20] found remarkable but largely overlooked experimental support in the early 1980s with the work of Clifton Meloan,[82,83] and thereafter died a lingering death until revived in 1998.

Even though for several times, similar theories have been placed recently throughout olfactory science, this theory has come into prominence again due to the pioneering works of Luca Turin, who applied the idea of electron tunneling, inelastic in nature, to be the main cause behind which vibrations are detected by the olfactory receptors within the nose. In this respect, vibrational theory of olfaction has attempted to describe a possible mechanism for olfaction so that researchers have been provided with a set of self explanatory principle, henceforth, predictions have been provided allowing also for the structure-odor relations.

In the following proposition and the corresponding development of the theory, emphasis has been on the fact that in the vibrational theory, receptors are assumed to consider not only the shape but also vibrational frequency in order to identify odourants. The multitude of receptors in flies, (upwards of 500 in humans, a thousand in mice, etc) point out to this important fact. Not only that the 'cuvette' of the spectrometer (the receptor binding pocket) has been the order of the size of the odourant and so cannot be wholly non-selective. Not only that, absolutely, it is needed to besure so that in catching an odourant one must have dozens of receptors and analyse its spectrum. However, the theory asserts that shape is not enough, but, on the contrary, in final role, indeed, the purpose of the system is revealed by the *fact* that most receptors, even at the individual level, appear to be relatively nonspecific.

Thus, Luca Turin and his group[43,45] stressed that receptors in the vibrational theory, use shape as well as vibrational frequency for the identification of odorants. They also noticed another important facts, i.e., this particular characteristic is quite prevalent in the multitude of receptors (63 in flies, upwards of 500 in humans, a thousand in mice, etc). However, in practicality, it can be asserted from this theory which states the shape not to be enough for the purpose of the system to be revealed, because of the fact that most receptors, even at the individual level, appear to be relatively nonspecific.

4.4 Inter-Chain or Intra-Chain Charge Transfer: Tunneling Process

Turin's mechanism can specifically be described as a quantum idea, partly due to the presence of tunneling, especially in which the quantum oscillator plays the crucial role by receiving or giving energy as quanta of specific energy. Of course, classical oscillator can give or receive any amount of energy. Even though, the phenomena of inelastic electron tunneling has been familiar, both in the physical sciences and biological context, it appeared as completely new to the biological signaling processing, leading to misunderstandings. On the contrary, Turin proposed certain kind of basic idea which leads directly to possible experimental and theoretical tests. For a biological inelastic tunneling process, the moving charge via the receptor helices could be proposed as two optimal routes.

Though, it might become clear that vibration frequencies do appear important, but there are still limits to what can be understood, primarily, in terms of odorant shape and vibrations. Thus to some extent shape and the weak interactions between the odorant and the receptor must matter. However, Turin's assertion is that vibrational modes of the odorant matter as well.[43,45] As is elaborated below, the conjecture is that the receptor exploits inelastic electron tunneling to detect the molecule's vibration frequency. Not only that, we note that vibrational frequencies are of course strongly dependent on the odorant geometry.

Usually, the inelastic tunneling transition has been described as — taking the electron from donor D on one of the olfactory receptor's polypeptide chains through the odorant to an acceptor A on another polypeptide chain. In this inter-strand picture, it is *not necessary* that the electron must passes through the molecule. Rather, the word 'through' means the odorant wave function to be a significant part of the transition matrix element. Not only sudden change in electric field at the odorant is sufficient to cause it to change vibrational state but also, an intra-strand charge transfer transition from the same, can possibly be an only satisfactory alternative.

Mathematically, it can be demonstrated that for both of the inter- and intra-chain charge transfers, the couplings between the odorant vibrations and the electron transition can easily posses the same size. One possibliliy of having such an advantage is that the intra-chain charge transfer could be observed in this process, i.e., there is no need for any long-range motion of charge for re-setting the donor and acceptor to their original states. Or, in other words, the original intra-chain electronic states will be recovered, simply by the olfactant, leaving the receptor. But the difficulty arises for this being purely a speculative proposition. Also, in the case of inter-chain charge transfer, possibly, it could be a single electron current, starting the next stage in the series of local processes. Thus, in short, the case could possibly be like this: in the case of intra-chain charge transfer, we can not claim in the same sense that there may be no significant current, but, might possibly be short-lived, electrical dipole moment. One conjecture could be mentioned, as an example, i.e., probably, the electric field from this transient dipole initiates the next stage.

4.4.1 *Dissipative Quantum Model of Olfaction Following Turin*

The basic ground behind the objections raised against the vibration-based theories of olfaction, specifically, lies on some crucially basic arguments, stating that enantiomers of chiral odorants have the same vibrational spectra but different smells.[13] But, the scientists, advocating such theories suggest that other molecular features, say — structural flexibility, — should be included in the model so that it accounts for this behavior,

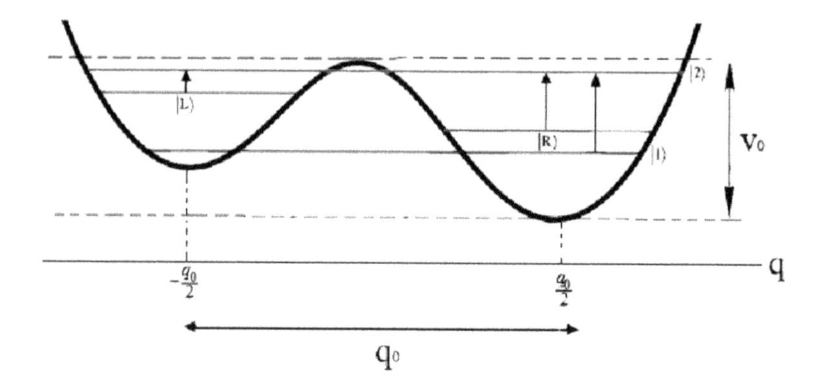

Figure 4.8. Possible transitions of the odorant described as a particle in an asymmetric double-well potential.

so to say, counter-intuitive.[14] A proposal, introduced by Brook *et al.*[18] (2012) proposed in his work by introducing this problem as follows: the chiral recognition in olfaction, basically, relies on the detection of energy difference between two enantiomers of the chiral odorant As a result, this result is possible to obtain from chiral interactions between the odorant and the receptor.

In fact, two enantiomers of a chiral molecule, can basically be considered as two localized states of an asymmetric double-well potential at sufficiently low temperatures. Added to this, it is supposed to portrays the energy difference between two enantiomers. It is well known that the biological environment of olfactory system can basically be represented as a collection of harmonic oscillators which, when coupled linearly to the double-well. Not only that, it produces the results which has claimed the possibility of crucial analysis related to the Spin-Boson model. But, it should be noted that this kind of model, i.e., in order to establish the validity of dissipative quantum model of olfaction, even after being applied extensively, particularly by Leggett and co-workers,[86] has been lacking the solid experimental support. Tirandaz *et al.*[76] in their pioneering works, have examined the physical plausibility of the odorant-mediated inelastic ET model of olfaction. The approach adopted by them examined the dynamics of olfactory ET by applying the Spin-Boson model, as a result of which, they obtained both the elastic and inelastic ET rates for each enantiomer, having a typical chiral odorant. Importantly, they also showed that the inelastic ET rates can be different for two enantiomers.

With the aim of obtaining the above mentioned results, firstly, they[76] examined the olfactory discrimination of left- and right-handed enantiomers of chiral odorants, embodied in a biological environment, based on the odorant-mediated electron transport from a donor to an acceptor of the olfactory receptors. It is an we known fact that within a laboratory environment, an analyte molecule is deposited upon a metal surface in close proximity to another metal surface. A tunable electrostatic potential is generated, as a resul, across the insulating gap between the plates, as a next, *driving these electrons from one side of the gap to the other via evanescent tunneling* Thus, during the elastic tunneling processes, electrons are driven from the donor (D) to the acceptor (A) while obeying strict energy conservation; if this is the only process, the junction becomes obviously of Ohmic nature. As depicted, the chiral odorant, can then effectively be expressed as an asymmetric double-well potential where minima are associated, only to the left- and right-handed enantiomers. This kind of asymmetry have been considered in order to measure an overall chiral interactions. Thus the biological environment is conveniently modeled as a bath of harmonic oscillators in such a way so that the resulting spin-boson model is expressed by a polaron transformation in order to derive the corresponding Born-Markov master equation to obtain the elastic and inelastic electron tunneling rates.

In their detail study of this problem, Tirandaz *et al.*,[76] gave importance, particularly, on the role of a typical vibrational degree of freedom of odorant molecule, known as contorsional vibration. In this, they focused on an atom or a group of atoms oscillating in between the two wells of the potential energy surface. Such kind of vibration can typically be modeled by considering the motion of a particle in a double-well potential. The corresponding associated biological environment is conveniently represented as a collection of harmonic oscillators. As a next, the time dependent perturbation theory has been developed by examining the dynamics of the odorant. Finally, the corresponding elastic and inelastic ET rates have been obtained for all possible transitions of the odorant. Next, with the aim of examining the physical limitations of the model, the rates in different limits of molecular and environmental variables have been analyzed, taking all through $\hbar = 1$. To do this, following three parts of their model have been mainly focused as the main component of the olfaction model, i.e.,

- The odorant,
- The *electron* tunneling through the odorant and lastly
- The surrounding *environment.*

4.4.2 *Odorant-Mediated Inelastic ET Model of Olfaction*

Hereby, we will try to follow inelastic ET model of olfaction which is odorant-mediated, keeping our focus particularly, on a vibrational degree of freedom of odorant molecule, called as contorsional vibration. In this theory, an atom or a group of atoms is considered as keeping oscillating, in between the two wells, containing the potential energy surfaces. Not only that, modeled by the motion, this occurs, typically, by a particle in a double-well potential. In contrast to the harmonic mode, the contorsional mode can be used to characterize the olfactory chiral recognition.[76] Essentially, the eigenstates of the double-well potential have been found as doplets. For most of the droplets, first doplet is energetically available for most molecules at room temperature.[86]

As an well known fact, the biological environment can conveniently be represented by as a collection of harmonic oscillators. By adopting this principle, Tirandaz *et al.*[76] applied the perturbation theory with time dependency for obtaining the corresponding elastic and inelastic ET rates for all possible transitions of the odorant and also to check the physical plausibility of the odorant-mediated inelastic ET model of olfaction. In order to test the viability of such physical limitations of the adopted model, analysis of the rates, considering different limitations of molecular and environmental variables, have also been followed by the above group throughout their conjecture, though in a short form. Even though the detailed biological origin of the electron tunneling through the odorant has not yet been known clearly, it was thought to occur, may be, due to redox agents in the cell fluid.[9] According to the original model there have been a conjecture that donor (D) and acceptor (A) sites of traveling electron as single molecular orbitals with energies ϵD and ϵA, coupled to each other by a weak hopping integral$^{\Delta}$. But, in order to satisfy energy conservation, — during the tunneling process, the electron's parameters should be consistent with the odorant's parameters. As a result, they cannot be considered as variables.

The model of biological environment has been constructed as a bath of harmonic nature having an ohmic spectral density following the original model. The environment, here, is characterized by its microscopic parameters (e.g., coupling frequency J_0 and cut-off frequency λ) and macroscopic parameters (e.g. temperature and pressure). Unlike the microscopic parameters, the macroscopic parameters can be controlled in experiment and thus they are considered as variables. The free Hamiltonian for the total system consisting of the electron at the receptor, the chiral odorant and the surrounding environment can be written as

$$\hat{H}_{\text{od}} = \hat{H}_e + \hat{H}_{\text{zero}} + \hat{H}_E \tag{4.1}$$

Here electron state at donor and acceptor sites of receptor are represented by $|D\rangle$ with energy ϵ_D and $|A_i\rangle$ having energy ϵ_A, respectively. We consider that $\epsilon_D \geq \epsilon_A$. Hence, here the electron tunneling corresponds to a vibrational absorption in the odorant. Then, we can describe the electron at the receptor with Hamiltonian

$$\hat{H}_e = \epsilon_D|D\rangle\langle D| + \epsilon_A|A\rangle\langle A| \tag{4.2}$$

A chiral odorant can occur at least as two chiral enantiomers through the inversion at molecule's center of mass, by a long-amplitude vibration known as contortional vibration.[86,87] This effectively can be described by an asymmetric double-well potential. The minima, associated to two chiral states $|L_i\rangle$ and $|R_i\rangle$, are separated by barrier $\dot{V}$. In the limit, $V_\circ \gg \omega_\circ$. Here, $\omega \gg k_B T(\omega^\circ)$ being the vibration frequency in each well, the state space of the molecule are effectively confined in two-dimensional Hilbert space, spanned by two chiral states. For most chiral molecules, this limit has been found to hold up to room temperature.[87,88] Hence, as the left- and right-handed enantiomers of the chiral odorant are associated to states $|L\rangle$ and $|R\rangle$, these are localized at left and right wells of the potential, respectively. Now, when the barrier becomes high enough to prevent the tunneling process, then molecule remains only in its initial chiral state. The corresponding Hamiltonian of the molecule in such 'chiral basis' can be expressed then as

$$\hat{H}_\circ = -\frac{\delta}{2}\hat{\sigma}_x - \frac{\omega_z}{2}\hat{\sigma}_z \tag{4.3}$$

Here, δ is meant for the frequency of tunneling between two chiral states and ω_z called as localization frequency giving the overall measure of chiral interactions.

As a next, the odorant's states of energy can be described as a superposition of chiral states, i.e.,

$$|1\rangle = \sin\left(\frac{\theta}{2}\right)|L\rangle + \cos\left(\frac{\theta}{2}\right)|R\rangle \tag{4.4}$$

$$|2\rangle = \cos\left(\frac{\theta}{2}\right)|L\rangle - \sin\left(\frac{\theta}{2}\right)|R\rangle \tag{4.5}$$

Hereby, $\theta = \arctan(\delta/\omega_z)$ and energies corresponding to these states can be expressed as

$$\pm\sqrt{\delta^2 + {\omega_z}^2}$$

Importantly, in the present case, the biological environment has been considered as a kind of **condensed bath**, so that conveniently this can be modeled as a collection of harmonic oscillators, the corresponding Hamiltonian being as

$$\hat{H}_E = \sum_i \omega_i \hat{b}_i^\dagger \hat{b}_i \tag{4.6}$$

4.4.3 *The Interaction Hamiltonian*

In order to avoid the undesired transitions from donor to acceptor, i.e., in case of the absence of the odorant, the coupling strength Δ should be small compared to other energy scales in the above expressions for modeling the interaction so that, conveniently, in this regime, it is possible to move into a polaron transformed reference frame. The corresponding three parts of the system can be stated as the main components of the olfaction model, i.e., (1) the odorant, (2) the corresponding electron, tunneling through the odorant, together with (3) the surrounding environment. A more realistic vibrational mode of non-planer odorant at this stage has been considered, known as *contorsional* mode, where, an atom or a group of atoms oscillates between the left and right wells of a double-well potential. Thus, unlike the harmonic mode, the contorsional mode is able to characterize the olfactory chiral recognition. Now, we model the odorant as an asymmetric double-well potential,

Tirandaz *et al.*[75] modeled the odorant as an asymmetric double-well potential, where, the minima of the potential correspond, importantly, to the left- and right-handed states, $|L_i\rangle$ and $|R_i\rangle$ of the odorant. However, in their original vibrational model of olfaction, the relevant vibrational mode has been represented only by a simple harmonic oscillator. At this point, it is quite important to remember that, the handed states can be inter-converted by the *quantum tunneling*, through the barrier V_0 only. In the limit, i.e., for $V_0 \gg \omega_0 \gg k_B T$, (ω_0) is the vibration frequency at the bottom of each well and the state space of the odorant is considered, effectively as confined in a two-dimensional Hilbert space, spanned by two handed states. Practically, such an approximation have been found working properly for a large class of odorants, even in the high-temperature limit.[87]

At this point, the odorant's Hamiltonian can be expanded by the handed states, as

$$\hat{H}_{od} = -\omega_z \hat{\sigma}_z + \omega_x \hat{\sigma}_x$$

Here, $\hat{\sigma}_i$ is the i-th component of Pauli operator, ω_x and ω_z being the tunneling and asymmetry frequencies, respectively. As a next, then, the tunneling frequency ω_x can be calculated by applying the WKB method as given below, i.e.,

$$\omega_x = A q_0 \sqrt{M \omega_0} \exp(-B V_0 / \omega_0)$$

In the above expression, M depicts the molecular mass,[76] and q_0, the distance between two minima of the potential. The value of the parameters A and that of B is explicitly dependent on the mathematical form of the potential considered which, however, can usually be approximated by 1.[76] The asymmetry found here, is caused by the fundamental parity-violating interactions[90, 92] and the chiral interactions (caused by interactions that are transformed as pseudo scalars[93]) between the odorant and environmental molecules. Interestingly, former is typically small among them. But on the contrary, the latter can be significant, especially, between a chiral odorant and ORs. Finally, the eigenstates of the odorant's Hamiltonian can be written as the superposition of the handed states, given by as

$$|E_1\rangle = \sin\theta |L\rangle + \cos\theta |R\rangle$$
$$|E_2\rangle = \cos\theta |L\rangle - \sin\theta |R\rangle$$

Here,

$$\theta = \frac{1}{2} \arctan\left(\omega_x / \omega_z\right)$$

In the next step, the physics working behind tunneling of electron through the odorant from a donor state $|D\rangle$, having an energy ϵ_D to the acceptor state ϵ_A with an energy ϵ_A, can expressed by an Hamiltonian of the electron, as:

$$\hat{H}_e = \epsilon_A |A\rangle\langle A| + \epsilon_D |D\rangle\langle D|$$

Importantly, in the present case, the biological environment associated with this condition, can typically be described as a kind of condensed bath, conveniently modeled, as collection of harmonic oscillators, having corresponding Hamiltonian,

$$\hat{H}_{\text{env}} = \sum_i \omega_i \hat{b}_i^\dagger \hat{b}_i \tag{4.7}$$

Here, $b_i^\dagger$ and b_i denote the creation and annihilation operators respectively, for the modes of frequency ω_i in the corresponding environment around it. As stated already, the interaction Hamiltonian has three contributions between donor and acceptor of the receptor with tunneling strength:

(1) Δ between the donor (acceptor)
(2) the odorant with coupling frequency $\gamma_D(\gamma_A)$ between the donor (acceptor) sites and
(3) i-th harmonic oscillator of the environment with coupling frequency $\gamma_{iD}(\gamma_{iA})$.

Under such circumstances, it is possible to describe the dynamics of the reduced density matrix of the receptor together with the odorant (hereafter, receptor + odorant). Hence, it is possible to describe the interaction Hamiltonian of the total system by the so-called Redfield equation,[94] i.e.,:

$$\partial_t \hat{\rho}(t) = -\int_0^t d\acute{t}\, \mathrm{Tr}_E \left[\tilde{H}_{\text{int}}(t), \left[\tilde{H}_{\text{int}}(\acute{t}), \hat{\rho}_{\text{tot}}(\acute{t})\right]\right] \tag{4.8}$$

Now, assuming the interaction between the receptor together with odorant and the environment to be sufficiently weak, the total density matrix remains in an approximate product form at all times, i.e.,

$$\hat{\rho}_{\text{tot}}(t) \approx \hat{\rho}(t) \otimes \hat{\rho}_E(t).$$

Now, in case, i.e., when there will be sufficiently high temperature and large environment, it could be assumed that the environment quickly forgets whether there have been any kind of internal self-correlation established during the course of the interactions with the receptor + odorant (i.e., Markov approximation) which effectively means that the temporal change of the environment density matrix can be neglected. Under such condition,

$$\partial_t \hat{\rho}(t) = -i \left[\hat{H}_{\text{zero}}^{RO}, \hat{\rho}(t)\right] - L_\rho(t) \tag{4.9}$$

Tirandaz *et al.*[76] dealt in details the dynamics of the reduced density matrix by relating the receptor and showed that it is possible to obtain the odorant (hereafter receptor+odorant) by applying Redfield equation.

Thus after solving some mathematics, i.e., using Born — Markov master equation, the following expressions have been calculated for the noise and dissipation kernals respectively (details can be obtained from[76]) as,

$$\nu(\tau) = \int_0^\infty d\omega \frac{J(\omega)}{\omega^2} \left[\sin^2\left(\frac{\omega\tau}{2}\right) \coth\left(\frac{\omega}{2k_BT}\right)\right] \tag{4.10}$$

$$\eta(\tau) = \int_0^\infty d\omega \frac{J(\omega)}{\omega^2} \sin(\omega\tau) \tag{4.11}$$

$J(\omega)$, here, depicts the spectral density of a continuous spectrum having the environmental frequencies, ω, encapsulating almost all of the physical properties of the environment surrounding it. Now, writing this in ohmic way, i.e., the spectral density can be expressed as an exponential cut-off which is as follows:

$$J(\omega) = \sum_i (\gamma_{iD} - \gamma_{iA})^2 \delta(\omega - \omega_i) \equiv J_0 \omega e^{-\omega/\Lambda} \tag{4.12}$$

In the above expression, J_0 denotes, in fact, measures the coupling strength of the system-environment whereas Λ is meant for a cut off at high frequency. However, in a particular situation, i.e., when the elastic ET coincides with that of the chiral odorant being present in the receptor but without undergoing any kind of transition. Importantly, the inelastic ET through left-and right-handed enantiomers, at this stage, are accompanied by a vibrational transition which aims left and right-handed states to the first excited state, respectively. Besides this, quite importantly, for the present case, the interaction Hamiltonians, especially, has three contributions, i.e.,

- Firstly, between donor and acceptor with tunneling strength Δ,
- Secondly, between the donor (acceptor) and the odorant with coupling frequency $\gamma_D(\gamma_A)$ and also
- Between the donor (acceptor) sites and the i-th harmonic oscillator of the environment having the coupling frequency $\gamma_{iD}(\gamma_{iA})$.

Finally, the interaction Hamiltonian of the whole system can be expressed as,[76]

$$\begin{aligned}\hat{H}_{\text{int}} &= \Delta(|A\rangle\langle A| + |D\rangle\langle D|) + (\gamma_D|D\rangle\langle D| + \gamma_A|A\rangle\langle A|) \otimes \hat{\sigma}_x \\ &\quad + \sum_i (\gamma_{i,D}|D\rangle\langle D| + \gamma_{i,A}|A\rangle\langle A|) \otimes (\hat{b}_i^\dagger + \hat{b}_i)\end{aligned} \tag{4.13}$$

Now, in absence of the odorant, to avoid the undesired trasitions from donor to acceptor, the coupling strength Δ should be small compared to other energy scale for conveniently moinge into a polaron transformed reference frame for this regime. Importantly, the above equation characterizes the time evolution of the whole system with the help of which the ET rates are calculated. Here, in order to achieve the polaron transformation, the free Hamiltonian of the total system which consists of the electron at the receptor, the chiral odorant and the surrounding environment, can be expressed as the unperturbed Hamiltonian:

$$\hat{H}_\circ = \hat{H}_{od} + \hat{H}_e + \hat{H}_{env} \tag{4.14}$$

In the present case, the state of the electrons of the receptor at donor and acceptor sites has been represented by $|D\rangle$ with energy ϵ_D and $|A\rangle$ with energy ϵA, respectively, assuming that $\epsilon_D > \epsilon_A$. In that case, the electron tunneling corresponds to a vibrational absorption in the odorant and in that case, the hamiltonian with electron at the receptor, can be described as

$$\hat{H}_e = \epsilon_D |D\rangle\langle D| + \epsilon_A |A\rangle\langle A| \tag{4.15}$$

as designated due to polaron transformation. In the present situation, occurence of a chiral odorant can be possible only when two chiral enantiomers-can be described in a effective way by an asymmetric double-well potential, at least, through the inversion at molecule's center of mass which is caused by a long-amplitude vibration, typically, called as a contortional vibration.[87,88] The associated two chiral states $|L\rangle$ and $|R\rangle$ have minima, separable by a barrier V_0 which, in the limit, becomes $V_0 \gg \omega_0 \gg k_B T$, having ω_0 as the vibration frequency in each well.

Also, state space of the molecule remains effectively confined in two-dimensional Hilbert space having two chiral states. Thus, for most of chiral molecules, this limit remains valid for the room temperature. The left and right-handed enantiomers of the chiral odorant are associated to two states, i.e., $|L\rangle$ and $|R\rangle$, localized at left and right wells of potentials respectively. In such situations, the tunneling process is prevented, caused by high enough barrier. As a result, in such case, the molecules remain in the initial state and the corresponding unperturbed Hamiltonian of the molecule in the chiral basis is diagonalized by a polaron transformation as[89]

$$\hat{H}'_\circ = \sum_{R=A,D} \left(\epsilon_R + \eta_R \hat{\sigma}_z\right) |R\rangle\langle R| + \sum_i \omega_i \hat{b}_i^\dagger \hat{b}_i \tag{4.16}$$

where,

$$\eta_A = -\frac{\omega_z}{2}\cos\left(\tan^{-1}\left(\frac{\omega_x+\gamma_A}{\omega_z}\right)\right)$$

$$\eta_D = -\frac{\omega_z}{2}\cos\left(\tan^{-1}\left(\frac{\omega_x-\gamma_D}{\omega_z}\right)\right)$$

In a similar way, we can write the interaction Hamiltonian which then has been transformed to

$$\hat{H}'_{\text{int}} = \Delta|A\rangle\langle D|\exp\left(\frac{i}{2}\left[\tan^{-1}\left(\frac{\omega_x+\gamma_A}{\omega_z}\right)+\tan^{-1}\left(\frac{\omega_x-\gamma_D}{\omega_z}\right)\right]\hat{\sigma}_y\right)$$
$$\times\exp\left(\sigma_i\left(\frac{\gamma_{iD}-\gamma_{iA}}{\omega_i}\right)(\hat{b}_i^\dagger-\hat{b}_i)\right)+\text{h.c.} \quad (4.17)$$

4.4.4 *Tunneling Rates*

It is now an well known fact that it has been Luca Turin, who proposed inelastic electron tunneling to be the method by which vibrations are detected by the olfactory receptors within the hose.

In order to characterize vibrationally assisted tunneling, the tunneling rate has been calculated by Tirandaz *et al.*[76] as a sum over individual rates considering all vibrational modes of the odorants. These have been obtained

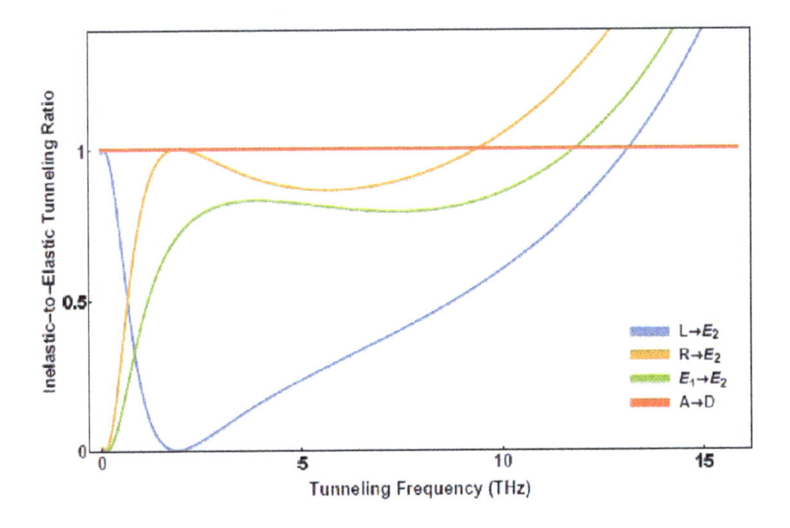

Figure 4.9. Tunneling frequency vs. Inelastic to elastic ratio.

from the corresponding probabilities by using $\Gamma_{i \to j} = dP_{r_{i \to j}}/dt$. The total electron tunneling rate is calculated as a function of the energy difference $\Delta\epsilon$ and the reorganization energy λ for acetophenone, citronellyl nitrile and octanol. The typical times for intrinsic electron tunneling in biological systems occurring on a length scale of (5–15) Angstroems are (10–100) nsec to obtain the tunneling rates. This means when the electron tunnels trough the odorant, there might be three kind of possibility of transitions which might take place in the odorant:

(i) $|L\rangle \to |E_2\rangle$,
(ii) $|R\rangle \to |E_2\rangle$ and might be also through and
(iii) $|E_1\rangle \to |E_2\rangle$.

In such kind of a situation, the electron transfer, i.e., ET (elctron tunneling) rates can be calculated from the corresponding probabilities, i.e., by using $\Gamma_{i \to j} = dP_{r_{\to j}}/dt$ [Details can be found in reference 73.

Interestingly, the rates for these three odorants behaving in a similar way, have also been studied quantitatively in detail which have been found,[76] behaving in a similar way, even though the maximal tunneling rates found as slightly different. However, for reorganization energies $\lambda \geq 0.5$ eV, the tunneling rate demonstrates only an weak dependence on the energy difference $\Delta\epsilon$, having a single peak at $\Delta\epsilon \approx 0.5$ eV for all studied odorants. As a result, this range of λ — renders the vibrationally assisted mechanism, appearing to be impractical for resolving the different odorant compounds as all of them will act similarly on the receptor

4.4.5 *Physical Parameters*

As a next, in order to calculate ET (Electron Tunneling) rates quantitatively, the parameters of the model requires to be finalized. Tirandaz *et al.*'s group calculated in details all such parameters in terms of controllable variables. This includes odorant's parameters, (for example, tunneling frequency ω_x and the asymmetry frequency ω_z) including the thermodynamical parameters of the environment (pressure and temperature). It should be noted that in the present analysis, the parameters with characteristic interaction properties, naturally depend on the odorant's parameters also. Not only that, the energy conservation requires that the energy gap between the donor and acceptor sites ϵ be close to the mean value of energy gap between odorant's states.

This crucial reason makes necessary, in assuming that ϵ be close to the mean value of energy gap, i.e., between the donor and acceptor sites. It could be assumed that

$$\epsilon \cong \sqrt{(\omega_x^2 + \omega_z^2)}.$$

In the present consideration, the coupling between donor and acceptor sites of OR(s) has been considered as 'weak' in comparison with the natural frequency of the odorant so that we can estimate $\Delta \approx 0.01\sqrt{\omega_x^2 + \omega_z^2}H_z$.[76]

As a next, the coupling frequency between the DA pair and odorant, calculated from the Huang-Rhys factor[62b] can be approximated as $\gamma_D = -\gamma_A \approx 0.1\sqrt{\omega_x^2 + \omega_z^2}$.[76] As is well known, microscopically, the biological environment is quite uncontrollable which, practically, makes the parameters of the corresponding spectral density as mere parameters to be considered. For example, taking the most common biological environment, i.e., water, the parameters of an aqueous environment can be estimated as $J_0 \approx 1$ and $\lambda \approx 10^{12}$ Hz,[76,93] In the tunneling-dominant limit, i.e., $\omega_x \geq \omega_z$, however, for all transitions, the inelastic ET is dominant. In an explicit way, for each transition, there is a threshold of tunneling frequency ω_x in the bottom limit of which, the olfactory system cannot recognize the odorant. This fact can be considered in judging the validity of the model in the corresponding experiment.

At this point, from the further studies it has been noted now that the energy dependence of the tunneling rate reflects the key IR active vibrations for the three odorants. This has been studied by the above group[76] and, obviously, recognition of the odorants through the vibrationally assisted electron tunneling mechanism have been possible to be feasible.

4.4.6 *Role of Odorant in Electron Tunneling*

The ET rates for different odorants have been analyzed by the above group[76] in terms of the corresponding molecular parameters, ω_x and ω_z. Thus, it is now possible to measure the magnitude of tunneling frequency ω_x, which could have a range from the inverse of the lifetime of the universe to millions of hertz which can be extracted from the spectroscopic data.[89] The asymmetry frequency of the odorant, i.e., ω_z represents an overall measure of all chiral interactions involved which, for the system followed in these interactions, are primarily due to the intermolecular interactions between the odorant and ORs. The magnitude of such intermolecular interactions, in principle, is possible to determine, taking the help of

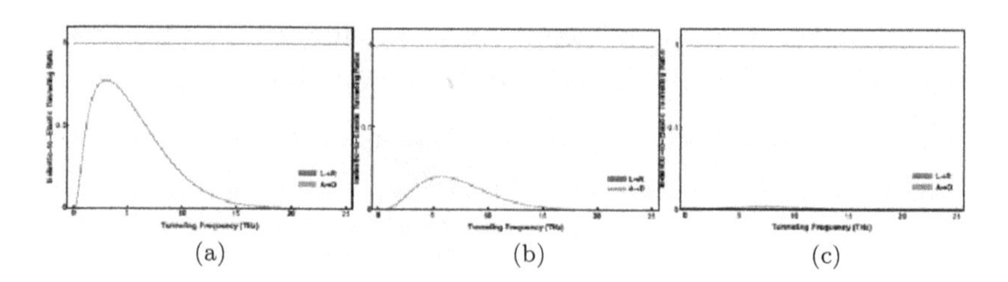

Figure 4.10. Tunneling frequency vs. Inelastic to elastic tunneling ratio.

quantum chemistry. But, it becomes crucially important to note that the dependence of the ET rates to ω_x and ω_z for different transitions of the odorant when plotted, clearly indicates towards a crucial decision, (at least for the present state of research), i.e., the vibrational model based on odorant-mediated inelastic ET is *improbable* for a wide range of odorants. In an alternate tunneling-dominant limit, i.e., $\omega_x \geq \omega_z$, the inelastic ET becomes dominant for all transitions. In other words, importantly, for each transition, a threshold of tunneling frequency ω_x in the bottom limit can be found, so that the olfactory system becomes unable of recognizing the odorant. This fact could be an crucial in examining validity of the model while doing experiment.

4.4.6.1 *Temperature Dependence*

The temperature dependency during different transitions of the odorant are found to be essentially similar. The Figure 4.9(a) clealry indicates it. At a fixed high magnitude of asymmetry parameter ω_z, for the transitions $L \rightarrow E2, R \rightarrow E2$, and $E1 \rightarrow E2$, the inelastic-to-elastic ratio versus the tunneling frequency ω_x when plotted for different temperatures of the environment, for each odorant (with a fixed tunneling frequency ω_x) a threshold for temperature in the bottom can be found in which limit, the olfactory system cannot recognize the scent. This fact could also be used in examining the validity of the model to be applied in experiment. However, at present, it has been observed that odor descriptions in the olfaction literature correlate more strongly with their vibrational frequencies than with their molecular shape.

4.4.7 *Effect of Pressure*

Tirandaz *et al.*[76] also did the detailed studies about the effect of pressure on the concentration of the odorant in the condensed environment and found it to be proportional to its pressure. We already know that the tunneling frequency of odorants is related to the pressure of the environment. In order to illustrate the pressure dependency of the odorant's dynamics, the above mentioned group focused on the ammonia molecule NH_3 as odorant. In the low-pressure limit, the tunneling frequency of ammonia, known as inversion frequency, their calculated estimation has been as $\omega_x \approx 2.4 \times 10^{10}$ Hz,[95] and at $P \approx 2$ atm, ω_x has been found to shift to zero which can be demonstrated theoretically, within the context of the mean-field theory.[96] The pressure dependency of the inversion frequency can be expressed as $\omega'_x = \omega_x\sqrt{1 - P/P_{cr}}$. Here, the critical pressure $P_{cr} \approx 1.6$ atm at room temperature. Now, for $P \to P_{cr}$, we have, $\omega' \to 0$ handed states become eigenstates of the molecular Hamiltonian. At the high-pressure limit, we have the relevant transition in the odorant $|R\rangle \to |L\rangle$. Then, the inelastic ET rate according to this transition is as follows:

$$\Gamma_{D,R\leftarrow A,L} = \Delta^2\sqrt{\frac{\pi}{k_BTJ_0\lambda}}\sin^2\nu \exp\left(\frac{-(\epsilon - J_0\lambda + (\eta_1+\eta_2))^2}{4k_BTJ_0\lambda}\right) \quad (4.18)$$

But, in the high pressure limit, the inelastic ET have been found **always ineffective**. Interestingly, this could provide another empirical test in order to probe the validity of the tunneling model of olfaction. Tirandaz *et al.*, specifically predict that at $P > P_cr$ the olfactory system becomes unable to recognize the smell i.e., when there will be no tunneling mechanism at work.

4.4.8 *Chiral Recognition*

The existence of enantiomers, responsible for different smells, originally have been applied in rejecting the quantum model of olfaction.[72] As these enantiomers have the same vibrational spectrum, it seems that the shape-based parameters should be included in the quantum model to distinguish them from each other. Importantly, the model presented here can be used to generalize the vibrational model for chiral recognition. It has been already proposed that chiral molecules can effectively be modeled by a double-well potential.[88] Since such enantiomers have the same vibrational spectrum, it seems that the shape-based parameters should be included in the quantum model to distinguish them from each other. The model presented here can be applied in generalizing, vibrational model for chiral recognition in which chiral molecules can effectively be modeled by a double-well potential.

Again, as the shape factor is known to be the configuration of the chiral odorant, the model, applied by Tirandaz *et al.* predicts that the ET rates of inelastic ET are different for the two enantiomers for transitions $|L\rangle \rightarrow |E_2\rangle$ and $|R\rangle \rightarrow |E_2\rangle$. The result obtained has been found in agreement with the Born-Markov master equation approach. From the results, it has been observed that the enantiomers with similar smells lie in the limit $\omega_x \ll \omega_z$. But the inelastic ET is ineffective in such limit. The enantiomers have different smells in the limit $\omega_x \approx \omega_z$, whereas inelastic ET is still ineffective. The inelastic ET has been found to be effective for all transitions in the limit $\omega_x > \omega_z = \lambda$. Importantly, in this limit, the ratio of the inelastic ET rate for the left-handed enantiomer to that of the right-handed one, increases with the ratio of the tunneling frequency to the asymmetry frequency. Typical times for electron transfer in proteins are of order 10^{-15}–10^{-12} s.[97] Even though, the difference between inelastic ET rates of transitions may appear insignificant, however, comparing similar processes in biology it can be possible for the system in discriminating between two enantiomers under quantum constraints.

4.4.9 *Conclusion*

Among many of discrepancies, one important consequence of going beyond these discrimination, based on shape alone, is that quantum phenomena, by time, become much more evident due to which, we consequently see the rise and gradual spread of the applications of Quantum Biology. Of course, shape already has implicitly invoked the quantum nature of chemical bonding. It should be noted here that, inelastic electron transitions of the sort, discussed in one of the conjectures, presently mentioned, involve a coherent quantum electron transfer event. Again, using the vibrational frequencies as a discriminant, this aspect relies on the quantum behavior of the odorant vibrational modes, since energy can only be given to an oscillator in units of its vibrational quantum. Some other quantum aspects, such as the role of zero-point motion might also be possible, but these are not evident at this stage and hence needed further extensive research.

The lock and key paradigm was one of the earliest attempts to rationalize remarkably selective responses to different molecules. For large molecules, it is still a key concept. However, for small molecules, the underlying idea that shape is the sole critical factor fails apart quite badly. Our sense of smell allows us to discriminate between small molecules in very low concentrations via scent molecules interacting with receptors in the nose. Presently, the biomolecular processes of olfaction are not yet fully understood, and some evidence suggests that a mechanism based solely on the size and shape of odorant molecules is inadequate. For example, it has been noted that molecules with very similar shapes and sizes have a remarkably different scent.[43,45] Thus traditional models of a "docking"-type mechanism, where the size and shape of an odorant molecule actuates the receptor in some way, are thought to be insufficient. The swipe card paradigm, on the other hand, whether at this stage definitive as a model or not, introduces perhaps more productive ways of thinking that confront interesting observations in nature. For this reason alone, it has the power to eliminate thinking based on theories that do not work and as a result, road-block progress. In fact, shape is not necessarily the actuating factor in smell; we must determine what factors are crucial for reasons of phenomenological interest. But also, it is again possible that the mechanisms and underlying processes of olfaction have parallels in the operation of a range of receptors activated by small molecules such as neurotransmitters, hormones, steroids and so on.

Turin proposed a mechanism which, in addition to "docking", gives a further level of selectivity (or sensitivity) by a process of inelastic electron tunneling. In this case the odorant molecule both docks with a receptor and then mediates phonon-assisted inelastic tunneling of an electron from a donor to an acceptor (i.e., donor and acceptor electronic states differ in energy by $\approx \omega$, and hence, transport only occurs whenever energy is conserved by emitting an odorant phonon, the vibrational degree of freedom of the molecule one is "smelling", of the right energy). A model proposed by Brookes[17] expanded on this idea, and presented evidence that such a mechanism fits the observed features of smell, and is at least "physically" credible. However, whether such a mechanism ultimately exists in nature has yet to be determined. While not specifically requiring "coherence" to function, this mechanism requires inelastic phonon-assisted tunneling of electrons, and is certainly more "microscopic", and sensitive, than previously thought. The details will never be precisely the same, but there is a clear grand challenge in the understanding of the responses of receptors to small molecules and linking them to their biomedical impacts.

References

[1] Zarzo, M. The sense of smell: molecular basis of odorant recognition, Biol. Rev. 82, 455, 2007.

[2] Amoore, John E.; 1964; "Current status of the steric theory of odor." Annals of the New York Academy of Sciences 116, no. 2: 457–476.

[3] Amoore, John E.; 1977; "Specific anosmia and the concept of primary odors." Chemical Senses 2, no. 3: 267–281.; J.E Amoore, Pelosi, P. & Forrester, L.J.; 1977; Specific anosmias to 5alpha-androst-16-en-3-one and omega-pentadecalactone: the urinous and musky primary odors. Chemical Senses, 2(4), pp. 401–425.

[4] J.E. Amoore, J.; 1963; The stereochemical theory of olfaction, Nature 199, 912.

[5] K. Mori & G. Shepard; 1994; Emerging principles of molecular signal processing by mitral/tufted cells in the olfactory bulb, Semin. Cell. Biol. 5. 65.

[6] Buck L, Axel R.; 1991; A novel multigene family may encode odorant receptors: a molecular basis for odor recognition. Cell (1991) 65:175–87].

[7] L. B. Buck, L.B.; 2005; Unraveling the Sense of Smell (Nobel Lecture), Angew. Chem. Int. Ed. 44, 6128.

[8] D. J. Rowe, Chemistry and technology of flavors and fragrances, Blackwell, Oxford, 2005.

[9] R. Axel, Scents and Sensibility: A Molecular Logic of Olfactory Perception (Nobel Lecture), Angew. Chem. Int. Ed. 44, 6110, 2005.

[10] S. H. Lee *et al.* Mimicking the human smell sensing mechanism with an artificial nose platform, Biomaterials 33, 1722, 2012.9.

[11] R. H. Farahi, A. Passian, L. Tetard and T. Thundat, Critical Issues in Sensor Science To Aid Food and Water Safety, ACS Nano 6, 4548, 2012.

[12] Klopping HL.; 1971; Olfactory theories and the odors of small molecules. J Agricult Food Chem.; 19:999–1004. doi: 10.1021/jf60177a002.

[13] F. Yoshii, S. Hirono and I. Moriguchi, Relations between the odor of (r) ethyl citronellyl oxalate and its stable conformations, Quant. Struc-Act Rel. 13 144–147, 1994.

[14] R. C. Araneda, A. D. Kini and S. Firestein. The molecular receptive range of an odorant receptor, Nature Neuroscience, 3, 1248, 2000.

[15] L. Turin and F. Yoshii, Struture-odor relations: a modern perspective, in Handbook of Olfaction and Gustaion, R. Doty, Marcel Dekker, New York, 2003.

[16] R. Bentley, The nose as a stereochemist: Enantiomers and odor, Chem. Rev. 106, 4099, 2006.

[17] J. C. Brookes, A. P. Horsfield, and A. M. Stoneham, Odour character differences for enantiomers correlate with molecular flexibility, J. R. Soc. Interface 6, 75, 2009.

[18] Brookes *et al.*; Swipe Card Model of Odor Recognition; Sensors; 2012, 12, 15709–15749.
[19] G. Dyson, The scientific basis of odour, Chem. Ind. 57, 647, 1938.
[20] R. Wright; 1977; Odor and molecular vibrations: neural coding of olfactory information, J. Theor. Biol. 64, 473.
[21] Wright, R.H.; 1982; The Sense of Smell, CRC Press, Boca Raton, FL.
[22] J. Lambe and R. C. Jaklevic; 1968; Molecular Vibration Spectra by Inelastic Electron Tunneling, Phys. Rev. 165, 821.
[23] Barwich AS.; 2015; What is so special about smell? Olfaction as a model system in neurobiology. Postgrad Med J.; 92:27–33.; doi: 10.1136/postgradmedj-2015-133249.
[24] Pauling Linas; 1946; Molecular architecture and Biological Reactions; Chem. Eng. News, 24, 1375; referenced by Ohloff, G., Scent and Fragrances, Springer-Verlag, Berlin Heidelberg, 1994.
[25] Moncrieff, R.W.; 1949; What is Odor. A New Theory, Am. Perfumer, 54: 453.
[26] Moncrieff, R.W.; 1954; The characterization of odours. The Journal of physiology, 125(3), p. 453.
[27] E.A. Hallem and J.R. Carlson, Cell 125 (2006) p. 143.
[28] D. Münch and C.G. Galizia, Sci. Rep. 6 (2016) p. 21841.
[29] Miao, Y. & J. McCammon; 2016; Curr. Opin. Struct. Biol. 41; p. 83.
[30] S. Jang and C. Hyeon; 2017; J. Phys. Chem. B 121; p. 1304. doi:10.1021/acs.jpcb.7b00486.
[31] I. Kufareva, M. Rueda, V. Katritch, G. Dock, R.C. Stevens & R. Abagyan; 2011; Structure; 19; p. 1108.
[32] Zhang Z, Wu J, Yu J, Xiao J.; 2012; A brief review on the evolution of GPCR: conservation and diversification. Open J Genet; 2:11–7. doi: 10.4236/ojgen.2012.24B003].
[33] Rosenbaum DM, Cherezov V, Hanson MA, Rasmussen SGF, Thian FS, Kobilka TS, *et al.* GPCR engineering yields high-resolution structural insights into 2 adrenergic receptor function. Science (2007) 318:1266–73. doi: 10.1126/science.1150609.
[34] Manglik A, Kim TH, Masureel M, Altenbach C, Yang Z, Hilger D, *et al.* Structural insights into the dynamic process of β_2-adrenergic receptor signaling. Cell (2015) 161:1101–11. doi: 10.1016/j.cell.2015.04.043.
[35] Sounier R, Mas C, Steyaert J, Laeremans T, Manglik A, Huang W, *et al.* Propagation of conformational changes during Mu-opioid receptor activation. Nature (2015) 524:375–8. doi: 10.1038/nature14680.

[36] Huang W, Manglik A, Venkatakrishnan AJ, Laeremans T, Feinberg EN, Sanborn AL, *et al.* Structural insights into Muopioid receptor activation. Nature (2015) 524:315–21. doi: 10.1038/nature14886.
[37] Vosshall LB, Amrein H, Morozov PS, Rzhetsky A, Axel R. A spatial map of olfactory receptor expression in the Drosophila antenna. Cell (1999) 96:725–36. doi: 10.1016/S0092-8674(00)80582-6.
[38] Vosshall LB (2015). "Laying a controversial smell theory to rest". Proc. Natl. Acad. Sci. USA. 112 (21): 6525–6526. Bibcode:2015 PNAS..112.6525V. doi:10.1073/pnas.1507103112. PMC 4450429. PMID 26015552.
[39] O. Man, Y. Gilad and D. Lancet, Protein Sci. 13 (2004) p. 140.
[40] C.D. Hanlon and D.J. Andrew, J. Cell Sci. 128 (2015) p. 3533.
[41] R. Strotmann, K. Schröck, I. Böselt, C. Stäubert, A. Russ and T. Schöneberg, Mol. Cell. Endocrinol. 331 (2011) p. 170.]
[42] Turin, L.; 2016; Inference 2.
[43] Turin, L.& F. Yoshii; 2003; Handbook of Olfaction and Gustation, 2nd ed., CRC Press Boca, Boca Raton, FL, 2003, p. 275.
[44] Turin, L.; 1996; A spectroscopic mechanism for primary Olfactory reception; Chem Senses; Vol. 21; p. 773.
[45] Horsfield, A. P.; Haase, A.; Turin, L. (2017). "Molecular recognition in olfaction". Advances in Physics: X. 2 (3): 937–977. doi:10.1080/23746149.2017.1378594.
[46] Block E. (2018). "Molecular basis of mammalian odor discrimination: A status report". Journal of Agricultural and Food Chemistry. 66 (51): 13346–13366. doi:10.1021/acs.jafc.8b04471. PMID 30453735.
[47] Block E, *et al.*; 2015; "Implausibility of the Vibrational Theory of Olfaction". Proc. Natl. Acad. Sci. USA. 112 (21): E2766–E2774. Bibcode:2015PNAS..112E2766B. doi:10.1073/pnas.1503054112. PMC 4450420. PMID 25901328.
[48] Block, E.; Batista, V.S.; Matsunami, H.; Zhuang, H.; Ahmed, L.; 2017; "The role of metals in mammalian olfaction of low molecular weight organosulfur compounds". Natural Product Reports. 34 (5): 529–557. doi:10.1039/c7np00016b. PMC 5542778. PMID 28471462.
[49] Keller A; Vosshall LB (2004). "A psychophysical test of the vibration theory of olfaction". Nature Neuroscience. 7 (4): 337–338. doi:10.1038/nn1215. PMID 15034588.
[50] Saberi M, Seyed-allaei (2016). "Odorant receptors of Drosophila are sensitive to the molecular volume of odorants". Scientific Reports. 6: 25103. Bibcode:2016NatSR...625103S. doi:10.1038/srep25103. PMC 4844992. PMID 27112241.

[51] Everts S (2015). "Receptor Research Reignites A Smelly Debate". Chem. Eng. News. 93 (18): 29–30.
[52] Sell, CS (2006). "On the Unpredictability of Odor". Angew. Chem. Int. Ed. 45 (38): 6254–6261. doi:10.1002/anie.200600782. PMID 16983730.
[53] Doszczak, L; Kraft, P; Weber, H-P; Bertermann, R; Triller, A; Hatt, H; Reinhold Tacke, R; 2007. "Prediction of Perception: Probing the hOR17-4 Olfactory Receptor Model with Silicon Analogues of Bourgeonal and Lilial". Angew. Chem. Int. Ed. 46 (18): 3367–3371. doi:10.1002/anie.200605002. PMID 17397127.
[54] Na, M.; Liu, M. T.; Nguyen, M. Q.; Ryan, K.; 2019; "Single-neuron comparison of the olfactory receptor response to deuterated and nondeuterated odorants". ACS Chem. Neurosci. 10 (1): 552–562. doi:10.1021/acschemneuro.8b00416. PMID 30343564.
[55] Crabtree, R.H.; 1978; "Copper(I) — Possible Olfactory Binding-Site". J. Inorg. Nucl. Chem. 1978 (40): 1453. doi:10.1016/0022-1902(78)80071-2.
[56] Duan, Xufang; Block, Eric; Li, Zhen; Connelly; Zhuang, Hanyi *et al.* 2012; "Crucial role of copper in detection of metal-coordinating odorants". Proc. Natl. Acad. Sci. U.S.A. 109 (9): 3492–3497.
[57] Li, S.; Ahmed, L.; Zhang, R.; Pan, Y.; 2016; "Smelling sulfur: Copper and silver regulate the response of human odorant receptor OR2T11 to low molecular weight thiols"; Journal of the American Chemical Society; 138 (40): 13281–13288. doi:10.1021/jacs.6b06983. PMID 27659093.
[58] Lewis, A.; Del Prioire, L.V.; 1988; The biophysics of visual photoreception. Phys. Today, 41, 38–46.
[59] Kaiser, L.; Graveland-Bikker, J.; Steuerwald, D.; Vanberghem, M.; Herlihy, K.; Zhang, S.; 2008; Efficient cell-free production of olfactory receptors: Detergent optimization, structure, and ligand binding analyses. Proc. Nat. Acad. Sci. 2008, 105, 15726–15731.
[60] Cook, B.L.; Ernberg, K.E.; Chung, H.; Zhang, S.; 2008; Study of a synthetic human olfactory receptor 17-4: Expression and purification from an inducible mammalian cell line. PLoS One, 3, e2920.
[61] Jacquier, V.; Pick, H.; Vogel, H. Characterization of an extended receptive ligand repertoire of the human olfactory receptor or 17-40 comprising structurally related compounds. J. Neurochem. 2006, 97, 537–544].
[62] D. Wrobel, U. Wannagat, and U. Harder, Chemical Monthly 113, 381 (1982).

[63] Warshel, A. Bicycle-pedal model for the first step in the vision process. Nature 1976, 260, 679–683.
[64] Fleischmann, A.; Shykind, B.; Sosulski, D.; Franks, K.; Glinka, M.; Mei, D.; Sun, Y.; Kirkland, J.; Mendelsohn, M.; Albers, M. Mice with a "monoclonal nose": Perturbations in an olfactory map impair odor discrimination. Neuron 2008, 60, 1068–1081.
[65] Jennifer C. Brookes; Andrew P. Horsfield; and A. Marshall Stoneham 2012; The Swipe Card Model of Odorant Recognition; vol. 12; 15709–15749.
[66] Brookes, J.C.; Hartoutsiou, F.; Horsfield, A.P.; Stoneham, A.M.; Could humans recognize odor by phonon assisted tunneling? Phys. Rev. Lett. 2007, 98, 038101.
[67] Lambe, J.; Jaklevic, R.C. Molecular vibration spectra by inelastic electron tunneling. Phys. Rev. 1968, 165, 821–832.
[68] Galperin, M.; Ratner, M.A.; Nitzan, A. Inelastic electron tunneling spectroscopy in molecular junctions: Peaks and dips. J. Chem. Phys. 2004, 121, 11965–11979.
[69] Troisi, A.; Beebe, J.M.; Picraux, L.B.; van Zee, R.D.; Stewart, D.R.; Ratner, M.A.; Kushmerick, J.G. Tracing electronic pathways in molecules by using inelastic tunneling spectroscopy. Prod. Nat. Acad. Sci. 2007, 104, 14255–14259.
[70] C. J. Adkins and W. A. Phillips; 1985; Frequency shifts in inelastic electron tunnelling spectroscopy of adsorbed species, J. Phys.C 18, 1313.
[71] I. A. Solov'yov, P.-Y. Chang and K. Schulten: 2012: Vibrationally assisted electron transfer mechanism of olfaction: myth or reality?, Phys. Chem. Chem. Phys. 14, 13861.
[72] E. R. Bittner *et al.* Quantum Origins of molecular recognition and olfaction in drosophila, J. Chem. Phys. 137, 22A551, 2012.
[73] D. J. Rowe; 2005; Chemistry and technology of flavors and fragrances, Blackwell, Oxford, 2005.
[74] A. Chęcińska *et al.*, Dissipation enhanced vibrational sensing in an olfactory molecular switch, J. Chem. Phys. 142, 025102, 2015.
[75] Arash Tirandaz, Farhad Taher Ghahramani, & Afshin Shafiee; 2017; Dissipative Vibrational Model for Chiral Recognition in Olfaction; arXiv 1704.08129[physics.chem-ph].
[76] A. Tirandaz, F. Taher Ghahramani and A. Shafiee, Dissipative vibrational model for chiral recognition in olfaction, Phys. Rev. E 92, 032724, 2015.-]
[77] W. Gronenberg *et al.*; 2014; Honeybees (Apis mellifera) learn to discriminate the smell of organic compounds from their respective deuterated isotopomers, Proc. Biol. Sci. 281, 20133089.

[78] M. I. Franco *et al.* Molecular vibration-sensing component in Drosophila melanogaster olfaction, Proc. Natl. Acad. Sci.USA 108, 3797, 2011.
[79] L. J. W. Haffenden, V. A. Yaylayan and J. Fortin, Investigation of vibrational theory of olfaction with variously labelled benzaldehydes, Food Chem. 73, 67, 2001.
[80] S. Gane *et al.* Molecular Vibration-Sensing Component in Human Olfaction, PLoS One 8, e55780, 2013.
[81] W. Ogle, Med. Chir. Trans. 53 (1870) p. 263.
[82] B.R. Havens and C.E. Meloan, Developments in Food Science, Vol. 37, Elsevier, 1995, p. 497.
[83] C.E. Meloan, V.S. Wang, R. Scriven and C.K. Kuo, Developments in Food Science, Elsevier, 1988.
[84] Jang, S. & Hyeon, C.; 2017; J. Phys. Chem. B 121; p. 1304. doi:10.1021/acs.jpcb.7b00486.
[85] C.S. Sell, Angew. Chem. Int. Ed. 45 (2006) p. 6254.
[86] A. J. Leggett *et al.*; 1987; Dynamics of the dissipative two-state system, Rev. Mod. Phys. 59, 1 (1987).
[87] C. H. Townes and A. L. Schawlow, Microwave Spectroscopy, McGraw-Hill, New York, 1955.
[88] Herzberg, G.; Molecular Spectra and Molecular Structure. Electronic Spectra and Electronic Structure of Polyatomic Molecules, Krieger, Malabar (1991).
[89] U. Weiss, U.; Quantum Dissipative Systems, World Scientific, Singapore 2008.
[90] (a) M. Quack, How important is parity violation for molecular and biomolecular chirality?, Angew. Chem. Intl. Ed. 41, 4618, 2002.
[91] M. Quack, J. Stohner and M. Willeke, High-resolution spectroscopic studies and theory of parity violation in chiral molecules, Annu. Rev. Phys. Chem. 59, 741, 2008.
[92] T.D. Lee & C. N. Yang; 1956; Question of Parity Conservation in Weak Interactions; Phys. Rev. 104, 254.
[93] J. Gilmore and R. H. McKenzie, Spin boson models for quantum decoherence of electronic excitations of biomolecules and quantum dots in a solvent, J. Phys.: Condens. Matter 17, 1735 (2005).
[94] M. Schlosshauer, Decoherence and the Quantum to Classical Transition, Springer, Berlin (2007).
[95] B. Bleaney and J. H. Loubster, Collision broadening of the ammonia inversion spectrum at high pressures, Nature 161 522, 1948.
[96] G. Jona-Lasinio, C. Presilla and C. Toninelli, Interaction induced localization in a gas of pyramidal molecules, Phys. Rev.Lett 88, 123001, 2002.

[97] K. Brettel and M. Byrdin, Reaction mechanisms of DNA photolyase, Curr. Opin. Struct. Biol. 20, 693, 2010.

[98] Yao, X.; Parnot, C.; Deupi, X.; Ratnala, V.R.P.; Swaminath, G.; Farrens, D.; Kobilka, B. Coupling ligand structure to specific conformational switches in the beta2-adrenoceptor. Nat. Chem. Biol. Lett. 2006, 2, 417–422.

[99] Fried, H.U.; Fuss, S.H.; Korsching, S.I.; 2002; Selective imaging of presynaptic activity in the mouse dependence of odor responses in identifies glomeruli. Proc. Nat. Acad. Sci., 99, 3222–3227.

Chapter 5

Navigation of Bird & Inclination Magnetic Compass by Long Distance Migratory Insects

"A perplexing property of quantum mechanics could be allowing birds to see and navigate the planet's magnetic fields. Birds also use quantum mechanics to navigate". —Sherwin, F.: 'Bird Brains and Quantum Mechanics' (ICR News; Posted on icr.org May 4, 2012)

5.1 Introduction

Quite recently several groups of researchers have raised a few viable intriguing possibility that living systems may use non-trivial quantum effects to optimize some tasks of this field. Studies range from the role of quantum physics in photosynthesis[1–5] and in natural selection itself,[6] through to the observation that 'warm and wet' living systems can embody entanglement given a suitable cyclic driving.[8,48] Thus, Quantum effects in biology are not just a quirk of plants and other organisms that do the peculiar job of turning sunlight into fuel. They may also provide an answer to a scientific puzzle that has been around since the 19th Century: how migratory birds know which way to fly. Bird navigation is a complex enterprise, requiring birds to make repeated and varying orientation decisions based on directional and positional information. Birds are aided by multiple physiological compass systems, among them a physiological magnetic compass. Thus a perplexing property of quantum effects in biology might explain how it could be allowing migratory birds to see and navigate, knowing which way to fly. In a journey, thousands of kilometer long, a migratory bird, such as the European robin will often fly to southern Europe or North Africa to escape particularly cold winters.

This journey over an unfamiliar landscape would be dangerous, if not impossible, without a compass. Start the journey in the wrong direction and a robin setting off from Poland might end up in Siberia rather than Morocco.

Amount of the chemical found to be present in the bird's nerve cells is a source of information, generating signals in the bird's nerve cells. As part of many other different environmental cues, this information will inform the bird about whether it is pointing towards Siberia or Morocco. This kind of novel ability of a migratory bird to orient itself relative to the Earth's magnetic field is at once a familiar feature of everyday life and a puzzling problem of quantum mechanics as well. The hypothesis that migrating birds can make use of the earth's magnetic field for orientation was first proposed by von Middendorff[9] though the precise mechanism by which an organism may sense the orientation of the weak geomagnetic field still remained somewhat unclear and theoretically problematic.

Magnetoreception has been observed to be one of many unique animal abilities to detect either the polarity or inclination of the earths magnetic field as a navigation tool. With the capability of electro-reception, i.e., detecting electrical fields, (for example, sharks hunting their prey) it represents a sense, however, slightly alien to human experience. Or so, it was thought until recently, several experiments have been undergoing in exploring and on manipulating a specific gene, for example, in flies and monarch butterflies, indicating that this gene plays a significant role in magnetoreception. As and since humans share this gene in a dormant form it suggests, — might be, even we may harbor this unique ability.[10,12] However, some evidence suggests it is not driven by the radical-pair mechanism.

Not only that the ability of many animal species, for example, insects, and mammals to sense the geomagnetic field for the purpose of orientation and navigation has led to huge interest in the field of biophysics.[5] Since long, scientists have speculated that certain animals are making use of the planet's magnetic fields to find their way, but biologists are mystified as to how they might do it. Now some answers might be coming from one of the most perplexing interactions in physics, related to the planet's magnetic fields. There are currently two leading hypotheses to explain this remarkable ability: (a) the magnetite-based mechanism, and (b) the radical pair mechanism.

There are a several mechanisms by which this sense may operate.[12] In certain species, including certain birds,[14,43] fruit flies[15,82] and even plants,[77] the evidence supports a so-called Radical Pair (RP) mechanism.

This process involves the quantum evolution of a spatially-separated pair of electron spins and such a model is supported by several results from the field of spin chemistry.[19–23] An artificial chemical compass operating according to this principle has been demonstrated experimentally,[24] and a very recent theoretical studies examines the presence of entanglement within such a system.

5.1.1 *Presence of Cryptochrome & Avian Compass*

Earth's 10,000 migratory bird species, navigate over great distances, i.e., almost around one-fifth of Earth, between their breeding and wintering grounds, and, seemingly, that also crossing insurmountable obstacles with remarkable accuracy. Billions of birds fly thousand of kilometers every year as they follow the seasons between their breeding and wintering grounds with the help of their astounding capabilities in detecting the direction of the Earth's magnetic field with an error of 5^0 or less. For example, Demoiselle Cranes are used to fly by negotiating the altitude in excess of 20,000 feet as and when they need to pass over Himalayan mountains. The Arctic Tern travels from pole to pole in pursuit of an endless summer, a distance of some 40,000 miles with the help of their incredible capacities. The biophysical sensory mechanism at the heart of this compass is thought to rely on magnetically sensitive, light-dependent chemical reactions in cryptochrome proteins in the eye. Thus far, theoretical models have been in great difficulty in accounting for the $<5^{\cdot}$ precision with which migratory birds are able to detect the geomagnetic field vector.

It was Schulten[34,150] who in 1978, first hypothesized out of his pioneering observations that some sort of biochemical reaction took place in bird's eyes, most likely producing electrons whose spin was affected by subtle magnetic gradients. For decades, scientists have adopted the idea that the navigational skill employed by bird, is an ability to detect variations in the earth's magnetic field. But which way this 'magnetic sense' works, however, has been frustratingly difficult to figure out. When birds migrate over long distances — sometimes thousands of miles, usually they end up in exactly the same place year after year. Such accurate feats of navigation, accomplished by millions of birds every year, have long made scientists wonder how they do it. Now a group of scientists in Germany has experimental evidence that reveals an important part of the birds' so called secrets in navigational success.

As per views of Jason S. Birdy, — birds navigate in part by orienting themselves with the sun, following physical landmarks. But these strategies alone have not been found enough as birds must be able to navigate even on cloudy days including the ways across huge swaths of ocean without existence of, may be, even a single recognizable landmark. This kind of astonishing capabilities found in bird's flying constraints led scientists, for years, to suspect birds having an "innate" ability to sense the Earth's magnetic field and adjust their paths accordingly, but still without success. Another group of scientists have hypothesized the mechanism to be rooted in bird's beak where iron-based minerals, acting as magnetic sensors, detect the bird's orientation feeding this information to its brain via a special nerve. Again, disputing this idea, another group of scientists proposed instead the magnetic sensors being placed, actually, in a bird's eyes where light receptors, sensitive to magnetic fields, feed data to the brain through optic nerves.

In 2000, Schulten refined this model by suggesting that the compass contain a photoreceptor protein called cryptochrome which, reacting with an as-yet-unidentified molecule, produces pairs of electrons existing in a state of quantum entanglement, i.e., spatially separated, but each still being able to affect the other. According to this model, when a photon hits the compass, "entangled electrons" are scattered to different parts of the molecule. Variations, present in Earth's magnetic field, cause them to spin in different ways, each leaving the compass in a slightly different chemical state, altering the flow of cellular signals through a bird's visual pathways, thus resulting ultimately in a perception of magnetism. Far-fetched as it sounds, subsequent research from multiple groups has found cellular evidence of such a system. Molecular experiments suggest that it's indeed sensitive to Earth's geomagnetic computational models thus suggesting a level of quantum entanglement, dreamed only by physicists, who hope to use entangled electrons to store information in quantum computers.

Peter Hore[25] proposed that such a chemical compass would work with the help of molecules with excitable lone electrons, known as radicals, and a quantum property known as spin. It is an well established fact that electrons in molecules usually come in pairs, spinning in opposite directions, effectively canceling out each other's spin. However, A "lone" electron spinning on its own, isn't cancelled out. This means it is free to interact with its environment — including magnetic fields. Or, more precisely," A "lone" electron spinning on its own is free to interact with its environment — including magnetic fields". Hiscock *et al.*[23] (2016), using computer simulations, established that genuinely quantum mechanical long-lived spin coherences in realistic models of cryptochrome, is able to explain the mechanism with necessary precision for which the crucial structural and dynamical molecular properties have been identified.

At present, the navigation of a bird is more or less already established and considered as a complex enterprise, requiring birds to make repeated and varying orientation decisions based on directional and positional information. Although commonly referred to as the "*Avian compass*", an ability to sense the local magnetic field orientation has been observed in every major group of vertebrates, as well as crustaceans, insects, including a species of mollusc.[27,29] For the majority of such species, the primary compass mechanism appears to be light-activated, with a few exceptions such as the sea turtle or the subterranean mole rat.[28] In addition to a light-activated compass, located in the eye, migratory birds are believed to possess a separate mechanism involving magnetite, with possible receptors identified in the beak,[27] the middle ear[36] and the brain stem,[37] even though the existence of a receptor in the beak has been challenged in a recent study.[38]

5.1.2 *Magnetoreception*

Also, since for a long time, it has been known from the extensive observations followed by both the observations and theoretical research showing that this is, in part, due to their ability to detect the direction of the Earth's magnetic field, i.e., some birds can detect the direction of field lines with an error of 5 degrees or less. Thus the phenomena of Magnetoreception turns out to be a unique animal ability to detect either the polarity or inclination of the earth's magnetic field as a navigation tool. As with electro-reception (the ability to detect electrical fields, which fact has been established already in sharks for hunting their prey), it represents a sense slightly alien to human experience. Or so it was thought, until recently.

When European robins were found utilizing a magnetic compass when migrating,[5] many experimental tests have been performed since then, related to the behavior of migrating birds. In particular, for some species, their magnetic sense has been found acting as an inclination compass, but insensitive to parity,[27] i.e., they can detect the relative angle of the local earth's magnetic field lines, but incapable of following the direction they follow, say, whether they are going north or south. There is some 'benefit' to

Figure 5.1. Picture of migratory bird European Robin.

this in that the compass would not be effected by geomagnetic reversal. This way, typically speaking, most models of magnetite-based magnetoreception act as a *parity compass.* However, scientists do not yet have a complete, clear & complete understanding of the biological mechanisms that make this magnetic sensing possible. Although commonly referred to as the "*avian compass,*" — an ability to sense the local magnetic field orientation has been observed in every major group of vertebrates, as well as crustaceans, insects, and a species of mollusc.[27,28]

For the majority of species, the primary compass mechanism appears to be light-activated, with a few exceptions such as the sea turtle or the subterranean mole rat.[28] In addition to a light-activated compass located in the eye, migratory birds are believed to possess a separate mechanism involving magnetite, with possible receptors identified in the beak,[27] the middle ear[36] and the brain stem,[32b] although the existence of a receptor in the beak has been challenged in a recent study.[38] However, many of such light-activated mechanisms being widespread, is well studied by a long series of behavioral experiments reviewed in.[28,29,31,32] However, among many, only two basic mechanisms are considered theoretically viable in terrestrial animals: iron-mineral-based magnetoreception and radical-pair based magnetoreception.

On the basis of current scientific evidence, *both the iron-mineral-based and radical-pair based magnetoreception mechanisms seem to exist in birds,* but seem to be used for different purposes. For each of these two types of magnetic sensing, plausible primary sensory molecules and a few brain areas involved in processing magnetic information have been identified in birds, even though, we are still far away from understanding the detailed

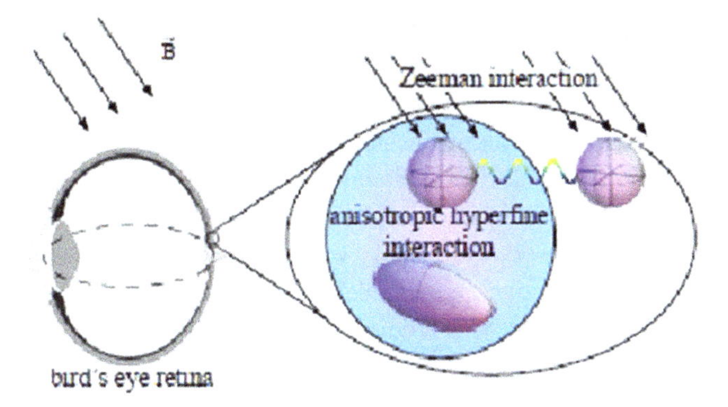

Figure 5.2. According to the RP model, the back of the bird's eye contains numerous molecules for magnetoreception. In the simplest variant, each such molecule involves three crucial components (see inset): there are two electrons, initially photo-excited to a singlet state, and a nuclear spin that couples to one of the electrons. This coupling is anisotropic, so that the molecule has a directionality to it.

function of any or, say, at least such two different magnetic senses existing in some, if not all bird species. At present, no primary sensory structure has been identified beyond reasonable doubt to claim as the source of avian magnetoreception. Walters[33] in his paper, represented avian compass as a unique example of a quantum mechanical process which not only survives but is actually sustained by interaction with a surrounding bath. However, even though the precise identity of the receptor/receptors involved in the avian compass still remains quite ambiguous, being undecided and unknown, simple geometrical assumptions allow sufficient information to compare numerically with experiment, derivable from first principles. Their[33] proposed mechanism requires neither unique properties nor elaborate manipulation of the radical pair state. Added to this, the biologically observable signal appears to be quite distinctive and easy to interpret, making the avian compass to be represented as a compatible model system in exploring the emerging and still largely unexplored role of quantum mechanics in biological processes.

Very recently, in many systems, especially, artificial systems, the quantum superposition and entanglement have already been observed to posses a very typical characteristics, i.e., decaying quite rapidly unless cryogenic temperatures are used. Now, the question may arise whether the life could have been evolved to that stage, i.e., it have been able to exploit such delicate phenomena? Quite surprisingly, as already stated, certain migratory birds have already been observed to posses the ability in sensing very subtle variations in the Earth's magnetic field. Ritz,[43] proposed first the model for quantum compass with cryptochrome molecule in robin's eye. Recently several scientists, also have raised the same interesting but somewhat intriguing possibility that living systems might be able to use some kind of non-trivial quantum effects in optimizing some tasks for the purpose of survival as well for their nourishment. The related studies range from the role of quantum physics in photosynthesis and in natural selection itself following the observation that 'warm and wet' living systems can embody entanglement, given a suitable cyclic driving.[48]

Gauger *et al.* together with Vedral[46] applied the quantum information theory and the widely accepted 'radical pair' model to analyze recent experimental observations of the avian compass. They examined the different phenomena related to superposition and entanglement in the process of magnetoreception, i.e., — the ability to sense characteristics of the surrounding magnetic field. They found that superposition and entanglement are sustained in this living system for at least tens of microseconds which exceeds the durations achieved in the best comparable man-made molecular systems. Interestingly, this they do, defying starkly at variance with the still existing view, i.e., life is too 'warm and wet' for such quantum phenomena to endure.

5.2 Radical Pair Mechanism (RPM)

The radical pair model gives an excellent phenomenological description of the avian compass, predicting disorientation, caused by an on-resonance oscillatory field. According to this model, the ability of migratory birds to orient relative to the Earth's magnetic field is believed to involve a coherent superposition of two spin states of a radical electron pair. However, the mechanism by which this coherence can be maintained in the face of strong interactions with the cellular environment has remained quite unclear and theoretically problematic, requiring the coherence to be maintained between different spin states for very long times, despite the presence of an environment, very hostile to this. As observed in[46] the slow spin flip time $(\pi/\Omega)Rabi = 3ms$) implies that the process it disrupts, must be slower still. Both the groups[46,47] use similar methods in inferring coherence times of $(10^{-6} - 10^{-4})$s. However, the proteins and water molecules, present in a cellular environment, possess large numbers of hydrogen nuclei, each of which interact with the radical pair via the hyperfine interaction, requiring necessary quantum information must survive such interactions long enough in order to give a biologically useful signal. Walter's detail works[33] considering the radical pair compass in the presence of decoherence, includes the treating effects of rapid singlet and triplet reaction rates on the evolution of the density matrix. Decoherence due to hyperfine interactions has been treated in terms of an effective magnetic field in,[67] while[48] and collaborators in their consecutive works, consider a radical pair interacting with a small number of nuclei. The related problem of decoherence in a singlet/triplet quantum dot has been treated in detail by Cai *et al.*, and Johnson *et al.* in some of their crucial works.[49]

Walter[33] in his paper, gives an exclusive detailed analytical treatment about the preservation and the decay of coherence for a radical pair, interacting with a bath of spin 1/2 nuclei. A long lived component of the quantum information has been identified, thus showing the yield as simple and robust compass mechanism. Design considerations for an efficient compass also have been identified showing the coherence lifetime to be consistent with lifetimes, inferred from behavioral experiments. Dynamics of corresponding radical pair density matrix have been derived showing to yield a simple mechanism for sensing magnetic field orientation. Rates of dephasing and decoherence have also been calculated ab initio and found to yield millisecond coherence times, consistent with behavioral experiments.

Another crucially important fact related to Radical pair mechanism (RPM) hypothesis of magnetoreception, first posited in the late 1970s, follows the discovery that electron transfer and related processes can generate a pair of radicals with properties (singlet and triplet spin states) that can be affected by exposure to a magnetic field (MF).[34] Importantly, Peter Hore & his group[25, 26] showed that birds are aided by light-dependent, multiple physiological compass systems, i.e., a physiological light-dependent magnetic compass together with chemical compass, relying on quantum phenomena to asses the perfect direction. This makes possible for migratory birds to be capable of knowing which way to fly. The mechanism behind this acts are thought to involve radical pairs (RP), formed photochemically in cryptochrome proteins in the retina, enabling them to rely on quantum phenomena to make correct directional choices during their flights. They applied coherent spin dynamics simulations to explore the behavior of realistic models of cryptochrome-based radical pairs, showing how the spin coherence persists for longer than a few microseconds, resulting to an output of the sensor containing a sharp feature, referred to as a *spike*. These spikes are explained as arising from avoided crossings of the quantum mechanical spin energy-levels of radicals, formed in cryptochromes. Importantly, *such a feature could deliver a heading precision, sufficient to explain the navigational behavior of migratory birds in the wild.* Anyway, the biophysical as well as the theoretical descriptions of such compass basis have thus far been unable to account for the high precision with which birds are observed, being able to detect the direction of the Earth's magnetic field, as the clear idea considered.

The basic parameters of the compass mechanism may be probed by confining a bird in a conical cage during its preferred migration period.[69] The restless nocturnal hopping behavior or *Zugunruhe*, will tend to orient in the preferred migration direction and the effects of environmental parameters can be judged by whether they affect the bird's ability to orient. Such experiments have established that the compass being light activated with an abrupt cutoff between wavelengths 560.5 nm and 567.5 nm[35], and importantly, birds are sensitive to the orientation of magnetic field lines but not their polarity — they *cannot distinguish* magnetic north from south.[27] Provocatively, a recent experiment has found that an oscillatory magnetic field, oriented transverse to the static field can cause disorientation when it is narrowly tuned to the Larmor frequency for an electron in the static field to flip its spin. On resonance, an oscillatory field strength of 15 nT, (Rabi frequency Ω-Rabi = 1250 Hz) has been found to be sufficient to cause disorientation. Interestingly, qualitatively such experiments are well explained by a "radical pair" model of the avian compass[44,45] in which the magnetic field drives coherent oscillations between the $|s, m_s\rangle = |0, 0\rangle$ singlet state and the $|s, m_s\rangle = |1, 0\rangle$ triplet state of an electron radical pair, formed by absorption of a photon. If the singlet and triplet states react to form distinguishable byproducts, or can be otherwise distinguished,[46] monitoring the ratio of the byproducts probing the time spent in each state, and thus the oscillation frequency is possible to know.

As far as the findings of behavioral studies are concerned, final obvious output indicates that a successful model should contain a light-dependent factor. Another important factor points to the fact that as the geomagnetic field is very weak (50 μT), the magnetic compass mechanism must be very sensitive, i.e., this mechanism, in addition, must transduce the response to the Earth's field into a biologically detectable signal. Finally, as mentioned earlier, in some species, this compass acts only as an inclination compass. In order to satisfy some of these criteria, the radical-pair mechanism (RPM) for magnetoreception has been proposed first by Schulten *et al.* in 1978. Within this radical-pair model, there exists the possibility of a photo-sensitive element,[47] i.e., absorption of light, triggering an electron transfer from a donor to an acceptor molecule, forming the initial radical pair (biologically this is proposed to occur within the cryptochrome proteins inside the eye). This process can be made to satisfy most of the frequency and intensity dependencies observed in experiments on Robins.

But next question arises demanding the sensitivity of this donor-accepter system towards the Earth's field? The answer lies in the fact that most known radicals contain atoms (hydrogen and nitrogen) with nuclear spins acting like an effective internal magnetic field, affecting the electron spin via the nuclear hyperfine interaction. One of these two major hypotheses postulates that birds use a light-induced radical pair reaction involving coherent spin evolution of two electrons as the foundation of their magnetic compass sensor. The existence of a magnetic compass was discovered in orientation experiment with birds in cages by Wolfgang Wiltschko,[27] in the late 1960's. In the 1970's, further studies indicated that weak magnetic fields can influence chemical processes involving photo-activated radical pair intermediates, i.e. a transient pair of molecules with an unpaired electron spin each. The underlying mechanism was shown to be based on the effects of magnetic fields on the electron spin evolution in each of the radical pairs. Discovery of such effects ultimately opened the now matured field of spin chemistry.

To sense the weak magnetic field of the Earth, the combination of anisotropic internal nuclear hyperfine fields and external geomagnetic fields causes a mixing of the electron singlet and triplet states, the strength of which is dependent on the angle of the external geomagnetic field. In order to deal with this problem, Ritz *et al.*[41] assumed a simplistic internal magnetic environment considering only one anisotropic nucleus (i.e., one radical is devoid of internal magnetic fields whereas the other has very strong internal magnetic fields) and showed that this maximizes the singlet-triplet mixing compared with other designs.[41,70] Whether this occurs in nature is unknown, but their extreme example helps to present a simple model of how this mechanism works. For the last stage, as mentioned, this radical pair (RP) can decay into different reaction products at different rates[47] depending on the spin state (singlet or triplet). If this radical-pair is located in the cryptochrome proteins within the eye, this could lead to a vision-based compass where the bird actually sees a directional signal.[47] Finally, this model requires that the many molecules forming the radical-pair-based magnetic compass must be ordered in some way so that the directional effects will not be averaged out which may require that the radical-pairs exist in a regular ordered biological structure, or lattice.

As such the radical pair model gives an excellent phenomenological description of the avian compass, and predicts disorientation by an on-resonance oscillatory field. But, some of theoretical problems still remains, requiring that coherence be maintained between different spin states for very long times, despite the presence of an environment, quite hostile to this. As observed in,[48] the slow spin flip time ($\pi/|\Omega_{Rabi}| = 3$ms) implies that the process it disrupts must be slower still. Gauger *et al.*[48, 49] applied similar methods to infer coherence times of $(10^{-6} - 10^{-4})$s. However, the proteins and water molecules present in a cellular environment, possess large numbers of hydrogen nuclei, each of which interact with the radical pair via the hyperfine interaction. Somehow, the necessary quantum information must survive such interactions long enough to give a biologically useful signal.

Most of us are now aware of the presence of two major postulates of the hypotheses that birds use a light-induced radical pair reaction involving coherent spin evolution of two electrons as the foundation of their magnetic compass sensor. But similar kind of the existence in magnetic compass was also discovered in orientation experiment with birds in cages by Wolfgang Wiltschko, in the late 1960's.[30] In the 1970's, their studies indicated that the presence of weak magnetic fields can influence chemical processes involving photo-activated radical pair intermediates, i.e. a transient pair of molecules with an unpaired electron spin each. Importantly, it was Klaus Schulten[34, 150] who first suggested that this radical pair mechanism might operate in the compass of migratory birds.

5.3 Application of Avian Compass in the Navigation of Major Group of Vertebrates, i.e., Birds

At present, the navigation of a bird is more or less already established and considered as a complex enterprise, requiring birds to make repeated and varying orientation decisions based on directional and positional information. Although commonly referred to as the "Avian compass", an ability to sense the local magnetic field orientation has been observed in every major group of vertebrates, as well as crustaceans, insects, including a species of mollusc.[27, 29] For the majority of such species, the primary compass mechanism appears to be light-activated, with a few exceptions such as the sea turtle or the subterranean mole rat.[28] In addition to a light-activated compass, located in the eye, migratory birds are believed to possess a separate mechanism involving magnetite, with possible receptors identified in the beak,[27] the middle ear[36] and the brain stem,[37] although the existence of a receptor in the beak has been challenged in a recent study.[38]

Even though commonly referred to as the "avian compass," an ability to sense the local magnetic field orientation has been observed in every major group of vertebrates, as well as crustaceans, insects, and a species of mollusc. In this context, the existence of a magnetic compass system that makes it possible to select and maintain the seasonally appropriate direction of migratory movements, as stated earlier, was first shown in birds.[27] However, at that period, the precise mechanism by which an organism may sense the orientation of the weak geomagnetic field remained unclear and theoretically problematic. Although commonly referred to as the "avian compass," an ability to sense the local magnetic field orientation has been observed in every major group of vertebrates, as well as crustaceans, insects, including several species of mollusc.[27,40] For the majority of species, the primary compass mechanism appears to be light-activated, with a few exceptions such as the sea turtle or the subterranean mole rat.[35]

Subsequently different authors reported the presence of the magnetic compass in many animals, including rodents,[49,50] Deutschlander *et al.*,[51] Phillips *et al.*;[52] Malewski *et al.*[53] and bats[54] among mammals; anuran amphibians;[55] by Diego-Rasilla *et al.*; Shakhparonov & Ogurtsov,[56] bony fishes,[57–59]sea turtles[60] and of course, in different types of birds.[62,63] The sensory basis of a magnetic map in fishes and turtles is currently not known, whereas for the birds there is evidence that positioning-related magnetic information from an unidentified receptor is carried to the brain by the ophthalmic branch of the trigeminal nerve.[64]

However, the physical bases of sensory systems behind the magnetic compass and magnetic map of birds have been observed to be different. Quite interestingly, it has been observed that unlike the magnetic map, *the magnetic compass of European robins (Erithacus rubecula) is not dependent on trigeminal-mediated information*, on the contrary, the birds with bilaterally sectioned trigeminal nerve *continue to select* the seasonally appropriate migratory directions on the basis of magnetic information.[65] It strongly suggests that *the work of the magnetic compass system of birds is based on a different receptor*. Currently, the most developed concept of the biophysical basis of the sensory system behind the magnetic compass sense of birds is the so-called model of radical pairs (RP), which assumes the existence of a chemical magnetoreceptor based on reversible biradical reactions.[66] A number of characteristics of the avian magnetic compass, known from behavioral experiments, point to this possibility, e.g. the fact that the compass is based on the inclination of the magnetic field rather than on its polarity, and light-dependence of its mechanism.[27,29] Important issues for constructing an adequate model of magnetoreception point to the ultimate necessity for the identification of the receptor molecule, and type(s) of magnetosensory cells participating in signal transduction. The radical pair (RP) model currently provides only hypothetical candidates for these roles that remain-still to be experimentally confirmed.

5.3.1 *Magnetoreception*

A new general approach for evaluating the "quantumness" of biological processes such as the ability of some birds to sense the Earth's magnetic field, has been developed by physicists, especially, in Switzerland and the USA. It involves describing the process as a "quantum meter" that uses quantum coherence to measure magnetic-field strength or light intensity. Atac Imamoglu of ETH Zürich and Birgitta Whaley of the University of California, Berkeley, have applied their framework to bird navigation and photosynthesis, and have concluded that only the 'former' is completely dependent on quantum coherence. In the long run, only two basic mechanisms are considered theoretically viable in terrestrial animals: They are,

- Iron-mineral-based magnetoreception and
- Radical-pair based magnetoreception.

On the basis of current scientific evidence, iron-mineral-based magnetoreception and radical-pair-based magnetoreception mechanisms seem to exist in birds, but they seem to be used for different purposes. Plausible primary sensory molecules and a few brain areas involved in processing magnetic information have been identified in birds for each of these two types of magnetic senses. Nevertheless, we are still far away, if not substantially enough from understanding the detailed function, i.e., about any kind or, at least two different magnetic senses existing in some, if not all, bird species. But at present, no primary sensory structure has not yet been identified beyond reasonable doubt to claim the source of avian magnetoreception.

Even though many of the theoretical studies explored the aspects of the radical pair (RP) hypothesis, the majority have concentrated especially on the magnitude of the anisotropic magnetic field effect, devoting little attention to a very crucial and determining factor in this matter: the precision of the compass bearing available from a radical pair sensor[26] which has been dealt in detail by Hore and his group. As pointed and suggested by Hore and his group,[66] the photochemistry of isolated cryptochromes in vitro has been found to support the approach, i.e., these isolated cryptochromes respond to applied magnetic fields in a manner that is quantitatively consistent with the radical pair mechanism.

Peter Hore proposed the presence of chemical compass and anisotropic magnetic interactions within the radicals which plays a crucial role by giving rise to intracellular levels of a cryptochrome signaling state which depends on the orientation of the bird's head in the Earth's magnetic field.[26] In their novel approach, Hore and his group[49] used coherent spin dynamics simulations to explore the behavior of realistic models of cryptochrome-based radical pairs. They showed that when the spin coherence persists for longer than a few microseconds, the output of the sensor contains a sharp feature, referred to as a *spike*. The spike arises from avoided crossings of the quantum mechanical spin energy-levels of radicals formed in cryptochromes. Such a feature could deliver a heading precision sufficient to explain the navigational behavior of migratory birds in the wild. Their results arrived at the following conclusive results, i.e.,

- afford new insights into radical pair magnetoreception,
- suggest ways in which the performance of the compass could have been optimized by evolution,
- may provide the beginnings of an explanation for the magnetic disorientation of migratory birds exposed to anthropogenic electromagnetic noise.
- Importantly, radical pair **magnetoreception may be more of a quantum biological phenomenon** than previously realized.

The locations of this primary sensory receptors are found in the eyes[65,67,69] of the bird, and in fact, directional information is processed bilaterally in a small part of its fore brain which is accessed via the thalamofugal visual pathway. Currently, the evidence collected, points to a chemical sensing mechanism based on photo-induced radical pairs in cryptochrome flavoproteins, present in the retina[150] of the bird's eye. Though scientists have long speculated that certain animals are making use of magnetic fields to find their way, biologists are mystified as to how they might do it. Now some answers might be coming from one of the most perplexing interactions in physics. Quantum entanglement dictates that if two electrons are created at the same time, the pair will be "entangled" so that whatever happens to one particle affects the other. Otherwise, it would violate fundamental laws of physics. The two particles remain entangled even when separated by vast distances. So if one particle is spin-up, the other must be spin-down, but what's mind-boggling is that neither will have a spin until they're measured. It means that not only will you not know what the spin of the electron is until you measure it, but that the actual act of measuring the spin will make it spin-up or spin-down. The suggestion that bird navigation is aided by the exploitation of coherent quantum effects, appeared at first sight rather far fetched and for many years, evidence supporting this idea was sparse. Its main impact was to trigger research into "light effects on the magnetic compass of birds". Long-lived spin coherence in proteins found in the eyes of migratory birds could explain how the creatures are able to navigate along the Earth's magnetic field with extraordinary precision.

An excellent phenomenological description of the avian compass can be obtained from the view point of radical pair (RP) model through which it predicts disorientation by an on-resonance oscillatory field. The basic parameters of the compass mechanism may be probed by confining a bird in a conical cage during its preferred migration period.[69] The restless nocturnal hopping behavior, or Zugunruhe, will tend to orient in the preferred migration direction, and the effects of environmental parameters can be judged by observing whether these factors affect the bird's ability to orient. Such experiments have established that the compass is light activated, with an abrupt cutoff between wavelengths 560.5 and 567.5 nm,[40] and importantly, the birds are sensitive to the orientation of magnetic field lines but not their polarity which points to their incapability in distinguishing magnetic north from south.[27] Not only that, a recent experiment has found an oscillatory magnetic field oriented transverse to the static field, causing disorientation when it is narrowly tuned to the Larmor frequency for an electron in the static field to flip its spin. On resonance, an oscillatory field strength of 15 nT. (Rabi frequency $\Omega_{Rabi} = 1250\text{Hz}$) is sufficient to cause disorientation.[41,70] Qualitatively, such experiments are well explained by a "radical pair" model of the avian compass,[43,70,150] in which the magnetic field drives coherent oscillations between the $|s, ms\rangle = |0, 0\rangle$ singlet state $|s\rangle$ and the $|s, ms\rangle = |1, 0\rangle$ triplet state $|t\rangle$ of an electron radical pair, formed by absorption of a photon. If the singlet and triplet states react to form distinguishable byproducts, or can be otherwise distinguished,[44] then it becomes possible in monitoring the ratio of the byproducts probes, the time spent in each state, and also the oscillation frequency.

But, it remains theoretically problematic as the coherence is to be maintained between different spin states for very long times, despite the presence of an environment, very hostile to this. As observed in,[46] the slow spin flip time[30] ($\pi/\Omega_{Rabi} = 3ms$) implies that the process it disrupts, must be slower still. Thus Gauger[46] and Bandopadhaya *et al.*[47] use similar methods to infer coherence times of $(10^{-6} - 10^{-4})$s. However, the proteins and water molecules present in a cellular environment possess large numbers of hydrogen nuclei, each of which interact with the radical pair via the hyperfine interaction. But the problem lies on the fact that somehow, the necessary quantum information must survive such interactions long enough to give a biologically useful signal. Many of previous works tried to explore this problem[60–65] from the related angle, considering the radical pair compass in the presence of decoherence. Not only that, it also includes the effects of rapid singlet and triplet reaction rates and have been treated taking into consideration the evolution of the density matrix. Decoherence due to hyperfine interactions has also been included and treated as well, in terms of an effective magnetic field in,[67] while they[7,8] considered a radical pair interacting with a small number of nuclei. However, the related problem of decoherence in a singlet/triplet quantum dot has been treated extensively in.[49]

Walters,[33] in his work, analytically explained 'the preservation and decay of coherence' for a radical pair interacting, considering a bath of spin 1/2 nuclei. Also, they identified a long lived component of the quantum information, yielding a simple and robust 'compass mechanism'. Design considerations for an efficient compass have also been considered and identified by this group. Corresponding coherence lifetime has been shown to be consistent with that expected from the inferred behavioral experiments. also, the avian compass, described in Walters[33] work, represents a unique example, i.e., it depicts an application of a quantum mechanical process, not only surviving but also actually being sustained by interaction with a surrounding bath. Although the precise identity of the receptor or receptors involved in the avian compass remains still to be known, a simple geometrical assumptions allow information collected to be sufficient for numerical comparison with that of experiment, derivable from first principles. But importantly, proposed mechanism by above group requires neither unique properties nor elaborate manipulation of the radical pair state. Not only that, the biologically observable signal has been found to be quite distinctive and easy enough in interpreting the functions of the avian compass, thus representing a simple model system for the emerging and still largely unexplored role of quantum mechanics in biological processes.

Walters,[33] applied quantum information theory and the widely accepted 'radical pair' model to analyze recent experimental observations of the avian compass. They found that superposition and entanglement are sustained in this living system for at least tens of microseconds, exceeding the durations achieved in the best comparable man-made molecular systems. They concluded this as starkly at variance with the view that life is too 'warm and wet' for such quantum phenomena to endure. It is an well known fact that in artificial systems, quantum superposition and entanglement decay quite rapidly in a typical way unless cryogenic temperatures are used. For a period, it had been really a continually puzzling question whether life could have evolved to exploit such kind of delicate phenomena. But, it already has been established as a true fact that *certain migratory birds have been observed to posses this ability to sense very subtle variations in the Earth's magnetic field.*

Applying quantum information theory and the widely accepted "radical pair" model to analyze recent experimental observations of the avian compass, Gauger *et al.*[46] and his group examined quantum phenomena in the process of magnetoreception — the ability to sense characteristics of the surrounding magnetic field. They observed several mechanisms viable which can be found by which this sense may operate.[31] According to their observations, superposition and entanglement are sustained in this living system for 'at least tens of microseconds', exceeding the durations achieved in the best comparable man-made molecular systems. Interestingly, their conclusion is also starkly at variance with the view that life is too "warm and wet" for such quantum phenomena to endure.

Mauritsen[75] proposed the role of earth's magnetic field and its effect on the process of bird's flying, i.e., the Earth's magnetic field provides potentially useful information, which birds could use for directional and/or positional information. It has clearly been demonstrated that birds are not only able to sense the compass direction of the Earth's magnetic field but also capable of using this information as part of a compass sense. Magnetic information could also be useful as part of a map sense as there is a growing body of evidence that birds are able to determine their approximate position on the Earth on the basis of geo-magnetic cues. In addition to direct uses for orientation and navigation, magnetic information also seems to be able to influence other physiological processes, such as fattening and migratory motivation, as a trigger for changes in bird's behavior. Although the behavioral responses to geomagnetic cues are relatively well understood, the physiological mechanisms enabling birds to sense the Earth's magnetic field have only started to be understood and understanding the magnetic sense(s) of animals, including birds, still remains one of the most significant but very little unsolved problems in biology. It appears quite challenging to sense magnetic fields as weak as that of the Earth using only biologically available materials.

5.4 Theoretical Approaches for the Source and Mechanical Aspects, Present in the Navigation of Bird

5.4.1 *Application of Quantum Coherence & Entanglement in the Navigation of Bird*

Since for a long time, even though scientists have speculated that certain animals make use of magnetic fields to find their way, biologists are mystified as to how they might do that. Now some answers might be coming from one of the most perplexing interactions in physics.

By now, we have been already aware of the fact and findings about Quantum entanglement which dictates that if two electrons are created at the same time, the pair will be "entangled" so that whatever happens to one particle affects the other. Otherwise, it would violate fundamental laws of physics. The two particles remain entangled even when separated by vast distances. So if one particle is spin-up, the other must be spin-down, but *what's mind-boggling is that* neither will have a spin until they're measured, meaning that it is not possible to know either the spin of the electron is, until you measure it or about what is the actual act of measuring the spin for making it spin-up or spin-down. As difficult as entanglement is to believe, as well as to understand, it is an well established property of quantum mechanics.

As has already been discussed, though briefly, some physicists have suggested that birds and other animals might be using this effect to see and navigate Earth's magnetic fields through the process which could work via light-triggered interactions on a chemical, present in bird's eyes. According to this suggestions, light would excite two electrons on a molecule in the bird's eye, switching one onto a second molecule, but the two would remain entangled even though they're separated. The Earth's magnetic field would alter the alignment of the electron's spins and in the process, also alter the chemical properties of the molecules. Physicists suspect that the reactions would leave varying concentrations of chemicals throughout the eye, possibly creating a picture of our planet's magnetic field that would allow birds *to orient* themselves. Even though the theory is still in its infancy, biophysicists already have their eyes on a few chemicals that might enable the birds to detect entanglement. One such chemical is called cryptochrome and its potential effects already being studied.

A group of physicists from the University of California at Irvine[70] also studied the European Robin's ability to sense small quantum changes by tampering with the magnetic field surrounding the birds. A robin was placed in a cage during migration season and then the physicists switched the polarity of the magnetic field around it. The test indicated that changes on the level of 'one-thousandth' the strength of Earth's magnetic field would impact the birds' abilities to orient themselves. Perhaps even more fascinating is that European Robins might do a better job of detecting quantum entanglement than physics labs currently can. A group of physicists from the University of Oxford have proposed that entanglement could last in a bird's retina for 100 microseconds, whereas physicists have only been able to make the interaction last for 80 microseconds — despite cooling their experiments to just above absolute zero. The studies have implications beyond birds as well. By extending this idea, a number of fish, reptiles, insects and even mammals are thought to use magnetic fields to navigate.

Bradlaugh *et al.*[86] discussed in his very recent work that the flavoprotein CRYPTOCHROME (CRY), now generally believed to be a magnetosensor, provide geomagnetic information via a quantum effect on a light-initiated radical pair reaction. Whilst there is considerable physical and behavioral data to support this view, the precise molecular basis of animal magnetosensitivity remains still frustratingly unknown. A key reason for this is the difficulty in combining 'molecular and behavioral biological' experiments with the sciences of magnetics and spin chemistry.

Thus, the phenomena of quantum entanglement has already been an well established fact by which it dictates that if two electrons are created at the same time, the pair will be "entangled" so that whatever happens to one particle affects the other. Otherwise, it would violate fundamental laws of physics, i.e., the two particles remain entangled even when separated by vast distances. So, if one particle is spin-up, the other must be spin-down, but what's mind-boggling is that neither will have a spin until they are measured, meaning that it is not possible to know either what the spin of the electron is, until it is measured about what is the actual act of measuring the spin for making it spin-up or spin -down. Though it appears quite tough and difficult as entanglement is to believe, as well as to understand, it is an well established property of quantum mechanics and this is the basic concept behind, which has stimulated the idea of scientists, especially physicists suggesting that birds and other animals might be using this very effect to see as well as navigating the Earth's magnetic fields.

The process thus far stated, could work via light-triggered interactions on a chemical, present in bird's eyes. Light would excite two electrons on a molecule in the bird's eye, switching one onto a second molecule, but the two would remain entangled even though they're separated. The Earth's magnetic field would alter the alignment of the electron's spins and in the process alter the chemical properties of the molecules. Physicists suspect that the reactions would leave varying concentrations of chemicals throughout the eye, possibly creating a picture of our planet's magnetic field that would allow birds to orient themselves. The theory might be still then could be called as in its infancy, but biophysicists already have their eyes on a few chemicals that might be responsible for enabling the birds to detect entanglement. One of such chemical is called cryptochrome and its potential effects have already been established by the famous works of Hore[58], followed by studies done by others[83] already being studied.

Very recently, in many systems, especially, artificial systems, it has already been observed that quantum superposition and entanglement posses a very typical characteristics, i.e., these two processes decay quite rapidly unless cryogenic temperatures are used. Now, the question may arise whether the life could have been evolved to that stage so that it could have been able to exploit such delicate phenomena? Quite surprisingly, certain migratory birds have been observed to posses the ability in sensing very subtle variations in the Earth's magnetic field. Ritz[43] in 2000, proposed the model with cryptochrome molecule for quantum compassin robin's eye. Recently, several scientists, working in this line of thought, also have raised the same interesting but somewhat intriguing possibility that living systems may use some kind of non-trivial quantum effects in optimizing some tasks for the purpose of survival as well for their nourishment.

Studies range from the role of quantum physics in photosynthesis and in natural selection itself through the observation that 'warm and wet' living systems can embody entanglement, given a suitable cyclic driving.[48] Gauger *et al.* together with Vedral[46] applied the quantum information theory and the widely accepted 'radical pair' model to analyze recent experimental observations of the avian compass. They examined the different phenomena related to superposition and entanglement in the process of magnetoreception, i.e., — the ability to sense characteristics of the surrounding magnetic field. Also they found that superposition and entanglement are sustained in this living system for *at least tens of microseconds* which exceeds the durations achieved in the best comparable man-made molecular systems. Interestingly, this they do, defying starkly at variance with the still existing view, i.e., life is too 'warm and wet' for such quantum phenomena to endure.

As already stated earlier, scientists have long speculated that certain animals are making use of magnetic fields to find their way but biologists are mystified as to how they might do it. In the case of artificial systems, quantum superposition and entanglement typically decay rapidly unless cryogenic temperatures are used. Now the question arise could life have evolved to exploit such delicate phenomena? Certainly, migratory birds have been found to posses such ability, i.e., to sense very subtle variations in the Earth's magnetic field. Intense research have brought now some answers which might be coming from one of the most perplexing interactions in physics.

As stated earlier, recently several researchers have raised the intriguing possibility that living systems may also use non-trivial quantum effects for the optimization of some tasks, just like in some artificial systems where quantum superposition and entanglement typically decay, i.e., rapidly unless cryogenic temperatures are used. The inevitable question of the probability arises out of this i.e., whether life could have evolved by exploiting such delicate phenomena in other sphere of life on earth. By now, it has already been noted that certain migratory birds posses the ability in sensing quite a very subtle variations in the Earth's magnetic field. At the present state of development, Gauger *et al.*[46] applied quantum information theory and the widely accepted 'radical pair' model in order to analyze recent observations of the avian compass, stating from the the role of quantum physics in photosynthesis as well as in natural selection itself and have already established through the observation that 'warm and wet' living systems can embody the principle of quantum coherence as well as entanglement, at least, tens of microseconds, given a suitable cyclic driving,[6] thus exceeding the durations, achieved in the best comparable man-made molecular system for at least tens of microseconds.

5.4.2 *Probable Role of Sustained Quantum Coherence & Entanglement Phenomena in Bird Navigation*

In many of the species, for example, certain birds,[14,43,70] fruit flies[15,41] and even plants,[78] the evidences support a process, so-called-Radical Pair (RP) mechanism involving the quantum evolution of a spatially-separated pair of electron spins.[43,70] Several mechanisms already been proposed and developed by which this sense may operate,[10] the model being supported by several results, available from the field of spin chemistry.[19–22] The fact that, following this principle, an artificial chemical compass operates — has been demonstrated experimentally,[24] followed by a very recent theoretical study which examined the presence of entanglement within such a system.[8]

Gauger *et al.*[46] considered the time scales for the persistence of full quantum coherence, and entanglement within a specific living system, i.e. of the European Robin. From the analysis of their data, they concluded that the RP model implies a decoherence time in the birds' compass which is extraordinarily long, i.e., beyond that of any artificial molecular system. Gauger *et al.* also addressed the question: Is there any noise model, which would cause rapid decoherence, consistent with experimental observations? Initially it seems that, might be, a particular type of such a form of noise might be viable to exist as the compass mechanism has been found to be almost immune to pure phase noise because, even starting from a fully dephased state, the compass operates well. Thus it might seem that strong phase noise could be present, rapidly degrading quantum coherence but permitting the compass to function. Importantly, they showed that for natural presence of at the level of or higher than that of decoherence rate in z-direction, i.e., $|\tau_z| \approx 10k$ or higher, it would render the bird to be immuned to the weak RF (Radio Frequency) magnetic fields.[41] After following the rigorous calculations related to systematic study of correlated noise processes, they finally concluded that the RF field immunity implied, unless, coherence is preserved on a time scale having the order of $100\mu s$.

As a next step, in order to characterize the duration of quantum entanglement which could be obtained from such living system, approximate values have been inferred for the key parameters, monitoring entanglement from the initial singlet generation up to the eventual decay. Dealing this problem in detail, Gauger *et al.* inferred approximate values for the key parameters so that the entanglement can be monitored from the initial singlet generation to the eventual decay and clarified how this negativity evolving under the generic noise model, proposed by them. Importantly, the initial singlet state has been found to be maximally entangled which importantly, under noise falls off at a faster rate than the decay of population from the excited state. Finally, Gauger et. al. summarized their conclusion out of their observations related to the sensitivity to RF fields. Quite importantly, they emphasized that both amplitude and phase (and thus entanglement) are indeed protected within the avian compass. According to their findings, the timescales are at least '*tens of microseconds*', even for a pure dephasing environment and *hundreds of microseconds* for the more general models. Though it is surprising and not yet clear why such remarkable protection occurs, but given the widely-accepted RP model together with the recent experimental data,[41] this conclusion follows.

Again, the inferences about the mechanism of the magnetic sensor can be made by manipulating a captive bird's magnetic environment and recording its response. Specifically, European Robins have been observed to be sensitive only to the 'inclination' but to the polarization of the magnetic field,[27] and this sensor is evidently activated by photons entering the bird's eye.[27,65] Importantly, a very small oscillating magnetic field can disrupt the bird's ability to orientate.[41,70] Not only that, birds have been found capable of being 'trained' significantly to different field strengths suggesting that their navigation sense is substantially robust.

Thus, the basics of the RP (Radical Pair) model could be stated as follows: there are molecular structures in the bird's eye each of which can absorb an optical photon giving rise to a spatially separated electron pair in a singlet spin state. A singlet-triplet evolution occurs due to the presence of differing local environments around the two electron spins, the evolution of which depends on the inclination of the molecule with respect to the Earth's magnetic field. Recombination occurs either from the singlet or triplet state, leading to different chemical end products, thus producing the concentration of these products, constituting a chemical signal, correlated to the Earth's field orientation. But importantly, however, the specific molecule involved has not been known yet at the time of Gauger's[46] work.

Figure 5.4 depicts the most basic form of the model: two electronic spins[43] and one nuclear spin. The nucleus interacts with only one of the electron spins, thus providing the asymmetry required for singlet-triplet oscillations. Similar to other models, they also employed the Hamiltonian corresponding to the system once the two electrons have become separated, corresponding to the moment of RP (Radical Pair) formation and measured the timescales for the persistence of full quantum coherence and entanglement. Long history of experimental evidence for magnetic field modulation of chemical reactions[34, 154] have already been recorded. Thus, the concept that the radical pair mechanism, by which a magnetic field can change the dynamics of a chemical reaction, arose out of investigations, made into the relevance of nuclear and electronic spin phenomena to chemical reactions, as long ago as late 1960s.[80] Since its introduction of such chemical reactions as a means to explain, the theory has been well documented.[80, 81]

Recent interests in magnetic field effects and the radical pair mechanism include a variety of subjects. For example, the emerging field of quantum biology, in particular, has been instrumental in raising interest in applying the RP mechanism to biological systems, for example, avian magnetoreception which started not long after the initial formalization of this mechanism. It has also been suggested that radical pairs (RP) in photosynthetic reaction centres play a number of diverse and important roles, starting from polarization to protection mechanisms. More recently, the observation of radical pairs in the flavoprotein cryptochrome has led to speculation that the mechanism controlling the circadian clock in some organisms, likely, could also be the radical pair mechanism.[82–84] But, substantial debate has also been there relating the role of magnetic field, for example, effects in enzyme reactions. Specifically, Flavin-dependent enzymes which are implicated in these processes, — as diverse as energy production, apoptosis, DNA repair and neural development,– have been less successfully investigated with respect to magnetic field effects, while researchers are trying to replicate the report that a magnetic field *can alter* also the enzymatic ATP synthesis.[85, 86, 94] The research is especially important in the context of our growing environmental exposure to electromagnetic fields and the lack of consensus on whether such weak fields might have an subtle but substantial effect on biological organisms in the long run.

It is important to mention that such continuing developments related to the radical pair mechanism has allowed for theoretical explanation of the fact that magnetic fields are observed to have an effect on chemical reactions. The mechanism describes how an external magnetic field can alter chemical yields by interacting with the spin state of a pair of radicals. Thus, in short, the radical pair can be described in the following rudimentary outline of the three important steps that constitute the mechanism:

- In the first step- a photon is incident on the donor molecule of the pair, causing an electron to be excited and donated to the acceptor molecule.

This results in a spin-correlated but spatially separated electron pair which is conventionally thought of as being in a singlet state.
- During the second step the radical pair oscillates between singlet and triplet state under the influence of the Zeeman or hyperfine interaction.
- The third step is then the recombination of the pair to form some sort of chemical product/signal. Importantly, the chemical product formed in this last step is dependent on whether the pair is in a singlet or triplet state and is thus dependent on the magnetic field, thus allowing the radical pair to function as a compass.

5.4.3 *Radical Pair (RP) Model of Gauger et al.*[46]*: Persistence of Full Quantum Coherence and Entanglement*

The process introduced by Gauger *et al.*,[46] deals with the idea of the quantum evolution of a spatially-separated pair of electron spins[44,70] dealing with problem of persistence of full quantum coherence and entanglement as well. After detailed analysis of the recent data obtained from experiments on live birds, they[46] finally arrived at a quite crucially important conclusion, i.e., *the implication of Radical Pair (RP) model points to a decoherence time in the birds' compass 'extraordinarily long'*, emphasizing the point, i.e., it lies quite beyond that of any artificial molecular system. However, it is possible to manipulate a captive bird's magnetic environment and its response as well so that one can make inferences about the mechanism of the magnetic sensor[29,30,41,82] within it. Specifically, out of another quite crucial observation, it has been found that the European Robins are sensitive only to the inclination, not the polarization of the magnetic field.[27] Then, obviously, this sensor is activated only by photons entering the bird's eye.[29] Importantly, a very small oscillating magnetic field have also been observed, enabled in disrupting the bird's ability to orientate.[36,37] Another significant characteristics of these birds have also been found, i.e., *they posses the ability to 'train' themselves to different field strengths*, suggesting their navigation sense to be quite robust, and *unlikely to depend* on very special values for the parameters in the model, supported by several results from the field of spin chemistry.[19–22]

That, an artificial chemical compass operates, following this principle, has been demonstrated experimentally[24] and very recent theoretical studies examined the presence of entanglement (Fig. 5.3 below), if any, within such a system.[48] However, the above group considered the timescales

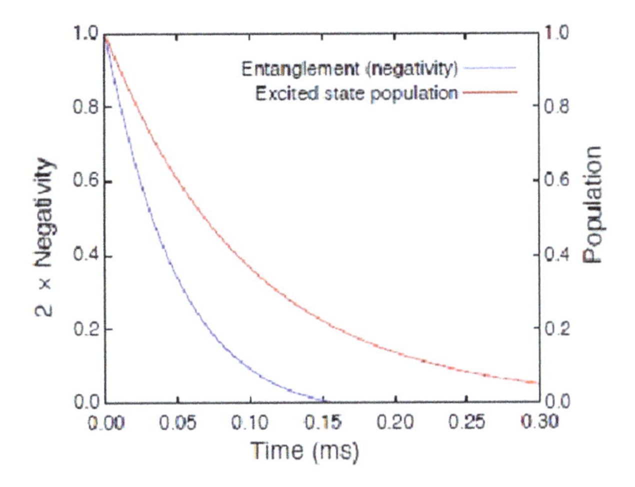

Figure 5.3. The decline and disappearance of entanglement.

for the persistence of full quantum coherence and entanglement within a *specific* living system: the *live European Robin.* Their conclusion has been as follows: the RP model implies a decoherence time in the bird's compass, extraordinarily long — beyond that of any artificial molecular system. However, it is quite possible in manipulating a captive bird's magnetic environment and recording its response, thus making inferences about the mechanism of the magnetic sensor.[27,41,70] Importantly, for the present analysis, a very small oscillating magnetic field has been found to be sufficient in disrupting the bird's stability to orientate.[41,70]

Haberkorn[87] applied a density matrix treatment of radical pair reactions and most subsequent detailed approaches in the field have since followed his lead, even though investigating the advantages and limitations of this approach.[87–89] In the context of avian migration, one of the limitations to the conventional approach is the increased computational complexity arising, as the radical pair interacts with increasing numbers of nuclei. It is estimated that the number of significant nuclei felt by each radical is approximately ten to fifteen.[24,48] Currently, the most likely mechanism of the magnetic compass sensing in migratory songbirds relies on the coherent spin dynamics of pairs of photochemically formed radicals in the retina. There are molecular structures in the bird's eye where each of which can absorb an optical photon, thus giving rise to a spatially separated electron pair in a singlet spin state.

Spin-conserving electron transfer reactions are thought to result in radical pairs whose near-degenerate electronic singlet and triplet states inter-convert coherently, as a result of hyperfine exchange, and dipolar couplings. Not only that, crucially, for a compass sensor, Zeeman interactions with the geomagnetic field is there, as a result of which, a singlet-triplet evolution occurs due to the presence of the differing local environments of the two electron spins. But, importantly, this evolution depends on the *inclination of the molecule with respect to the Earth's magnetic field*. In such case, recombination occurs either from the singlet or triplet state, leading to different chemical end products, the concentration of which constitutes a chemical signal correlated to the Earth's field orientation. However, according to their findings, the specific molecule involved remained unknown still then.

A few essential assumption had been undertaken, regarding the specific structure of the molecule, especially, in order to describe the basic idea of the RP model which can be described as follows: as we are well acquainted by now with the fact that there are molecular structures in the bird's eye, each of which can absorb an optical photon, giving rise to a spatially separated electron pair in a singlet spin state. Now, due to the presence of the differing local environment of the two electron spins, Gauger *et al.* proposed evolution of a singlet-triplet occurring whereas evolution is dependent on the inclination of the molecule with respect to the Earth's magnetic field. Recombination follows only either from the singlet or triplet state, leading to different chemical end products whose concentration products constitute a chemical signal, correlated to the Earth's field orientation. But, it should be noted that the specificity regarding the molecule as well as that of directionality condition applied, had been still unknown to the group. However, it has been possible in making the model only by following that of 'O. Efimova and Hore, P.J.[87]

As stated by the RP model,[46] the back of the bird's eye contains numerous molecules for magnetoreception[93] which give rise to a pattern, discernible to the bird, indicating the orientation of the field, implying the fact that the molecules involved are at least fixed in orientation, and possibly ordered with respect to one another.[43] In the simplest variant, each such molecule involves three crucial components: two electrons, initially photo-excited to a singlet state, and a nuclear spin that couples to one of the electrons. This coupling is anisotropic, so that the molecule has a directionality to it. Now, as stated in Fig. 5.1 (seems O.K.), starting with two electronic spins and one nuclear spin,[10] the nucleus is considered to interact with only one of the electron spins at a time, this way providing the necessary symmetry needed for singlet-triplet oscillations. In this model, Gauger and his group, have considered a particular case, i.e., in the system, once the two electrons become separated, at the time $t = 0$, the Hamiltonian considered to that condition, corresponds to the 'moment of RP (Radical Pair) formation'.

At this point, let us describe the anisotropic hyperfine tensor, coupling the nucleus and electron to be 1 so that it can conveniently be written as $A = diag(A_x, A_y, A_z)$, assuming an axially symmetric molecule with $A_z = 10^{-5}$ meV and $A_x = A_y = A_z/2$, i.e., this model has been considered by taking the general shape and magnitude of the tensor to be consistent with reference.[94]

The corresponding Hamiltonian can be defined as

$$H = \widehat{I} \cdot \mathbf{A} \cdot \widehat{S}_1 + \gamma \mathbf{B} \cdot \widehat{S}_1 + \widehat{S}_2$$

In the above equation, $\widehat{I}$ is the nuclear spin operator, $\widehat{S} = (\sigma_x, \sigma_y, \sigma_z)$ are the electron spin operators, $\mathbf{B}$ is meant for magnitude of magnetic field vector and $\gamma = 1/2\mu_0 g$ for the gyromagnetic ratio, taking account of the fact that a factor $1/2$ in the gyromagnetic ratio accounts for taking a spin one-half system. For this, we need to use Pauli matrices taking $\sigma_z = diag(1, -1)$ and so on. Even though only one electron is taken in the present model as coupled to one nucleus, but essentially, the remote electron is considered as free in practical sense as it possess quite weakly interacting nature.

Gauger and his group also considered a family of variants, for example, involving different hyperfine tensors, adding second nuclear spin, also replacing the nuclear asymmetry with an anisotropic electron g-factor. But, in essence, all of such variation of models produced the same qualitative behavior as that of basic model. Anyway, it is not at all surprising, because, all of such variations are based on the over all same basic principle, for example, one of the basic assumption stating that the electron spins of the RP must be protected from an irreversible loss of quantum coherence, because of their susceptibility towards the experimentally applied RF field. The extremely low strength of such applied field *dictates* the timescale over which quantum coherence must be preserved. As a result, the inference of extraordinarily long coherence times does not vary significantly over the various models.

5.4.4 *Quantum Master Equation Approach*

In order to model the dynamics of the system with a quantum master equation (ME) approach, we need to add two 'shelving states' to the 8 dimensional Hilbert space of the three spins. Employing operators to represent the spin-selective relaxation into the singlet shelf $|S\rangle$ from the electron singlet state, or the triplet shelf $|T\rangle$ from the triplet configurations, we arrive at the point where one of the two events will mean to occur, and the final populations of $|S\rangle$ and $|T\rangle$ give the singlet and triplet yield. With the usual definition of singlet $|s\rangle$ and triplet states $|t_i\rangle$ in the electronic subspace, while $|\uparrow\rangle$ and $|\downarrow\rangle$ describing the states of the nuclear spin, we are now able to define the following decay operators:

$$P_{S,\uparrow} = |S\rangle\langle s, \uparrow|; P_{T_{0,\uparrow}} = |T\rangle\langle t_0, \uparrow|,$$
$$P_{T_{+,\uparrow}} = |T\rangle\langle t_+, \uparrow|; P_{T_{-,\uparrow}} = |T\rangle\langle t, \uparrow|$$

The same way will follow for the 'down' nuclear states, giving two singlet and six triplet projectors in total, with the aim of discriminating the respective decays, related to the above standard Lindblad Master equation which can be written as

$$\dot{\rho} = -\frac{i}{\hbar}[H, \rho] + k\sum_{i=1}^{8}\left[P_i\rho P_i^\dagger - \frac{1}{2}P_i^\dagger P_i\rho + \rho P_i^\dagger P_i\right] \tag{5.1}$$

For simplicity, the choice adopted here, corresponds to the usual expression for the singlet yield and specifically, all eight projectors have been assigned for the same decay rate k. But, it is quite important to note that, for the above treatment has been dealt without the contribution of the effect caused by the presence of 'environmental noise', if any. Importantly, as Walter *et al.*[33] claimed, even if included, the estimation of k will not be effected.

5.4.5 *Noise: Proper value for the parameter k*

Though this approach goes at per with other contemporary groups who considered the same approach, Gauger *et al.* group claimed to introduce various kinds of noise operators. Here, the initial state of their model ρ_0 assigns a pure singlet state to the electrons, and a completely mixed state to the nucleus, so that

$$\rho(0) = |s, \downarrow\rangle\langle s, \downarrow| + |s, \uparrow\rangle\langle s, \uparrow|$$

from which an appropriate choice for the parameter i.e., the decay constant k can be achieved.

5.4.5.1 *To find the Proper Value of Parameter k'*

Ritz *et al.*, in their pioneering work[41] noted that, a perturbing magnetic field of frequency of 1.316 MHz (i.e. the resonance frequency of the 'remote' electron) can disrupt the avian compass, thus implying a bound on the decay rate constant k (since the field would appear static for sufficiently rapid decay) immediately. Following this result, Gauger *et al.* refined this bound on k by considering the oscillating magnetic field strength, sufficient enough to completely disorient the bird's compass, i.e. $150nT$. (practically, even a $15nT$ field was reported as being disruptive).

Again, to find out the exact amount of the effect, i.e., the effect of the oscillatory field component has been examined from the singlet yield which is a function of the angle between the Earth's field and the molecular axis. Consistent with the experimental work no effect at such weak fields could be traced when the oscillatory field is *parallel* to the Earth's field as a result of which the aforementioned group decided to put the oscillatory

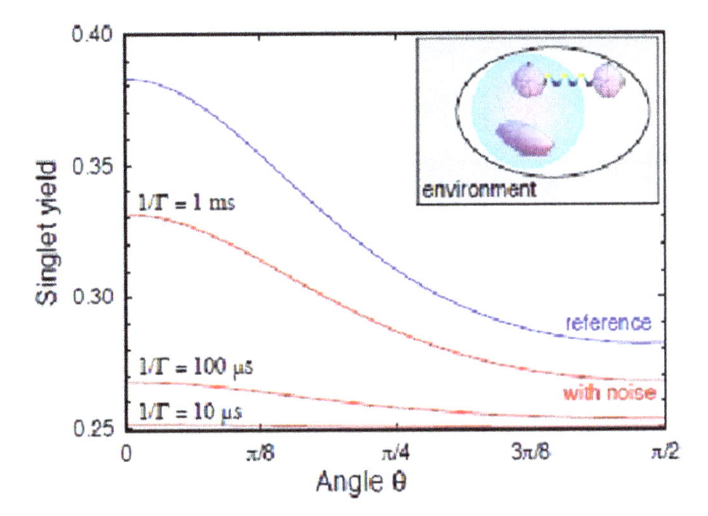

Figure 5.4. Angular dependence of the singlet yield in the presence of noise (for $k = 10^4$). The blue curve provides a reference in the absence of noise and the red curves show the singlet yield for different noise rates. As in apparent from the plot a noise rate $|Gamma\rangle 0.1k$ has a dramatic effect on the magnitude and contrast of the singlet yield. Inset: partitioning between compass and environment.

field to be *perpendicular.* The results are shown in Figure 5.2, out of which conclusion has been reached that if the oscillating field is to disorient the bird, as indicated by the experimental results, then the decay rate k should be approximately 10^4 s^{-1} or less and quite importantly, no higher values of k (shorter timescales for the overall process) is possible due to the absence of proper weak oscillatory field to significantly perturb the system. Instead, it relaxes before it has suffered any effect. However, such a value for the decay rate is consistent with the long RP lifetimes in certain candidate cryptochrome molecules found in migratory birds.[46]

5.4.5.2 *Role of Noise*

Taking the value $k = 10^4$ s^{-1}, next related question remains is to find out the *robustness* of this mechanism against environmental noise. There are several reasons for decoherence, for example, dipole interactions, electron-electron distance fluctuations and other particles' spin interactions with the electrons. All of such kind of environmental noise can be dealt with by extending Eqn. (5.1) with a standard Lindblad dissipator,[95] which can be expressed as

$$\dot{\rho} = -\frac{i}{\hbar}[H,\rho] + k\sum_{i=1}^{8}\left(P_i\rho P_i^\dagger - \frac{1}{2}P_i^\dagger P_i\rho - \frac{1}{2}\rho P_i^\dagger P_i\right)$$
$$+\sum_i \Gamma_i\left(L_i\rho L_i^\dagger - \frac{1}{2}L_i^\dagger L_i\rho - \frac{1}{2}\rho L_i^\dagger L_i\right) \tag{5.2}$$

A general formalism for Markovian noise as well as several noise models have been considered by Gauger *et al.* following above expression in which a few crucial points have been observed. A physically reasonable generic model, when considered, produced both of phase and amplitude being perturbed with equal probability. Here, $\sigma_x, \sigma_y,$ *and* σ_z for each electron spin individually (i.e. tensored with identity matrices for the nuclear spin and the other electron spin), give six different noise operators in total, i.e., L_i, where, the same decoherence rate Γ for all of them have been used.

As a next, the approximate level of noise is to be determined which the compass may suffer, by finding the magnitude of Γ for which the angular sensitivity fails. Conservatively it could be stated that when $\Gamma \geq k$, the angular sensitivity is highly degraded (Figure 5.3). This is remarkable as it implies the decoherence time of the two-electron compass system to be of order 100 μsec or more. It is quite demanding in providing proper context for this number. It could be mentioned that the best laboratory experiment involving preservation of a molecular electron spin state has accomplished a decoherence time of 80μs by Morton *et al.*[95]

Next inevitable serious query arising out of this context, is "Is there any noise model existing, consistent with experimental observations, which would cause rapid decoherence?"–answer of such query initially seems that really a form of noise can exist, i.e., the compass mechanism is almost immune to pure phase noise. Even starting from a fully dephased state $(\langle s| + |t_0\rangle\langle t_0|)/2$, the compass operates well. Thus it might seem that strong phase noise could be present, rapidly degrading quantum coherence but permitting the compass to function. But, the crucial points which could be shown that if such noise were naturally present at the level of $\Gamma_z \approx 10k$ or higher, it would render the bird immune to the weak RF magnetic fields, as considered by Ritz.[96] After studying systematically all such correlated noise processes related to the generic noise model, it has been found that these models implying RF field immunity unless coherence is preserved on a time scale, i.e., of the order of 100 μsec. This value directs us to avail an 'quite interesting query' in order to characterize the duration of quantum entanglement in the very process involved.

Thus, in this living system, the next important thing is to characterize the duration of quantum entanglement, taking the inferred approximate values for the key parameters, where starting from the initial singlet generation to the eventual decay can be monitored. For such aa case, the metric we use is negativity, i.e.,:

$$N(\rho) = ||\rho^{T_A/2}||$$

where $||\rho^{T_A}||$ is the trace norm of the partial transpose of the system's density matrix. The transpose is applied to the uncoupled electron, thus performing the natural partitioning between the electron, on one side, and the coupled electron plus its nucleus, on the other. Fig. 5.4] depicts how this negativity evolves under this generic noise model. Clearly, the initial singlet state becomes maximally entangled.

Interestingly, *this entanglement have been found to fall off in the presence of noise* and that also at a faster rate than the decay of population from the excited state. Thus, Gauger *et al.* suggested convincingly that the reported sensitivity to RF fields implies that both the amplitude and phase (and thus entanglement) are protected in reality, within the avian compass. They suggested the timescales to be at least tens of microseconds, even for a pure dephasing environment and hundreds of microseconds for the more general models. However, they remained inconclusive why such remarkable protection occurs, but given the widely-accepted RP model together with the recent experimental data,[41] their present conclusion follows.

Furthermore, in order to perform a systematic study of correlated noise processes, Gauger *et al.* took the basic RP model, consisting of three spins, making possible the existence of 64 combinations, having the form

$$L = S_i \otimes S_j \otimes S_l$$

for

$$S_i, j, l \in I_2, \sigma_x, \sigma_y, \sigma_z$$

Considering the generic noise model above, the authors also found these models implying RF field immunity unless coherence is preserved on a time scale $\approx 100\ \mu$ sec. Also, it appears quite interesting to calculate the duration of quantum entanglement in this kind of living system in which case, by considering the approximate values for the key parameters, it is possible to monitor entanglement from the initial singlet generation up to the eventual decay. For this, the criteria to be calculated, i.e., the negativity should be

$$N(\rho) = ||\rho^{T_A/2}||$$

$||\rho^{T_A}||$ being the trace norm of the partial transpose of the system's density matrix. The transpose is applied to the uncoupled electron, performing the natural partitioning between the electron, on one side, and the coupled electron plus its nucleus, on the other.

Fig. 5.4 shows how this "negativity" evolves under their[46] generic noise model which establishes the initial singlet state to be maximally entangled. Also, under noise, entanglement falls off at a faster rate than the decay of population from the excited state.

As a conclusion, we could state that implication of the reported sensitivity to RF fields is such that both amplitude and phase (and thus entanglement) are indeed protected within the avian compass. The timescales are at least tens of microseconds, even for a pure dephasing environment, and hundreds of microseconds for the more general model. However, it is not yet quite clear why such remarkable protection occurs, but given the widely-accepted RP model together with the recent experimental data,[8,41] this conclusion follows.

5.4.6 *Zachary Walter's Model*

The capability of a migratory bird in orienting itself relative to the Earth's magnetic field — even though a familiar every day feature of life — has been long since a puzzling problem of quantum mechanics. That birds have this ability is presently an well established fact, after following a long series of their behavioral patterns through experiments, believed to involve a coherent superposition of two spin states of a radical electron pair.[33] In addition to a light-activated compass located in their eyes, migratory birds are believed to possess a separate mechanism involving 'magnetite', possessing possible receptors, identified in the beak,[29] the middle ear[37] and the brain stem,[37] even though the existence of a receptor in the beak has been challenged in a recent study.[38] However, the mechanism by which this coherence can be maintained in the face of strong interactions with the cellular environment has still remained unclear. Walters addressed the problem of decoherence between two electron spins due to hyperfine interaction with a bath of spin 1/2 nuclei in which the dynamics of the radical pair density matrix have been derived by showing that it yields a simple mechanism for sensing magnetic field orientation and the rates of dephasing and decoherence have been calculated ab initio and found to yield millisecond coherence times. Light-activated mechanism has also been considered in addition to being widespread, well studied by a long series of behavioral experiments, reviewed in.[30–32]

The basic parameters of the compass mechanism may be probed by confining a bird in a conical cage during it's preferred migration period.[69] The restless nocturnal hopping behavior, or Zugunruhe, will tend to orient in the preferred migration direction, and the effects of environmental parameters can be judged by whether they affect the bird's ability to orient. In such experiments, the compass is light activated, with an abrupt cutoff between wavelengths 560.5 nm and 567.5 nm,[28] and that birds have been noticed to be sensitive to the orientation of magnetic field lines but not their polarity — they cannot distinguish magnetic north from south.[33] Provocatively, a recent experiment has found that an oscillatory magnetic field, oriented transverse to the static field, also can cause disorientation when, narrowly tuned to the Larmor frequency for an electron in the static field, to flip its spin. On resonance, an oscillatory field strength of 15 nT (Rabi frequency $\Omega_{Rabi} = 1250$ Hz) is sufficient to cause disorientation.[30,70] Qualitatively, such experiments are well explained by a "radical pair" model of the avian compass[34,70] in which the magnetic field drives coherent oscillations between the $|s, m_s\rangle = |0,0\rangle$ singlet state $|s\rangle$ and the $|s, m_s\rangle = |1,0\rangle$ triplet state $|t\rangle$ of an electron radical pair formed by absorption of a photon. If the singlet and triplet states react to form distinguishable byproducts, or can be otherwise distinguished,[44] monitoring the ratio of the byproducts probes, the time spent in each state, and thus the oscillation frequency.

The radical pair model gives an excellent phenomenological description of the avian compass, and predicts disorientation by an on-resonance oscillatory field. However, it remains theoretically problematic, requiring that coherence be maintained between different spin states for very long times despite the presence of an environment, very hostile to this. As observed in,[46] the slow spin flip time ($4\pi/\Omega_{Rabi} = 3ms$) implies that the process it disrupts must be slower still. Similar methods have also been used by the groups[46,47] to infer coherence times of $(10^{-6} - 10^{-4})$s. However, the proteins and water molecules present in a cellular environment possess large numbers of hydrogen nuclei, each of which interact with the radical pair via the hyperfine interaction. Somehow, the necessary quantum information must survive such interactions long enough to give a biologically useful signal.

Zachary[33] treated analytically the preservation and decay of coherence for a radical pair interacting with a bath of spin 1/2 nuclei, following which, a long lived component of the quantum information has been identified and shown to yield a simple and robust compass mechanism. Included to it, design considerations for an efficient compass have also been identified, and the corresponding coherence lifetime has been shown to be consistent with lifetimes, inferred from behavioral experiments with the calculations following atomic units throughout. Even though the radical pair model gives an excellent phenomenological description of the avian compass, and its prediction related to the disorientation by an on-resonance oscillatory field, anyway, the problem related to the theoretical treatment still remains problematic, as the coherence needed to be maintained between different spin states for sufficiently long times despite the presence of an environment, appearing very hostile to this. As observed in[30], the slow spin flip time ($\pi/\Omega_{Rabi} = 3ms$) implies that the process disrupted by this act must be slower still. Stoneham *et al.* (2012)[44] and Gauger *et al.* also applied similar methods to infer coherence times of $(10^{-6} - 10^{-4})$s. However, the proteins and water molecules present in a cellular environment possess large numbers of hydrogen nuclei, each of them interacting with the radical pair via the hyperfine interaction. Somehow, the necessary quantum information must survive such interactions long enough to give a biologically useful signal.

5.4.7 *Avian Compass*

Considering a radical pair interacting within in a Markovian bath, the evolution of the reduced density matrix ρ can be written as the Lindblad master equation, i.e.,

$$\frac{\partial \rho}{\partial t} = i[H_0^{rp}, \rho] + \sum_k \Gamma_k \mathcal{L}_k[\rho] \tag{5.3}$$

The above equation depicts that when a nonzero field drives states $|t^+\rangle$, having $|s, m_s\rangle = |1, 1\rangle$, $|t^{-1}\rangle$, and $|s, m_s\rangle = |1, -1\rangle$, due to the presence of degeneracy with the value of $m_s = 0$ states as a result. However, for the present, they can be omitted from the above equation. But, they will need to be considered while deriving the dephasing rates Γ_k. In the present case, the Zeeman Hamiltonian can be expressed as,

$$L_k[\rho] = -\rho L_k^\dagger L_k - L_k^\dagger L_k \rho + 2L_k \rho L_k^\dagger$$

The above expression is meant for the Lindblad super-operator corresponding to projection operator $L_k = |k\rangle\langle k|$. As both the receptor involved in the avian compass and the origin of $\Delta\mu$ have not been identified, in the absence of specific knowledge, $\Delta\mu$ has simply been taken as the order of 1 by the group, i.e., Zachary.[33]

$$H_0^{rp} = \frac{1}{2}\Delta\mu \vec{B} \cdot \vec{S} = B_z \frac{\Delta\mu}{2}(| \uparrow\downarrow\rangle\langle\uparrow\downarrow | - | \downarrow\uparrow\rangle\langle\downarrow\uparrow |) \tag{5.4}$$

The above equation describes the Zeeman Hamiltonian for the radical pair. Here, $\Delta\mu$ arises as a result of interactions between electrons together with their immediate surroundings, whereas $B_z = B\cos\theta$ describes the projection of the field onto some axis, considered by the interaction as preferable, together with $\Delta\vec{S} = \vec{S_1} - \vec{S_2}$.
According to Zachary's work, the dephasing induced by the hyperfine bath has been obtained from two parameters, the values of which can be obtained analytically, the details of which described in Zachary's original work.

For a short review this can be stated as follows: We know by now that

$$\Gamma_{|s\rangle} = \Gamma_{|t\rangle} = \bar{\Gamma}/2$$

and

$$\Gamma_{|\uparrow\downarrow\rangle} = \Gamma_{|\downarrow\uparrow\rangle} = \Delta\Gamma/2$$

Here, $\bar{\Gamma}$ is large enough for moderate field strengths and $\delta\Gamma$ becomes zero for some of orbital symmetries. Now, mapping the density matrix, to a Bloch sphere we get

$$(\rho_{ss} - \rho_{tt}) \Rightarrow (\rho_{01} + \rho_{10}) = x\sigma_x;$$
$$(\rho_{st} - \rho_{ts}) \Rightarrow (\rho_{10} - \rho_{01}) = iy\sigma_y,$$

and,

$$(\rho_{st} + \rho_{ts}) \Rightarrow (\rho_{00} - \rho_{11}) = z\sigma_z$$

The above, i.e., $\sigma_{x,y,z}$ are Pauli matrices where, z-component of Bloch vector $\vec{V} = (x, y, z)$ has been found to decay rapidly for large value of $(\bar{\Gamma})$, so that x and y components will behave like damped harmonic oscillators having the following nature, i.e.,

$$V_{x,y}(t) = Ae^{\lambda_+ t} + Be^{\lambda_- t} \tag{5.5}$$

Here,

$$\lambda_\pm = \frac{-\bar{\Gamma} \pm \sqrt{\bar{\Gamma}^2 - 4(B_z\Delta\mu)^2}}{2} \tag{5.6}$$

whenever, $\bar{\Gamma} \gg B\Delta\mu$, the system becomes over damped, resulting to a value of as $\lambda_+ \approx (B_z\Delta\mu)^2/\bar{\Gamma}$. Not only that, in such case, decay of x and y both becomes more slow with the growth of the dephasing. It should be mentioned mention that these dynamics are quite similar to quantum Zeno effect even though it has a different source of origin.[67,69] In the present case, one can find fast singlet or triplet reaction rates taking place out of rapid dephasing. Also interestingly, as observed by Cai *et al.* here, instead, because of rapid decay of z-component in the Bloch vector, the symmetry group of the long lived information is found[33] to be $U(1)$ rather than $SU(2)$.

Next, the value of $\bar{\tau} = \frac{5}{3}B\Delta\mu N_{sphere}$ can be obtained, considering a cellular environment by assuming a density of hydrogen nuclei equal to that of liquid water. This follows, for example, for $B = 50\mu T$, $N_{sphere} = 3300$, i.e., for the number of nuclei within radius $r_0 = 43$ Bohr, at which the hyperfine interaction equals to Zeeman interaction in magnitude, as a result of the corresponding dynamics becoming overdamped with a dephasing factor $\overline{\tau}^{-1} = 43$ ps and a coherence life time $\tau = \lambda_{+}^{-1} = 1.3$ ms. This is somewhat longer than the $(10^{-4}s - 10^{-6}s)$ as inferred in the works of Gauger *et al.*[46] and Bandyopadhyay *et al.*,[47] appearing quite crucial as this very slow rate in loss of coherence allows for a biologically useful signal.

Another important consequence is that, as the dynamics of the radical pair (RP) are strongly overdamped, as a result, the Bloch vector will not precess about the z-axis as in the original radical pair model. Rather, a Bloch vector in the equatorial plane will be frozen in place and evolve only due to decoherence. Thus, loss of coherence manifests itself as a transfer of population from singlet to triplet at a rate which varies as $B^2 \cos^2 \theta$. Identical logic can also be applied in the case of the triplet, taken as the initial state. For a compass, then, only requirement is the triplet reaction rate to be sufficiently large in order to prevent backwards population transfer. Assuming such a rate, the ratio of triplet to singlet byproducts becomes as

$$R_{ts}(\theta) = \frac{\lambda_{+} + \Delta\Gamma}{k_s} \approx \frac{|B\Delta\mu \cos\theta|^2}{\overline{\Gamma} k_s} + \frac{\Delta\Gamma}{k_s} \tag{5.7}$$

Here, k_s denotes the singlet reaction rate. Importantly, following the consistency of behavioral experiments, the compass signal has been found to be sensitive to the alignment of the magnetic field lines but **not** their polarity.

Another quite interesting fact needed to be mentioned here, i.e., whenever the static field is doubled in the absence of an oscillatory field, behavioral experiments[97] show temporary disorientation, lasting less than an hour. This important observation indicates that the biological signal is affected by the field strength, but the ability to orient the direction of bird's flight. The above approach, adopted by Walter Zachary[33] for the avian compass, represents an unique example of a quantum mechanical process which not only survives but, in fact, is sustained by interaction with a surrounding bath. Although the precise identity of the receptor or receptors involved in the avian compass still remains unknown even at that time, simple geometrical assumptions allowed information, sufficient for numerical comparison with experiment, to be derived from first principles. The proposed mechanism requires neither unique properties nor elaborate manipulation of the radical pair state, and the biologically observable signal is distinctive and easy to interpret. The avian compass thus represents a simple model system for the emerging and still largely unexplored role of quantum mechanics in biological processes.

5.4.7.1 *Quantum Coherence & Entanglement in the Avian Compass*

In the field of biophysics,[27] an huge amount of interests have been developed observing the ability of many animal species, such as birds, insects and even mammals to sense the geomagnetic field for the purpose of orientation and navigation, leading to huge interest. Among the two leading hypotheses in explaining this remarkable ability have been: the magnetite-based mechanism, and the radical pair mechanism[37,46] and references there in. Recently, several authors have raised the intriguing possibility of using nontrivial quantum effects in case of many living systems; especially, birds which may use nontrivial quantum effects in optimizing their orientation behavior.

This followed the suggestions stating that entanglement, rather than mere quantum coherence, is one of the important contributing factors, responsible in allowing the avian compass to achieve its high level of sensitivity.[48] But these suggestions immediately raised a few crucial question: if this is so, does the duration of the entanglement last sufficiently long enough to impact biological processes? Not only that, — is the entanglement sensitive enough to the inclination of the radical pair with respect to the Earth's magnetic field? In order to answer these questions, the lifetime of radical pair entanglement corresponding to different magnetic field strengths are examined and compared with the results with the candidate chemical reactions,[92] including the angular dependence of the radical pair entanglement within the geomagnetic field. A new model have been explored, the underlying details of which states that the presence of entanglement in their proposed model displays both directional sensitivity as well as a sufficiently long duration of entanglement.

The study of the lifetime of radical pair entanglement for the magnitude and direction of magnetic fields in detail showed that the entanglement lasts long enough in bird's eyes to be used for navigation. Quite importantly, they demonstrated that, due to a *lack of orientational sensitivity* of the entanglement in the geomagnetic field, the birds are not able to orient themselves by the mechanism based directly on radical-pair entanglement. Thus, with the aim of exploring the entanglement mechanism further, specifically, in which, the hyperfine interactions are replaced by local magnetic fields of similar strength, the entanglement of the radical pair in this model have been found to last longer and display an angular sensitivity in weak magnetic fields, which factors have not been found to be present in the previous models considered.

By now, it has been an established fact that the basic scheme of the Radical Pair Mechanism (RPM) involves three steps i.e.,

- the first step is light absorption,
- followed with the formation of the radical pair and its interconversion between the singlet and triplet for electron spins.
- The final step is the decay of the singlet and triplet states to chemical products, producing a chemical signal detectable through an biological pathway.

It has been already noted by now that typically, the RP reaction involves two kinds of molecules which play the roles of electron donor and acceptor. After absorbing light, the electronic state of the donor molecule, e.g. one of the electrons in the donor molecule is excited from the highest occupied molecular orbital (HOMO) to the molecular orbital (LUMO).[88] As a next step, whenever the two molecules are close enough to each other, the donor electron will be transferred to the acceptor molecule, thus both molecules contributing an unpaired electron to form a pair of radicals. These two unpaired electrons in the donor and acceptor molecules are initially bound in a singlet state before they are spatially separated. Finally, under the effect of the external magnetic field, i.e. the geomagnetic field and local nuclear spins, the electron pair transfers between the singlet and triplet states. Finally, the singlet and triplet states will produce different reaction products.[43,88] The underlying process has been demonstrated in the following figure.

To investigate the role of entanglement in this chemical compass, the second step of the RPM scheme [give the figure number of just above], has been taken into consideration because of the fact that while the electrons are separated, it is believed that geomagnetic field may influence inter system conversion. Considering the Zeeman and the hyper fine interaction, as well, in the Hamiltonian of the system, one gets

$$H = g\mu_B \sum_{i=1}^{2} \vec{S}_i \cdot \left(\vec{B} + \widehat{A}_i \cdot \vec{I}_i \right) \tag{5.8}$$

The first term, in the above equation, accounts for the Zeeman interaction and, the second term for the hyperfine interaction. Here, $\vec{I}_i$ represents the nuclear spin operator; $\vec{S}_i$ the electron spin operator, i.e., $\vec{S} = \vec{\sigma}/2$ with $\vec{\sigma}$ being the Pauli matrices; Here, g is the g-factor of the electron, in the present case, chosen as $g = 2$; μ_B the Bohr magneton

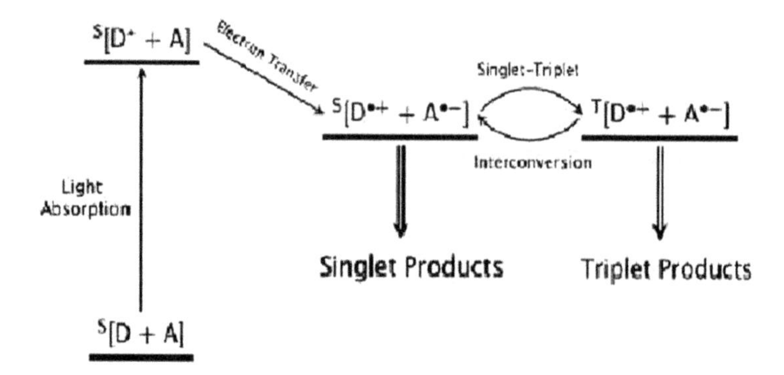

Figure 5.5. Scheme of RPM. After a light induced electron excitation, the donor transfers an electron to the acceptor, forming a radical pair with the accceptor molecule. The interconversion between singlet and triplet is affected by an external magnetic field. Finally the singlet and triplet decay into different products.

of the electron; and $\hat{A}_i$ is the hyperfine coupling tensor, a 3×3 matrix. Following the suggestions, from the work of Ritz *et al.*, as a next step, the model with radical pair dynamics with a Liouville equation, becomes as,

$$\dot{\rho}(t) = -\frac{i}{\hbar}[H, \rho(t)] - \frac{k_S}{2}\{Q^S, \rho(t)\} - \frac{k_T}{2}\{Q^T, \rho(t)\} \tag{5.9}$$

In the above equation (5.2), H is the Hamiltonian of the system, Q^S is the singlet projection operator, i.e., $Q^S = |S\rangle\langle S|$, and $Q^T = |T_+\rangle\langle T_+|\mathrm{T}_0\rangle\langle T_0| + |\mathrm{T}_-\rangle\langle T_-|$ is the triplet projection operator. $|S\rangle$ is for the singlet state and $(|T_+\rangle, |T_0\rangle, |T_-\rangle)$ stand for the triplet states.

$\rho(t)$ represents density matrix for the system; k_S, k_T are the decay rates for the singlet state and triplet states, respectively.

Since the radical pair must be very sensitive to different alignments of the magnetic field, it is necessary to assume that the hyperfine coupling tensors in Eq. (5.1) are anisotropic and hence, for the sake of simplicity, the hyperfine coupling has been employed as anisotropic. In order to determine what values of the decay rates are reasonable for biological systems, the influence of different decay rates has been calculated on the triplet yield, ΦT, for the variation of the external magnetic field.

5.4.7.2 *Assignment of Decay Rate 'k' & Effect of Earth's Magnetic Field*

In order to determine what values of the decay rates are reasonable for biological systems, the influence of different decay rates, following the variation of external magnetic field rates on the triplet yield, Φ_T, has been calculated by the above mentioned group. The triplet yield has been expressed as,[47]

$$\Phi_T = k \int_0^{\infty} \mathrm{Tr}[Q^T \rho(t)]\, dt \tag{5.10}$$

where, $Q^T = |T\rangle\langle T|$, and $|T\rangle = |T_+\rangle + |T_0\rangle + |T_-\rangle$.

It should be noted that the effect of the radical pair decay rates on the triplet yield has a twofold function. For a very high decay rate, i.e., larger than 10 μs^{-1}, the rapid decay of the radical pair prevents efficient singlet-triplet mixing which could be noticed by increasing the triplet yield in the weak magnetic field meaning that the weak magnetic field has very little effect on the triplet yields with respect to fast decay rates. However, for very slow decay rates, i.e., smaller than 0.1 μs^{-1}, the triplet yield increases up to its maximum, i.e., almost immediately when the magnetic field increases from zero, but remains essentially static as the magnetic field continues to increase. A decay rate of the order of 1 μs^{-1}, seems to be optimum for the detection of a weak magnetic field. Having fixed the decay rate 'k' to be 1 μs^{-1}, the above mentioned group studied the radical pair entanglement as a function of the magnitude of the geomagnetic field in detail.

In the above group's experiments, it has been observed that when the magnetic fields are weaker than the Earth's magnetic field, or as strong as $1G$, the entanglement curves are almost identical. There does not appear to be any unique behavior that distinguishes a field in the neighborhood of 0.5 Gauss. However, under the Earth's magnetic field, the entanglement will be robust periodically during the first 0.5 μs, which is longer than the suggested duration of radical pair separation.[92] A stronger magnetic field (e.g. 5G) will disturb this periodicity. It should be noted that previous research on the magnetic-field sensitivity of the chemical compass has already demonstrated that the entanglement is helpful only if nature allows birds to optimize this behavior.[8] On these grounds, one can say that the entanglement lasts long enough to play crucial role in the orientation of birds. Not only that, the present result obtained, showed that the dynamics of entanglement are nearly static for different angles under the symmetric hyperfine tensors. But, the next, immediate, question arises, i.e., what might happen in case of using an asymmetric hyperfine tensor. After going through several such cases, the above group arrived at the final conclusion, i.e., by using the asymmetric hyperfine coupling tensor pair of $\cap A_i^c$, a certain kind of intriguing result have been obtained, which means in a clear way that the dynamics of the entanglement is dependent on the system's orientation.

This crucial result inspired the above group, in developing a new model, in which only the external magnetic fields are considered. They proposed that each electron interacts with additional local magnetic fields, $\vec{B}$ rather than with the hyperfine fields. Changing the relative angles and strengths of the local magnetic fields, a dramatic impact has been noticed on the angular sensitivity. If indeed, a protein, such as cryptochrome, is in part responsible for magnetoreception, there must be some directional bias of the orientation of the protein, so that there will be a strong net signal. It is possible that this directional dependence could be provided by embedding within the 'membrane shelves' of the photoreceptor cells. This form of embedding leaves the protein free to rotate about one axis, but greatly restricts the rotations about its second axis.[92] Importantly, for this reason, it is necessary for the RP compass to be sensitive to rotation about one axis, while being virtually unaffected by rotation about the other axis. If the RP compass were to be sensitive to rotation in both θ and ϕ, the result of randomly oriented proteins about the θ axis would average out to create a back ground signal that could potentially reduce the contrast of the RP compass.

At this point, it should be pointed out that, unlike the previous model and its variants, new variant is not symmetric about 90°, but it is symmetric about 180°. While this might seem to contradict an inclination-only compass model, it is reasonable to assume that cryptochrome is either bound to both sides of the cell membrane, or embedded within the membrane in both up and down orientations, so that the net signal cannot discern the polarity of the geomagnetic field. However, another quite important point, needs to be mentioned, proposed by the above group, is that this model is not symmetric about 90°, but symmetric about 180°. Importantly, while this might seems to contradict an inclination-only compass model, but is quite reasonable to assume that cryptochrome is either bound to both sides of the cell membrane, or embedded within the membrane in both up and down orientations, so that the net signal cannot discern the polarity of the geomagnetic field. Finally, the above group decided that the entanglement decay endures long enough for living systems to conduct the entanglement-based reactions. However, finally and importantly, it should be noted that the dynamics of the entanglement is not sensitive to the change of angle between the z axis of the radical pair and the geomagnetic field vector in the hyperfine model. Therefore, if we still believe that entanglement plays a crucial role in the orientation of birds as demonstrated before, there must be an **indirect** mechanisms by which the entanglement can affect the birds' behavior.

Thus, finally the above mentioned group arrived at their conclusion that the entanglement endures long enough for living systems in order to conduct the entanglement-based reactions. However, importantly, they opined that the dynamics of the entanglement is not sensitive enough to the change of angle between the z axis of the radical pair and the geomagnetic field vector in the hyperfine model. Therefore, possible role entanglement, plays a crucial role in the orientation of birds as demonstrated here and before as well. Not only that there must be an indirect mechanisms by which the entanglement can affect the bird's behavior. However, adjustment of the decay rates, for example, i.e., using different values for the decay rates of the singlet and triplet state, might put new signature in this line of thought in improving the above mentioned model which might find the hidden bridge between the entanglement of the radical pair and that of the orientation in a magnetic field.

5.4.8 *An Open Quantum System Approach to the Radical Pair (RP) Mechanism: Collective Coupling Approach*

Adams *et al.*[98] adopted an mechanism, i.e. an open quantum systems approach to a model of the radical pair (RP) mechanism in order to derive a master equation by applying the Born-Markov approximation for the case of two electrons, each interacting with an environment of nuclear spins as well as that of external magnetic field, then fially placed in a dissipative bosonic bath. This model is used to investigate two different cases relating to radical pair dynamics:

- firstly, uses a collective coupling approach to simplify calculations for larger numbers of nuclei interacting with the radical pair, whereas
- secondly, looks at the effects of different hyperfine configurations of the radical pair model, considering a particular case, i.e., the case in which one of the electrons interact with two nuclei having different hyperfine coupling constants.

The results, thus obtained out of these investigations have been analyzed to notice, whether, they offer any insights into the biological application of the radical pair (RP) mechanism in avian magnetoreception. As we are well aware by now that, the radical pair can be described in the following rudimentary outline of the three important steps that constitute the mechanism, such as

- In the first step: a photon is incident on the donor molecule of the pair, causing an electron to be excited and donated to the acceptor molecule. This results in a spin-correlated but spatially separated electron pair which is conventionally thought of as being in a singlet state.
- During the second step: the radical pair oscillates between singlet and triplet state under the influence of the Zeeman or hyperfine interaction.
- The third step is then, responsible for the recombination of the pair to form some sort of chemical product/signal. Quite importantly, the chemical product formed in this last step is dependent on whether the pair is in a singlet or triplet state, thus dependent on the magnetic field. This allows the radical pair to function as a compass.

Importantly, Adams *et al.*[98] proposed an open quantum systems and applied to the second step of the mechanism. However, even though the theoretical approach outlined in their work is not restricted to the case of avian magnetoreception only, the programme has been simulated, specifically, using parameters relevant to this specific application. The results of these simulations have also been analyzed to see if they offer any insights into biological systems, particularly, the avian compass. To this end, the works of this group[92] presented two different investigations. Firstly, the novelty of this theoretical approach is to offer first theoretical approach to the radical pair mechanism in the form of the 'Collective Coupling Model'. Secondly, this novel approach looks more closely at the hyperfine environment of the radical pair (RP) to study whether altering hyperfine coupling strength might alter radical pair dynamics in any way. The motivation for such investigations can be described in a short form from the followings:

Haberkorn, in his 1976 paper,[87] described in detail a density matrix treatment of radical pair reactions and most subsequent approaches in this field have since followed his lead, however, with expected attempts to investigate both the advantages and limitations of such an approach. One of the monumental limitations faced by the conventional approach, in the context of avian migration, is the increased computational complexity arising, as the radical pair interacts with increasing numbers of nuclei. It is estimated that the number of significant nuclei felt by each radical is approximately ten to fifteen. As a means to simplify these calculations the group of Adam *et al.* outlined an extraordinary approach, i.e., they considered the Hamiltonian as a function of the collective spin operators of the nuclear spin environment so that the space of the total system can be decomposed as the sum of the subspaces.[99,100] This model, i.e., the collective model has been found to be quite useful whenever simplification is needed, i.e., in order to get a broad idea of the effects of increased numbers of nuclei and to look for potential patterns of radical pair behavior, needed to be considered. However, this simplification faces difficulty whenever, the context of the complex biological environment is to be considered.

To address this problem in their Collective Coupling model, the above mentioned group adapted two preliminary but important conditions, i.e., in having the predominant influence on radical pair dynamics for the two 'non-identical nuceli'. But, secondly, quite crucially, they specifically meant the nuclei, not being confined in having the exact same hyperfine coupling strength, but followed by the investigation of the effects of various arrangements, happened to these specific nuclei when interacting with the radical pair. It should be noted that there have been a number of reports detailing experimental results in which electromagnetic radiation has been noted in disrupting the avian compass,[43,101,102] the most recent of such reports indicating a broad band of radio-frequency radiation to be responsible for this kind of disruption.[102]

The master equation derived in their work[98] has been expressed in terms of the transition frequencies associated with the singlet-triplet oscillations, present in the radical pair which depends on the magnetic field strength as well as that of hyperfine coupling constants. By investigating different hyperfine parameters and the way the radical pair dynamics have been altered by them, they gained some insight into how such a broad band of electromagnetic radiation might interact with the radical pair. In these investigations, both for the case of the collective model as well as that for non-identical nuclei, Adam *et al.* did follow an open quantum systems master equation route for the present case. Generally, the Lindblad master equation derived, is trace-preserving (i.e., preserves physical probabilities), however, the approach, adopted by Adam's *et al.*, differs from the conventional approach, in two main aspects:

- Firstly, master equation is not restricted in demonstrating only the effects of the weak interaction.
- Secondly, their model focuses only on the dynamics of the radical pair prior to, but importantly, not including recombination.

This way, their model involves in calculating the coherence lifetime of the radical pair (RP) and also, the effects of different hyperfine environments associated within these dynamics. It is crucially important to clarify the fact that the rates under scrutiny are not the rates at which recombination happens, or product being formed, but rather the rates, governing singlet/triplet interconversion. However, so far further dynamics is related to the production of singlet or triplet, specific chemicals would still need to be added to the master equation as described below.

5.4.9 *Model adapted by Adams et al. (2018): Open Quantum System — Application of the Complete Spin Hamiltonian for the Radical Pair (RP)*

The model adopted by Adams *et al.* (2018) considered the complete Hamiltonian for the radical pair so that[88]

$$H = H_{zee} + H_{hf} + H_{dip} + H_{ex} + H_{nuc} \tag{5.11}$$

where H_{zee} and H_{hf} correspond to the Zeeman and hyperfine effects and are the only two terms retained. The dipole and exchange effects, H_{dip} and H_{ex} can be discounted either due to sufficient separation of the electrons or because of the separation being optimal for the effects to cancel.[92,93] Finally, H_{nuc}, i.e., the interaction of nuclear spins with the external field, has also been neglected as the gyromagnetic ratio is much smaller than in the electronic case.[88] Thus, the Hamiltonian for such system models, involves the two electrons, labeled (1) and (2), each of them interacting separately with their nuclear environment,

The Hilbert space of the system can be written in this case as,

$$H_s = H_e^{(1)} \otimes H_e^{(2)} \otimes H_n^{(1)} \otimes H_n^{(2)}$$

where, H depicts to the Hilbert space of either electron while H_n refers to the Hilbert space of the nuclear spin environment.

5.4.10 *Collective Coupling Approach for Identical Nuclei: Radical Pairs*

In order to describe the present model, the Hamiltonian needed for describing the system, consists of two electrons, labeled (1) and (2), each interacting separately with their nuclear environment, (see Fig. 5.1).

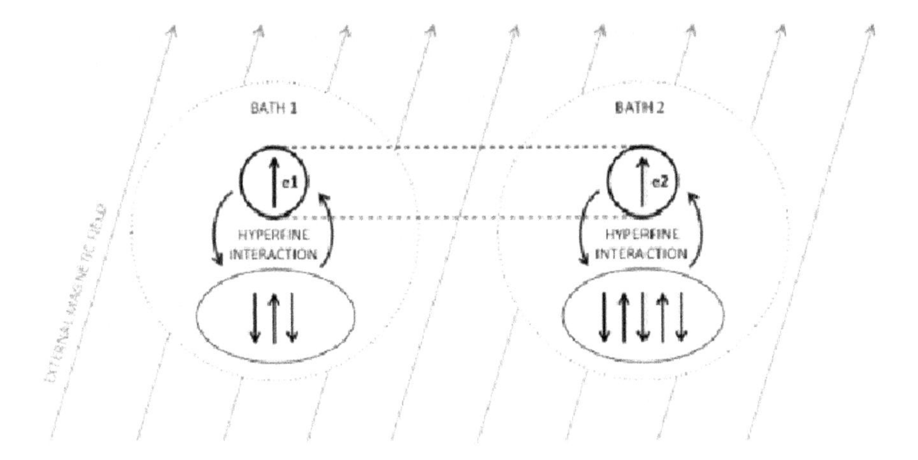

Figure 5.6. A diagram of the modl shows the two electrons e_1 and e_2 respectively. Both electrons experience the Zeeman Effect due to the external geomagnetic field. Each of the radicals in the pair (modelled as the system is then assumed to interact with an external heat bath, undergoing decoherence and dissipation. This allows for a mathematical description of the second step of the radical pair mechanism as outlined in the introduction.

The corresponding Hilbert space of the system can be written as

$$H_s = H_e^{(1)} \otimes H_e^{(2)} \otimes H_n^{(1)} \otimes H_n^{(2)}$$

Here, H_e refers to the Hilbert space of either electron. H refers to the Hilbert space of the nuclear spin environment.

As nuclei are added, complication starts with the number of the increasing degrees of freedom, meant for the nuclear spin environment, thus making the solution of the problem related to the model more ad more complicated. In tackling this problem, Adam *et al.* adopted the approach in which the Hamiltonian is considered as a function of the collective spin operators of the nuclear spin environment meant for each of the electrons. For convenience, it has been assumed that the two electrons of the radical pair can be treated separately. In order to investigate the dynamics of the radical pair itself, it is important in having the partial trace over the nuclear spin, the degrees of freedom being carried out within the subspaces, corresponding to the different values of the total angular momentum of the surrounding nuclei.[100, 101]

Now, with the aim of developing this model, a particle having a state of spin $\frac{1}{2}$ been considered so that it can be described by the complex inner product of space, i.e., $\mathbb{C}$. A basis of this space can be constructed using the eigenvectors $|1\rangle, |0\rangle$, having corresponding eigenvalues $\pm\frac{1}{2}$ of the z-component with the spin operator, given by $S_z = \frac{1}{2}\sigma_z$ with σ_z, being one of the Pauli operators. After making generalization, for the present system of N spin-half nuclei, the state space can be written as

$$\mathbb{C}^{2\otimes N}$$

and a basis

$$\otimes_i^N |\epsilon_i\rangle,$$

where $\epsilon_i = 1, 0$. This can be constructed from the eigen vectors of the collective spin operator I_z, with $\overrightarrow{I} = \frac{1}{2}\sum_{i-1}^{N} \overrightarrow{\sigma_i}$. When collective interaction is considered then $|j, m\rangle$ becomes common to both the operators, i.e., I^2 and I_z having corresponding eigenvalues $j(j+1)$ and also m. Thus, importantly, for a system having N nuclei, the quantization of angular momentum dictates this state so that for N even, $0 \le j \le \frac{N}{2}$ and for odd N, $12 \le j \le \frac{N}{2}$, having $-j \le m \le j$.

Now, as seen from the above process, for different values of j, the scalar product of the state vectors is zero and the space of the total system can be decomposed as the sum of the subspaces,[99, 101, 102] i.e.,

$$\bigotimes_{i=1}^{N} \mathbb{C}_i^2 \cong \bigotimes_{j=0}^{\frac{N}{2}} \nu(N, j)\mathbb{C}^{2j+1} \tag{5.12}$$

In the above expression, $\nu(N, j)$ denotes the degeneracy corresponding to a specific value of j so that the above expression can be expressed as

$$\nu(N, j) = \frac{2j+1}{(\frac{N}{2}+j+1)} \cdot \frac{N!}{(\frac{N}{2}-j)!(\frac{N}{2}+j)!} \tag{5.13}$$

At this stage, for $\hbar = 1$, the Hamiltonian for the system can be written as follows i.e.,

$$H_S = \gamma_e(\overrightarrow{B} \cdot \overrightarrow{\mathbf{S}}^{(1)}) + \overrightarrow{B} \cdot \overrightarrow{\mathbf{S}}^{(2)} + \lambda_h \sum_{k=1}^{2}\sum_{l=1}^{3} A_{nl}^{(k)} \mathbf{S}_n^{(k)} \cdot \mathbf{I}_l^{(k)} \tag{5.14}$$

Here, $\overrightarrow{B}$ is the magnetic field vector. By considering now, for example,

$$S_z^{(1)} = S_z^{(1)} \otimes \mathbb{I}_e^{(2)} \otimes \mathbb{I}_n^{(1)} \otimes \mathbb{I}_n^{(2)}$$

Here, $\mathbb{I}_e$ and $\mathbb{I}_n$ depicts the identity matrices for electron and nuclei respectively.

Hence, in the present case, the expression, for example, $S_+^{(1)} I^{(1)-}$ uses ordinary matrix multiplication.

Again, $\mathbf{S} = (S_x, S_y) = \frac{1}{2}\sigma_y \& S_z) = \frac{1}{2}\sigma_z$ is the vector of the spin operators for each of the electrons having $S_x = \frac{1}{2}\sigma_x, S_y = \frac{1}{2}\sigma_y \&, S_z = \frac{1}{2}\sigma_z$, whereas, σ_x, σ_y and σ_z are the Pauli matrices.

$$\sigma_x = \begin{pmatrix} 0 & 1 \\ 1 & 0 \end{pmatrix} \quad \sigma_y = \begin{pmatrix} 0 & -i \\ i & 0 \end{pmatrix} \quad \sigma_z = \begin{pmatrix} 1 & 0 \\ 0 & -1 \end{pmatrix} \tag{5.15}$$

where

$$\mathbf{I} = (I_x, I_y, I_z)$$

is the vector of the spin operators for nuclear spin.

In the above case, Zeeman interactions have been represented by $\gamma_e \mathbf{B} \cdot \mathbf{S}$ having electron gyromagnetic ratio $\gamma_e = -g\mu_B$ with $g = 2$ and μ_B, is meant for Bohr magneton. Here the sum runs over for both of k in order to include both the electrons. Here, the hyperfine coupling, as described by the hyperfine coupling tensor, must be anisotropic because the radical pair is expected to be sensitive to different magnetic field alignments. Following the convention of quantum biology and the approach of Ritz *et al.*[43] they have adopted the simplest case considering the first electron as anisotropically coupled to its spin environment while the second being isotropically coupled, with respective hyperfine coupling tensors, we get

$$A^{(1)} = \begin{pmatrix} 2 & 0 & 0 \\ 0 & 2 & 0 \\ 0 & 0 & 1 \end{pmatrix}, \quad A^{(2)} = \begin{pmatrix} 1 & 0 & 0 \\ 0 & 1 & 0 \\ 0 & 0 & 1 \end{pmatrix} \tag{5.16}$$

with having hyperfine coupling constant as λ_h. Next, the diagonalized system Hamiltonian can be expressed as

$$\begin{aligned} H_S &= (\gamma_e B_0 + \lambda_h j)|0, j, j\rangle\langle 0, j, j| - (\gamma_e B_0 - \lambda_h j)|1, j, -j\rangle\langle 1, j, -j| \\ &\quad + \sum_{m=-j+1}^{j} \left[\nu_1(j, m)|\lambda_{jm}\rangle\langle\lambda_{jm}| + \nu_2(j, m)|\lambda_{j,m}^i\rangle\langle\lambda_{jm}^i|\right] \\ &\quad + \sum_{m=-j+1}^{j} \left[u_1(j, m)|\phi_{jm}^-\rangle\langle\phi_{jm}^-| + u_2(j, m)|\phi_{jm}^+\rangle\langle\phi_{jm}^+|\right] \end{aligned} \tag{5.17}$$

The next part is to express the necessary Born-Markov master equation for this system which is as follows:

$$\frac{d\rho_I^S(t)}{dt} = \sum_{k=1}^{2} \gamma_D [V_0^{(k)}, \rho_I^S(t) V_0^{(k)}] - \frac{1}{2}\{V_0^{(k)} V_0^{(k)}, \rho_I^S(t)\}$$
$$+ \sum_{k=1}^{2}\sum_{q=1}^{N_T} \gamma_q \left((N(\omega_q^T) + 1)[V_q^{(k)}, \rho_I^S(t) V_q^{\dagger(k)}] - \frac{1}{2}\{V_q^{\dagger(k)} V_q^{(k)}, \rho_I^S(t)\} \right)$$
$$+ \sum_{k=1}^{2}\sum_{q=1}^{N_T} \gamma_q N(\omega_q^T) \left([V_q^{(k)}, \rho_I^S(t) V_q^{\dagger(k)}] - \frac{1}{2}\{V_q^{(k)} V_q^{\dagger(k)}, \rho_I^S(t)\} \right) \tag{5.18}$$

Here, the V_q represents the transition operators. The first term of the equation describes the effects of decoherence whereas, all other terms describe the dissipation present in the system and the summation runs here over $k = 2$ for the inclusion of the electrons, N_T is for number of operators whereas γ_D and γ_q describe the decoherence and dissipation rates respectively. Lastly, the Plank distribution is given by

$$N(\omega_q^T) = \frac{1}{e^{\beta\omega_q^T} - 1}$$

Here, $\beta = \frac{1}{k_B T}$, gives the number of thermal photons (bosons) in a mode of frequency ω_q^T at a given temperature T, k_B representing to Boltzmann constant. The group of Adams *et al.*,[98] considered their investigation keeping in mind the biologically relevant dynamics of the radical pairs so that the values adopted here, are the specific case of avian magnetoreception. To do this, Zeeman effect has also been taken into consideration as the results are to be affected by the presence of a geomagnetic field of $47\mu T$. Thus, they applied accurately the model to biological systems by taking the temperature to be ≈300 K. Next, the appropriate frequency-dependent rates of dissipation have been calculated from the following relation which can be expressed a

$$\gamma_q = \frac{\omega_q^3}{3\epsilon_0 \pi \hbar c^3} |d|^2 \tag{5.19}$$

The term, d, relates to transition dipole moment. For this range of transition frequencies, Adams *et al.* calculated a range of transition rates from $(10^{-9} - 10^{-3})\text{s}^{-1}$. The decoherence rate, taken by this group, has been at least ten times greater than the largest of the dissipation rates.

There exists a general acceptance that spin relaxation via spontaneous emission is very slow so that this has been considered balanced in this model by the inclusion of relaxation due to thermal fluctuations, as represented by $N(\omega_q^T)$. It has already been known fact that electron spins in radicals are relaxed by a number of other mechanisms which occur faster than spontaneous emission. Typically, it has been noticed that spin relaxation results from local magnetic field fluctuations due to motion of the radical or its surroundings[104] which effect has already been discussed by Adams *et al.* group, i.e., effects of rotational motion on the spin relaxation rate, demonstrating that the maximum relaxation rate can be up 100 times faster than that of the radical pair recombination.[104]

5.4.10.1 *Extension of the Collective Coupling Model*

By considering the collective coupling model, this group showed the usefulness of its application, i.e., this approach allows for increased numbers of nuclei without over complication of the matrix calculations. Added to that this model predicts a coherence lifetime of the order of seconds, appearing surprisingly long-lived, getting to the millisecond range whenever nuclei are added this, getting closer to the millisecond range. Finally, the effects of both the decoherence and dissipation have been calculated following the increase of the number of nuclei interacting with the radical pair, showing that these effects are represented as the probability over time of finding the radical pair in one of the four possible spin states. Also, a general trend has been noticed towards a decreased coherence lifetime caused by increasing the number of nuclei in the model which, i.e., increasing the number of possible transitions means allowing for a greater dissipative effect from the environment. Interestingly, this effect is not same for odd and even numbers of nuclei, as the radical pair decays, interestingly, slower for integer values of j, because, for integer values of j, the possibility exists for both radicals in the pair could in having $j = 0$, thus leading to a 'particularly' long life time. Again, an additional difference between integer and non-integer j is the slightly faster oscillation between singlet and triplet states for the case of integer spin.

Finally, if this model is to be viewed in the context of biological systems, recent transient absorption measurements involving the avian compass candidate, cryptochrome, indicate radical pairs with millisecond lifetimes,[46] even though, it should also be noted here that it is difficult to compare the lifetimes discussed in the literature due to the 'different' interpretation of lifetime. However, the group, mentined here, described the persistence of the coherence whereas elsewhere it is used to describe how long the radicals last before they recombine. It could be the case that the disparity between integer and half integer spins arises from the simplifications inherent in the model itself, which treats all the nuclei for each radical as identical spin-$\frac{1}{2}$ nuclei with identical hyperfine interactions, and hence, would no longer be relevant, should we consider a more realistic nuclear environment in which case, all hyperfine tensors are different and both spin $-\frac{1}{2}$ and spin 1 nuclei are present. However, the efficacy of the above mentioned model lies exactly to this treatment of the nuclei as identical, so that it has been possible in simplifying the increasing degrees of freedom for increasing numbers of nuclei by using the fact that the space of the total system could be decomposed as the sum of the subspaces. However, the present model of Adams *et al.* has allowed the present theory in identifying any patterns which might possibly emerge for increasing numbers of nuclei.

5.4.10.2 *Non-collective Case*

The approach by Adams *et al.* has given the results, after abandoning the collective coupling approach by adapting the model to a slightly more realistic case. Here, they did introduce the idea, where the radical pair interacts with its nuclear environment having a range of hyperfine coupling strengths. For such an particular case, they have investigated the effects of two different nuclei interacting anisotropically with, having one of the radicals in the pair while the other radical experiencing no hyperfine interaction. At this stage, discounting all other interactions except the Zeeman and hyperfine effects, the system Hamiltonian can be written as

$$H_s = \gamma_e(\overrightarrow{B}^{(1)}) + \lambda\hbar\sigma_{n=1}^{2}\sigma_{(i=x,y,z)}\omega_i(n)S_i^{(n)} \otimes \mathbf{I}_i^{(n)} \tag{5.20}$$

Here, $\overrightarrow{B}$ is the magnetic field vector and the second electron has no hyperfine interaction. The sum over n for the first electron, represents the two different nuclei whereas, ω_i represent the anisotropic hyperfine coupling strengths. Once again, $S = (S_x, S_y, S_z)$ is the vector of spin operator for each electron with $S_x = \frac{1}{2}\sigma_x; S_y = \frac{1}{2}\sigma_y; S_Z = \sigma_z$ and $\sigma_x, \sigma_y = \sigma_z$ are the Pauli matrices given by,

$$\sigma_x = \begin{pmatrix} 0 & 1 \\ 1 & 1 \end{pmatrix}, \quad \sigma_y = \begin{pmatrix} 0 & -i \\ i & 1 \end{pmatrix}, \quad \sigma_z = \begin{pmatrix} 1 & 0 \\ 0 & 1 \end{pmatrix} \tag{5.21}$$

where, $\mathbf{I} - (I_x, I_y, I_z)$ is the vector of spin operators for nuclear spin.

In the above case, the Zeeman interaction has been represented by $\gamma_e \vec{B} \cdot \mathbf{S}$ having the gyromagnetic ratio as $\gamma_e = -g\mu_B$ where, $g = 2$ and μ_B is the Bohr magneton. In the above case, the system Hamiltonian for the non-collective case, has been diagonalized by finding eigenvectors and eigenvalues.At this stage, the above collective model demonstrated explains some potentially interesting differences between integer and non-integer collective spins but it still falls short as a realistic model, as might be applied to complex biological systems, most obviously in the fact that it uses a single coupling strength for each nucleus. A more appropriate approach in this case would be to allow for a number of nuclei with differing coupling strengths to interact with the radical pair. Adams *et al.* investigated three different hyperfine configurations with the two nuclei interacting anisotropically with one of the electrons in the pair.

- In the first case, both nuclei had a comparable hyperfine coupling strength of the order of 107 Hz.
- In the second, the two nuclei again had a comparable but much smaller hyperfine coupling of the order of 104 Hz.
- In the third case, the hyperfine coupling strengths of the two nuclei differed, one nucleus had a hyperfine coupling constant of 107 Hz with the other set at 104 Hz. Though it might be expected that dissipation would be enhanced by larger hyperfine coupling constants, the results show 'otherwise'.

The coherence lifetime is comparably longer for the cases, i.e.,when the nuclei had similar hyperfine coupling magnitudes, being independent of whether these coupling constants were both 107 Hz or 104 Hz (although it has been found to be slightly shorter for the case in which the coupling constants are both large). However, the coherence lifetime, is much shorter here, even if each of the two nuclei posses a very different coupling strength. The reason behind this is caused by the dependence of the transition frequencies on various combinations of the hyperfine coupling constants and the magnetic field strength. Thus, in case of both nuclei having a different magnitude of coupling constant, this gives rise to a greater proportion of transition frequencies with correspondingly fast transition rates. Finally, it is perhaps possible to conclude that "the coherence lifetime is sensitive to a varied nuclear environment rather than being solely dependent on the strength of the hyperfine coupling". But, it should be noted that the magnitude of such coupling constants becomes an important factor, whenever taken with respect to the rate of singlet-triplet oscillation.

Adams *et al.* did observed in their experiment that when both coupling constants were too small, the singlet and triplet 1 states oscillations vary slowly but the triplet 2 and triplet 3 states do not oscillate at all. In contrast, whenever the radical pair of avian magnetoreception has a recombination lifetime, much smaller than milliseconds, then the singlet would not have time to oscillate at all which points out to the importance of the role of magnitude of the coupling constants in the functioning of the radical pair and hence it is sufficient to have one nucleus with larger coupling constant for the oscillation to be ensured.

The open quantum systems approach to the radical pair mechanism presented by Adams *et al.* demonstrates a novel theoretical approach (the collective coupling approach) as well as a novel numerical result (variation rather than hyperfine coupling strength magnifying dissipative effects). Finally, some tentative conclusions to be made with respect to the radical pair mechanism, developed by Adam's group is

- Firstly, the collective coupling simplify calculations for increasing numbers of nuclei and shows that when the radicals in the pair interact with even numbers of nuclei the coherence lifetime of the pair is extended. This suggests that for a hyperfine environment consisting of identically coupled nuclei, the radical pair coherence might be preserved by the case in which an even number of nuclei give the dominant hyperfine interaction.
- Secondly, the results of the non-collective approach suggest that the coherence lifetime does not simply depend on the magnitude of the hyperfine coupling constant but rather on a combination of coupling strengths which might be optimized to offset dissipation and extend the coherence lifetime.

As far as the biological context is concerned, one of the advantages of the above described model is that the master equation is formulated in terms of transition frequencies. If the model is to be applied to specific biological systems then it should be tailored to address the experimental evidence associated with these systems.

For instance, avian magnetoreception has been postulated to rely on the radical pair mechanism and it has been demonstrated that birds are disoriented in oscillating radio-frequency fields.[41, 101, 102] This would be the case if the frequency of the applied field corresponded to one of the various energy-level splittings of the hyperfine interaction. The most recent of such studies introduced two non-overlapping frequency bands from 20 kHz to 450 kHz or from 600 kHz to 3 MHz[102] and demonstrated the birds' disorientation under both, thus proving that this effect is unlikely to be the result of a single specific frequency or confined to one part of the radio frequency spectrum. That such a broad range of frequencies could cause resonance effects has already been suggested by the model of the radical pair developed by Adams *et al.*'s group. As discussed previously, the transition frequencies in the present case, follow from the external magnetic field and the hyperfine coupling constant. For example,

$$\omega_q^T = \gamma_e B_0 + \lambda_h j$$

Thus, choosing the hyperfine coupling constant to be $\lambda_h = 30M$ Hz, resulted in transition frequencies ranging from 1–100 MHz. Added to this, if, as it is estimated, the radical pair interacts with 10 nuclei, this would give 50 different transition frequencies across a range from kHz to MHz. This is the case if a single value for λ_h is chosen. If it is taken into consideration that each radical might interact with each of these 10 nuclei with differing strengths and a range of λ_h then the number of possible transition frequencies runs into the hundreds. Thus it might be conceivable that the complexity of the hyperfine environment does allow for effects across such a broad range of frequencies. If it is conceivable that birds employ quantum effects through the radical pair mechanism in something as integral to their survival as migration, then it is not too much of a leap to consider that other complex species might also make use of radical pairs. It has recently been shown that the molecule in which the radical pair mechanism manifests in birds is the flavoprotein cryptochrome.[106,107] Cryptochrome is also found in the human retina[108] and it stands to reason that it might be employed through a mechanism similar to that of avian cryptochrome. It would thus be extremely useful to develop a model by which this mechanism might be investigated, a model that is both simple enough to simulate for differing hyperfine environments but not so simple as to lose biological relevance.

5.4.11 *The Avian Compass: Role of Magnetic Field & Cryptochrome in the Navigation of Migratory Birds*

Historically, migratory birds appear to be the most studied class of animals which posses an intrinsic magnetic compass. The typical strength of the geomagnetic field is 0.5G ($50mT$), putting severe limitations on possible physical mechanisms of magnetoreception. The magnetic compass of birds is embedded in the visual system and it has been hypothesized that the primary sensory mechanism is based on a radical pair reaction. In the quest for explaining the origin of this sense, two models have attracted much attention, one involving iron mineral structures[109,110,113,114] in the bird's eye.

The idea behind the latter mechanism, referred to as the radical pair mechanism,[22,24,92,117] is that in the course of a photochemical reaction in the retina, a pair of reactive radicals is produced, the reaction yield of which is influenced by the orientation of the bird with respect to the geomagnetic field. This, in turn modulates visual perception.[43,65,75,105] It is thought that when the bird, by moving its head, changes the angle between its head and the Earth's magnetic field; it then generates a moving visual impression that reveals the external magnetic field.[43] Consistent with this suggestion, studies have found that at least some migratory birds use head-scanning behavior to detect magnetic compass information.[102] Most likely, both types of magnetic sensing mechanisms are realized side-by-side also in animals. Importantly, the avian compass is an inclination compass, sensitive only to the inclination of the Earth's magnetic field lines and not to their polarity.[27] Not only that the avian compass has been found to be highly sensitive to the strength of the ambient magnetic field, requiring a period of acclimation before orientation. Any kind of orientation can occur at intensities differing from that of the natural geomagnetic field.[97] Low-intensity radio frequency radiation affects orientational behavior of bird[27,70] as expected for radical pair processes.[118] The avian compass is light dependent and, normally requires light in the blue-green range for the proper function.[24c–e,35]

5.4.12 *Cryptochrome*

The flavoprotein CRYPTOCHROME (CRY) is now generally believed to be a magnetosensor, providing geomagnetic information via a quantum effect on a light-initiated radical pair reaction. Although there exists considerable physical and behavioral data to support this view, the precise molecular basis of animal magnetosensitivity remains still frustratingly unknown. A key reason for this is due to the difficulty in combining molecular and behavioral biological experiments together with the sciences of magnetics and spin chemistry. In 2000, Ritz first suggested that the blue-light (BL)-sensitive protein CRYPTOCHROME (CRY) might be the elusive magnetoreceptor in magnetically sensitive organisms.[43] The flavoprotein CRYPTOCHROME (CRY) is now generally believed to be

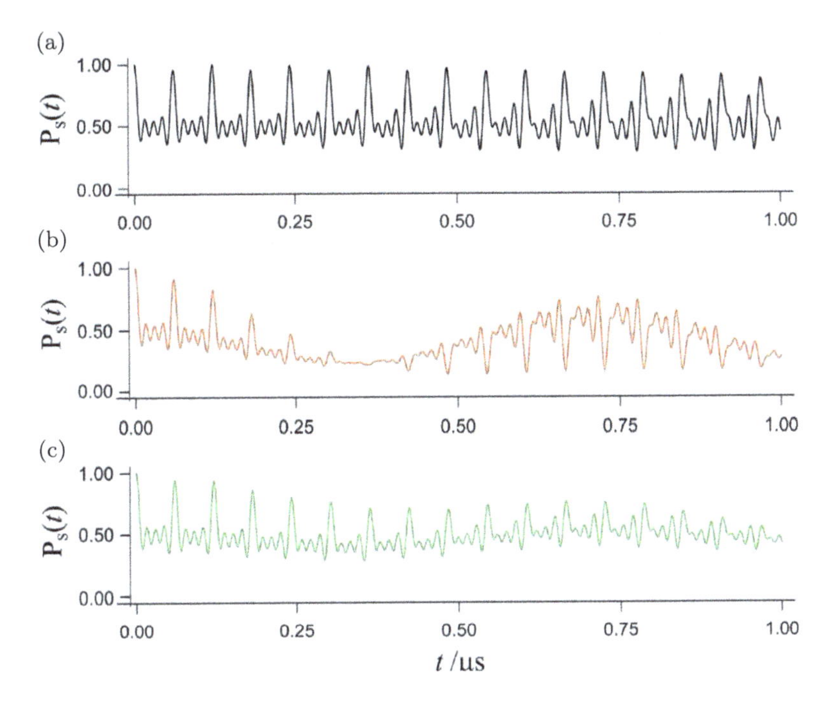

Figure 5.7. Interconversation of the singlet and triplet states of a model radical pair. The fraction of radical pairs n the singlet state, $P_S(t)$, has been calculated as a function; of time ($0 \leq t \leq 1$ μs). (a) In absence of an external magnetic field. (b) In the presence of a 50 μT external magnetic field. (c) A in panel (b) but with magnetic field rotated by 90°. This model radical pair contains two nuclear spins in one radical (with hyper fine tensors appropriate for N5 and N10 in $FAD^{\dot{-}}$ and no nuclei in the other radical. For (b), the magnetic field is at an angle of 60° to the symmetry axis of the hyperfine interactions. chemical reactions and spin relaxation were note included.

a magnetosensor, providing geomagnetic information via a quantum effect on a light-initiated radical pair reaction. The spin dynamics of which are fundamentally quantum mechanical in nature. Unless the electron spins are strongly exchange-coupled, the initial singlet state is a coherent superposition of the spin eigenstates of the pair resulting in **oscillatory time-dependence at frequencies corresponding to eigenvalue differences** (Figure 5.1).

Whilst there has been considerable amount of physical and behavioral data to support this view, the precise molecular basis of animal magnetosensitivity remains frustratingly unknown. A key reason for this is the difficulty in combining molecular and behavioral biological experiments with the sciences of magnetics and spin chemistry. The main requirement for a magnetic field as weak as the Earth's (≈50 mT) to be affective in the spin dynamics is that the coherence persists for at least one period (≈700 ns) of the electron Larmor frequency (≈1.4 MHz). This condition is not particularly restrictive: electron spin relaxation times of radicals in cryptochromes could be as long as ≈1 ms.[134–136] Field-induced changes in the instantaneous probability that the radical pair is singlet or triplet, determine the probability of the pair reacting along spin-selective pathways which translate into reaction product yields depending on the intensity and direction of the magnetic field. As the electron spins are not at thermal equilibrium, it is irrelevant that all the magnetic interactions are orders of magnitude weaker than kBT. Thus, the potential involvement of coherent spin dynamics in the "warm, wet and noisy" environment of a living cell has led to the inclusion of avian magnetic sensing in the currently fashionable field of "quantum biology". Thus, radical pair magnetoreception has been important and popular amongst those, interested in quantum information and quantum computation[46, 48, 88, 128, 142] and has prompted speculations about magnetic sensing devices inspired by the quantum physics of migratory birds.

The photochemistry of CRY is mediated by the photoexcitation of a bound cofactor, flavin adenine dinucleotide (FAD), and a subsequent electron transfer to FAD from a chain of neighbouring tryptophan residues, generating a radical pair (RP) consisting of a flavin semiquinone (FAD•−) and an oxidised Trp (TrpH•+).[119] Here, electron transfer has been proposed to be mediated by a triad of Trp residues in CRY, although, recently a fourth Trp residue has also been implicated, raising the idea of a Trp-tetrad and/or possible redundancy in the pathway. This radical pair (RP) initially forms with correlated spins. As previous work on similar systems has indicated, this could be influenced by an external magnetic field. The generally accepted mechanism requires an RP, generated by photo-reduction of FAD, undergoing interconversion between the singlet and triplet states whereas the relative population of each spin state is altered by exposure to an mgnetic field. But, in the canonical model, the reverse reaction in CRY (electron returning to TrpH•+ from FAD•−) can only occur whenever the RP is in the singlet state. Thus, exposure to an magnetic field (MF) is predicted to influence the probability of occurence of the reverse reaction and thus modulate the half-life of "active" CRY, correlating with the flavin radical.[120]

Formation of such protein harboring blue-light-dependent radical pair formation, cryptochrome, is found localized in the retinas of migratory birds,[105, 121, 122, 124] its effects being used in the visual neuronal pathway. During magnetic compass orientation, a visual brain region, named Cluster N[22, 23, 28, 38, 69] acts as the most active fore-brain region.[105, 123, 129] This activation region requires light perceived through the eyes for its neuronal activation.[68, 123] Importantly, the differences in activation between migratory and nonmigratory birds have been documented[68, 123] whereby European robins with bilateral lesions of Cluster N are found being unable to show oriented magnetic compass guided behavior, but their ability to perform sun-compass and star-compass orientation behavior is unaffected by Cluster N lesions;[65] in contrast, bilateral section of the ophthalmic branch of the trigeminal nerve, leading to the putative iron-mineral-based receptors in the upper beak of European robins did not influence the birds' ability in using their magnetic compass for orientation. A radical pair model in which a light-driven, magnetic field-dependent chemical reaction in the eye of a bird, modulates the visual sense, indeed predicts all of these properties.[34, 117]

In vertibrates, as cryptochromes are the only known class of proteins, currently found in photoreceptor molecules to form radical pairs (RP) upon photo-excitation,[43, 81, 92, 124, 125] representing these only suggested candidate molecules. In them, Cryptochrome 4 (Cry4) is particularly interesting as it has only been found in vertebrates which using a radical pair magnetic compass mechanism is able of functioning as the primary receptor molecules. But, up till now, its detailed structure and localization within the retina has remained not known precisely. However, by now, Cryptochromes are considered as flavo-proteins, sharing moderate amino acid similarity to photolyases but do not show photolyase activity[41, 112, 126] in plants and various animal species, involved in blue-light-dependent pathways and in the circadian clock.[41] Mammalian cryptochromes involved in the circadian rhythm are mainly localized in cell nuclei,[8, 46] whereas magneto-receptive cryptochromes should be located in the cytosol and be associated with membranes and/or the cytoskeleton.[105, 112, 122, 126] To date, four different kind of cryptochromes have been found in the retina of several bird species.[43, 65, 110, 126]

Mouritsen *et al.*[120] demonstrated the presence of the cryptochrome in noticeable quantities within the retinal ganglion cells and in the photoreceptor cells of the retina. Cryptochrome is activated via light-induced electron transfer, probably involving a chain of three tryptophan amino acids and a molecule called flavin adenine dinucleotide (FAD).[22,97,110,125,127] Cryptochrome internally binds FAD in its oxidized state before light activation. In the active (signaling) state, this FAD is transformed to the FADH form. The magnetic field could influence the photoactivation process of cryptochrome acting on the unpaired electron spins as described in detail by Solov'yov *et al.*[110] and Solov'yov & Schulten.[129] Cryptochrome's signaling state has a lifetime of 1–10 ms, as the FADH state slowly reverts to the FAD state.[97,105] In the back-reaction, this may involve the superoxide radical $\dot{O}_2^-$ and could be modulated by the Earth's magnetic field.[41,126] The notion that superoxide might play a role in avian magneto-reception arises principally from the observation that European robins display a resonant disorientation response to weak radio frequency fields at the electron paramagnetic resonance frequency.[41,70]

Importantly, in terms of the radical pair mechanism, the radiation effect can be understood, only if the radicals has no hyperfine interactions, i.e., $T \geq 0.2G$, a condition that excludes most biologically plausible paramagnetic molecules. But, by contrast, $O_2^{\cdot}$, contains no magnetic nuclei, and is ubiquitous in animal cells. Thus, to have a viable radical pair magnetoreceptor, other crucially important conditions are required to be satisfied[120]. For example, to account for the resonant response to a 1.3-MHz radiofrequency field, the radical in question must have a reasonably isotropic g - value, close to 2 and should not undergo electron spin relaxation faster than ≈1 ms. Not only that, the chemistry of the radicals must generate also a spin correlated initial state, permitting appropriate spin-selective reactions.

As discussed by Hogben *et al.*,[127] these constraints are quite stringent for $\dot{O}_2^-$ due to its orbital angular momentum so that in the case of $\dot{O}_2^-$, these requirements can only be satisfied only if the radical is complexed tightly enough to cryptochrome, in order to quench the majority of its orbital angular momentum, also preventing rapid reorientation. But this should be in such a way so that any hyperfine interactions with the cryptochrome are ≥ 0.2G. Another 'essential' requirement for cryptochromes to act as magnetic compass sensors is that they must be orientationally restricted. This might appear a challenging condition because intracellular structures, for example, membranes and cytoskeleton, are dynamic and wobbly. Solovyov *et al.*,[130] studied in detail how much disorder could be permitted for the cryptochrome-based magnetic compass in the eye to remain functional by presenting data which showed that a cryptochrome-based compass can be functional, also at night. However, it is not possible to conclude from their analysis whether constraints on the radical pair mechanism might help to explain why some bird species prefer night navigation over day navigation which uncertainty should be investigated further.

5.4.13 *Effect of Magnetic Field in Cryptochrome Activation-Reaction*

Cryptochrome is brought to its active (signaling) state via the photoreduction process whereas, this could revert to its non-active form if ever the unpaired electron on FADH back-transfers to one of the three tryptophans. Interestingly, this back-transfer process is spin-dependent and can only take place only if the spins of the two unpaired electrons on FADH and the tryptophan are in an overall singlet (antiparallel) state, rather than a triplet (parallel) state. The spins of the unpaired electrons precess about the local magnetic field, which consists of contributions from the surrounding nuclei together with that from the external magnetic field. As each of the electron spins precess, they change their orientation with respect to one another. For example, if the spins begin in a singlet (antiparallel) state, their precession will bring them out of alignment, introducing some triplet contribution. This way, the presence of the external magnetic field can influence not only the precession of the electron spins but also influence thereby the amount of time the spins spend in their singlet state. This, in turn, influences the probability for electron back-transfer and therefore the amount of time that cryptochrome spends in its signaling state.

Computational studies on a model of the photoreduction pathway in cryptochrome have shown that the magnetic field effect described above can have an effect on cryptochrome activation. The model using the cryptochrome's photoreduction pathway[110] makes use of realistic electron transfer rate constants and hyperfine coupling constants. Calculations involving this model predict that the magnetic field effect could alter cryptochrome's activation yield (the amount of time it spends in its active state) by approximately 10% over the range from 0 to 5G. This is of the same order of magnitude as the magnetic field effects observed by experimentalists in Arabidopsis thaliana. The calculations also predict an angular dependence which matches the observed inclination-only magnetic sense of birds. Also, the magnetic field effect has been observed to be highly sensitive to the hyperfine coupling constants for each nucleus; unfortunately, at that time, these hyperfine constants have not yet been known for cryptochrome (the values used in the calculation were those for the highly-similar photolyase). Also, computational constraints limit the number of nuclei that could be included in the calculation. Calculations as described here, strongly suggest that a radical-pair-based magnetic sense involving cryptochrome is feasible, and an important first step in explaining and understanding the magnetic sense of animals.

5.4.14 *Magnetic Field Effect in Cryptochrome Back-Reaction*

By now, it is now an well established fact that the external magnetic field and the hyperfine interaction affect the interconversion between the singlet and the triplet states of the radical pair, depending on the orientation in the Earth magnetic field. Once the FAD cofactor is reduced to the FADH-state, cryptochrome stops signaling, because the reaction $FADH + O_2 \rightarrow FADH^- + O_2$ is considered irreversible. However, before this reaction occurs, the radical pair may separate, namely, if the O_2- radical escapes from cryptochrome's molecular pocket, leaving cryptochrome, still in its signaling state. The escape reaction is governed by the rate constant kb occurring equally likely from either the singlet or the triplet state of the radical pair. Thus, cryptochrome remains in its signaling state until another O_2 radical arrives, and the FADH radical gets another chance to be reduced. The separation and re-encounter of O_2- delay the magnetic field-dependent reaction, shifting it to the millisecond time scale, i.e., the time scale relevant for biological signaling.

Computational studies of the radical pair-based back-reaction in cryptochrome in the presence of a weak (i.e., the Earth's) external magnetic field demonstrated that the duration of cryptochrome's $FADH + O_2^- \rightarrow FADH^- + O_2$ reaction can be changed significantly.[126] Moreover, it was shown that the suggested reaction can act as an inclination compass by demonstrating that a field of 0.5G produces effects that vary significantly during reorientation of cryptochrome. The involvement of the superoxide radical in avian magnetoreception, still has to be verified experimentally, as such involvement has been corroborated, so far only indirectly. The involvement of a radical pair in which one of the radical partners is devoid of the hyperfine interaction in the magnetoreception process is consistent with studies on the effects of weak radio-frequency oscillating magnetic fields on migratory bird orientation, following investigations that had been inspired by earlier theoretical works. Another argument that speaks for the suggested reaction, is its robustness. Indeed, the suggested back-reaction is much simpler than the magnetic field-dependent reaction in cryptochrome activation process[110] and, therefore, is expected to function more reliably in a weak magnetic field.

At a first glance, the involvement of superoxide, O_2- in the magnetic field dependent back-reaction of cryptochrome seems rather controversial. It seems odd that an organism should rely on a toxic substance for a sensory mechanism. However, one should note that superoxide arises naturally in organisms, and is well controlled by superoxide dismutase, which keeps the concentration of superoxide low. This low concentration level, though, is key to the suggested mechanism as the reaction back to the non-signaling state of cryptochrome should be slow, i.e., take about 10 ms, as corroborated by Liedvogel *et al.*[105] Such slow rate of diffusion-controlled encounter is ensured through the low O_2- concentration. At a concentration of $O_2- = 3$ nM, which is tolerable to an organism, the formation of FADH $+ O_2$ — is estimated to take about 1.1 ms, which is indeed the time needed for the suggested mechanism to function optimally.

The studied models of magnetoreception have assumed that the radical pair-forming molecules are rigidly fixed in space, and this *assumption has been a major objection* to the suggested hypothesis. In 2010 Solov'yov *et al.* investigated theoretically how much disorder is permitted for the radical pair-forming, protein-based magnetic compass in the eye to remain functional. Their studies showed that only one rotational degree of freedom of the radical pair-forming protein needs to be partially constrained while the other two rotational degrees of freedom do not impact the magnetoreceptive properties of the protein. The result implies, importantly, that any membrane-associated protein is sufficiently restricted in its motion to function as a radical pair-based magnetoreceptor. Not only that, signaling of cryptochromes may work in the eye by interfering with the normal rhodopsin-based visual process or independently from this process. However, Solov'yov *et al.* showed that for the principle results of the calculations, the exact signaling mechanism is irrelevant. They also discussed after assuming that the currently unknown cryptochrome activation cascade involves amplification steps resulting in a similar degree of amplification as known from the rhodopsin signaling cascade.

In summary, the recent studies[131] follows that the molecular movement of a radical pair through cryptochrome provides deeper insights into the origin of the mysterious magnetic sense. For example, animal cryptochromes have a slightly different structure than the plant protein, which means radical pairs in animal cryptochromes should behave slightly differently than those in the plant protein. However, a related electron transfer in the animal protein could create the stabilizing structural change needed for magnetic sensing. In vertebrates, as the cryptochromes are the only class of proteins that have been found to form radical pairs upon photo-excitation, they are currently the only candidate proteins for light-dependent magnetoreception. Cryptochrome 4 (Cry4) is particularly interesting because it has only been found in those vertebrates that use a 'magnetic compass'. However, its structure and localization within the retina has remained unclear still now. As is already known now, Cryptochromes are flavoproteins that share moderate amino acid having similarity to photolyases but do not show photolyase activity.[105,119,129] In plants and various animal species, they are involved in blue-light-dependent pathways and in the circadian clock.[132] Mammalian cryptochromes involved in the circadian rhythm are mainly localized in cell nuclei,[121,133] whereas magneto-receptive cryptochromes should be located in the cytosol and be associated with membranes and/or the cytoskeleton.[105,119,121,131]

As mentioned by the works of Anju Gunthur *et al.*,[119] according to findings put to date, four different kind of cryptochromes have been found in the retina of several bird species.[105,119,121,135–139] Among them, Cry1a from 'migratory garden warblers' has been shown to form long-lived radical pairs upon photo-excitation,[105] located in the inner and/or outer segments of UV-sensitive photoreceptor cells[121,136] whereas Cry1b was found in the cytosol of ganglion cells and in the photoreceptor inner segments.[87] Cry2 seems to be widespread and nucleic, i.e.,it can be stated as almost certainly a clock protein.[121] Especially, Cry4 has been reported to be located in more or less every cell within the retina of domestic chicken and feral pigeons.[137,162] Interestingly, it has been suggested that Cry4 is either a 'common housekeeping gene' that does not play a highly specific role in a sensory system, or that the 'antibodies used, might have been unspecific'. In any case, 'no information on Cry4 location within the retina' of any migratory bird exists. However, Cry4 has been found to be a particularly interesting magnetoreceptor candidate molecule, since

- It has so far only been found in birds, amphibians, and fish,[107] i.e., three of the four animal classes in which magnetically guided behavior is particularly well documented,[107]
- Cry4 is the only cryptochrome that seems to show no clear endogenous circadian oscillation,[107] and

- It undergoes a light-triggered photocycle, involving the formation of a flavosemiquinone radical in chicken.[107]

Mauritzson and his group[122] have documented Cry4 as very strongly expressed in the outer segments of double cones and long wavelength single cones, all over the retina of European robin and chicken. Also, these have been found to be more strongly expressed during the migratory season than in the non-migratory season, especially, in the case of night-migratory European robins. Even though there is no clear evidence showing which of the four cryptochromes — if any — is the primary magnetoreceptor in night-migratory songbirds, several properties of 'erCry4' indicate it's difference from other cryptochromes. Among them, Cryptochrome 4 (Cry4) has been proved itself particularly important as it has been found only in vertebrates that use a magnetic compass. However, its structure and localization within the retina has still remained unknown.

Importantly, so far, Cry4 has only been found in species showing magnetoreceptive behavior even though, there has been no informative indications that it operates as a clock protein in the control of circadian rhythms. However, after sequencing night-migratory European robin (Erithacus rubecula) Cry4 from the retina, Anju Gunthur *et al.* predicted the currently unresolved structure of the erCry4 protein, suggesting that erCry4 should bind Flavin and, most probably, is going to bind FAD and in this way, is expressed at the seemingly most suitable location, to become primarily, a light-dependent, radical pair-based magnetoreceptor in birds. They also found that Cry1a, Cry1b, and Cry2 mRNA display robust circadian oscillation patterns, but Cry4 on the contrary, shows only a weak circadian oscillation. This observation leads this group to opine that Cry4, magnetoreceptor, is most likely the candidate among the cryptochromes. Cry4 protein has been found, expressed specifically in the outer segments of the double cones and long wavelength single cones in European Robins and chickens. Thus, the localization of Cry4 in double cones seems to be ideal for light-dependent magnetoreception.

5.4.15 *Possible Role of Cryptochrome-4 in the processes of Magnetoreception: Double-Cone Localization and Seasonal Expression Pattern in European Robin*

Cryptochrome 4 (Cry4) is particularly interesting because it has only been found in vertebrates that use a magnetic compass. However, its structure and localization within the retina has not yet been known in a conclusive way. Anja Gunthur *et al.*, sequenced night-migratory European robin (Erithacus rubecula) Cry4 from the retina and predicted the currently unresolved structure of the erCry4 protein, suggesting that erCry4 should bind Flavin, besides which Cry1a, Cry1b, and Cry2 mRNA display robust circadian oscillation patterns, except Cry4, showing a weak circadian oscillation only. After extensive studies regarding the comparison of the relative mRNA expression levels of the cryptochromes in the spring and autumn migratory seasons with that of the non-migratory seasons in European robins and domestic chickens (Gallus gallus), they did found the Cry4 mRNA expression level in European robin retinae, but not in chicken retinae, to be significantly higher during the migratory season, compared to that in non-migratory seasons. Also, importantly, Cry4 has been found to be specifically expressed in the outer segments of the double cones and long wavelength single cones, in the case of European robins and chickens.

Thus, a localization of Cry4 in double cones seems to be ideal for light-dependent magnetoreception. Considering all of the data, in their work, especially including its localization within the European robin retina, Gunthur *et al.* opined that it is likely binding of Flavin, and its increased expression during the migratory season in the migratory bird but not in chicken, Cry4 could be the *magneto-receptive protein.* The retinal localization pattern, as discovered by the above mentioned group, suggests that Cry4 fulfills a specialized function in a specific sensory pathway. Not only that, as the cryptochromes are currently the only class of vertebrate proteins known to form radical pairs upon photo-excitation,[24, 43, 105, 119, 131] Cry4 could be a magnetoreceptive protein and its location in the outer segments of the double cones makes it an even likelier magnetoreception candidate, because the hundreds of parallel cell membranes in photoreceptor outer segments would facilitate an, at least, partial alignment of the cryptochromes, relative to one another,[104, 105, 119, 136] and a magnetic compass sensor most likely require the correlated responses of many radical pairs.

Gunthur *et al.* have also documented Cry4 as very strongly expressed in the outer segments of double cones and long wavelength single cones all over the retina of European robin and chicken and in the night-migratory European robins, more strongly expressed during the migratory season than during the non-migratory season. Even though, especially, in the case of night-migratory song birds, up till now, there is no clear evidence as such, showing which of the four cryptochromes — if any — is the primary magnetoreceptor, however, several properties of **erCry4** indicate

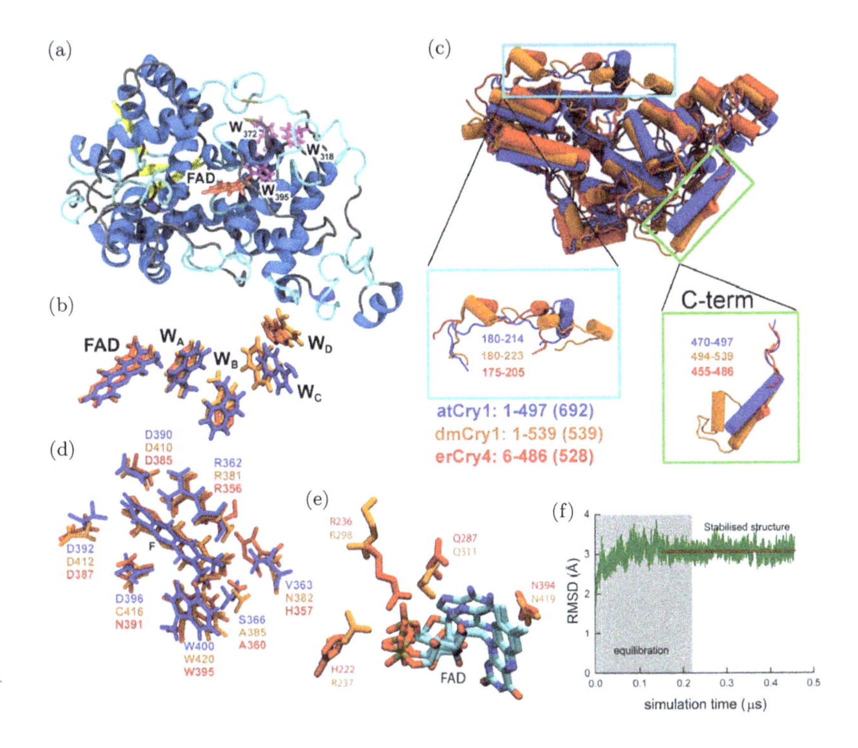

Figure 5.8. Equilibrated Structural Model of erCrys (a) The secondary structure of the erCry4 model obtained after extensive MD equilibration, with the isoalloxazine part of the FAD (red) and the conserved tryptophan triad (magenta). (b) Superimposed molecular structures of the FAD isoalloxazine part and the tryptophan triad from at Cry1 (blue), tetrad from dmCry1 (orange), and the tetrad from erCry4 model (red). (Structural alignment of atCry1 (blue), dmCry1 (orange), and the erCry4 model (red), suggesting two major structural differences in the three proteins: (1) the surface-exposed parts of the proteins feature one large helix in atCry1, whereas dmCry1 and the erCry model feature three and two smaller helices respectively (blue box), and (2) atCry1 and dmCry1 feature one helix at the end of the PHR domain, while the erCry-4 model features a shorter helix in the same structural area (green box). One reason for this discrepancy might be that dmCry1 is the only cryptochrome where the C terminus is resolved in the crystal structure, whereas the at Cry1 structure and the erCry 4 model only include the PHR domain. To include the complete structure of atCry1 and erCry4 would change the C-terminal part, which, unfortunately, cannot be predicted reliably yet. (d and e) Comparison of the amino acid residues surrounding the FAD in the binding pocket of atCry1 (d) and dmCry1 and erCry4 (d and e). (f) Homology models of proteins obtained directly from web servers can rarely be used as reliable structures for further analysis since these models are usually not stabilized. The stability of the homology model can be established and probed by MD simulations, as done here for the model of erCry4. (f) shows the root- mean-square deviation (RMSD) of the protein backbone atoms computed relatively to the initial structure of the model for erCry during the entire 0.5-us MD simulation performed on our erry-4 model. The gray area indicates the equilibration interval, while the remaining simulation was used to justify the stability of the model (production simulation) See also Figures S1 and S2.

its differences from other cryptochromes. In fact, so far, Cry4 has only been found in species showing magnetoreceptive behavior but they did not observe any indication whether it operates as a clock protein in the control of circadian rhythms or not. After detailed studies, *Mauritzson's group opined that erCry4 is probably going to bind FAD, expressed at the seemingly most suitable location, to be a primary, light-dependent, radical-pair-based magnetoreceptor in birds.* They concluded out of their experiments by suggesting a structural predictions that erCry4 will bind FAD.

5.4.16 *Application of Quantum Physics in Bird Navigation Based on Radical Pair (RP): Hore's Approach*

By now, Radical pair recombination reactions are known to be sensitive to extremely weak magnetic fields and can therefore be said to function as molecular magnetoreceptors. The recently developed classic example is a carotenoid-porphyrin-fullerene (C•+PF•−) radical pair that has been shown to provide a "proof-of-principle" for the operation of a chemical compass.[24] Previous simulations of such radical pair have employed semi-classical approximations, routinely applicable to its 47 coupled electronic and nuclear spins. However, calculating the exact quantum mechanical spin dynamics, presents a significant challenge and has not been successful until now in solving that challenge quite successfully. Recently, Lachlan P. Lindoy *et al.*[200] applied a developed method to perform numerically converged simulations of the C•+PF•− quantum mechanical spin dynamics, including all coupled spins. Importantly, a comparison of these quantum mechanical simulations with various semi-classical approximations reveals that while it is not perfect, the best semi-classical approximation does capture essentially all of the relevant physics in this problem.

The radical pair of magnetoreception has found a place in the emerging field of Quantum Biology[147, 189, 190] on the strength of the absolute requirement that the radical pair must be in a coherent superposition of the quantum states of the two electron spins. In fact, the initial electronic singlet state of the radical pair is quantum mechanically entangled [although the entanglement, as such, confers no advantage in terms of the general operation of the compass,[159] nor is it essential for the existence of the spike]. Recently it has been showed by Hishcock *et al.* that the spin dynamics of long-lived radical pairs in weak magnetic fields can be described by a semiclassical approximation that becomes increasingly accurate as the number of nuclear spins is increased.[153,174] They did put in many more nuclear spins, with realistic couplings to the electron spins, specifically, assuming the spin coherence of the paired radicals to be much longer lived than in previous models. As a result, they observed a marked "spike" in the yield of the signaling state whenever the spin coherence times in their model were set to be longer than a few microseconds, produced by spin-selective reactions of the radical pairs. Interestingly, as the spin lifetime is prolonged from 1 μsec toward 100 μsec, the spike 'emerges, strengthens and narrows'.

If the behavior of a realistic radical pair magnetoreceptor can satisfactorily be modeled in terms of classical rather than quantum oscillations, then arguably it does not belong under the quantum biological umbrella. However, the spike discussed here is undeniably a quantum effect, arising from the mixing of states associated with avoided energy-level crossings, not captured by the semiclassical theory. In this sense, radical pair magnetoreception may be more of a quantum phenomenon than hitherto realized. Such a crucial feature could provide birds with directional information in explaining their navigational behavior with sufficient precision. Hore *et al.* pointed to a crucial observation, i.e., while the exact chemical signal produced in birds, is unknown, it is almost certainly a form of the cryptochrome protein, having a different shape, allowing the protein to have different interactions with other proteins, thus starting the 'neural signaling process'. The long-lived spin coherences provide enough time "*for the interconversion of the singlet and triplet states of the radical pairs, and therefore the yield of the signaling state, to respond sensitively to the direction of the geomagnetic field*" whereas, as stated before, radical pairs, not going forward to the signaling state,- revert to the original protein structure.

As has already been explained, the process of magnetoreception in radical pair mechanism has been established itself quite a strongly emerging field in Quantum Biology as the absolute requirement, stating that the radical pair must be in a coherent superposition of the quantum states of the two electron spins. In fact, the initial electronic singlet state of the radical pair is quantum mechanically entangled [even though the entanglement, as such, confers no advantage in terms of the general operation of the compass, nor is it essential for the existence of the spike]. The group led by Hore showed that the spin dynamics of long-lived radical pairs in weak magnetic fields can be described by a semiclassical approximation that becomes increasingly accurate as the number of nuclear spins is increased.[153,154] If the behavior of a realistic radical pair magnetoreceptor can be satisfactorily modeled in terms of classical rather than quantum oscillations, then arguably it does not belong under the quantum biological umbrella. But, the spike discussed here is undeniably a quantum effect, as they arise from the mixing of states associated with avoided energy-level crossings, not captured by the semiclassical theory. In that sense, radical pair magnetoreception may be more of a quantum phenomenon than hitherto realized.

Finally, — to reveal the inner working of the avian compass, a novel aspect, i.e., another *crucially important and significant step* in this direction, has been pointed out and explained by Hore and his group,[26] i.e., though coherence phenomena are not expected to be long-lived in most biological systems, birds could have evolved a molecular system that do not rapidly lose coherence, i.e., the source of the spin relaxation that would destroy the coherence, is the stochastic fluctuations of the radicals in their binding sites in the protein so that, at this juncture, if mutations in the proteins make those motions both fast and relatively of low amplitude, then the loss of coherence should be inefficient, thus allowing the spin-correlated state to persist, perhaps, for $(5–10)\mu$m, which would be sufficient to produce the spikes in the yield of the signaling state.

As proposed and developed by the very recent developments,[143] the most likely mechanism of the magnetic compass sense in migratory songbirds relies on the **coherent** spin dynamics of pairs of photochemically formed radicals in the retina. The question arising here is whether it is possible to account only for the coherent spin dynamics using quantum mechanics. Importantly, Hore and his collaborators[135] established that semiclassical approximations to the spin dynamics of radical pairs can only provide a satisfactory description of the anisotropic product yields in absence of no electron spin–spin coupling. However, having such a situation is unlikely to be consistent with a magnetic sensing function. Although these methods perform reasonably well for shorter-lived radical pairs with stronger electron-spin coupling, the accurate simulation of anisotropic magnetic field effects relevant to magneto-reception seems to demand for the full quantum mechanical calculations. Spin-conserving electron transfer reactions are thought to result in radical pairs whose near-degenerate electronic singlet and triplet states inter-convert coherently as a result of hyperfine exchange and dipolar couplings and crucially for a compass sensor, i.e., Zeeman interactions with the geomagnetic field. Thus, the yields of the reaction products can be influenced by magnetic interactions, even a million times smaller than $k_B T$.

Another crucially important and already an established fact is that the magnetic compass sense of migratory birds is thought to rely on magnetically sensitive radical pairs formed photochemically in cryptochrome proteins in the retina. An important requirement of such hypothesis needs electron spin relaxation to be slow enough for the Earth's magnetic field in having a significant effect on the coherent spin dynamics of the radicals. General assumption in this regard is that the evolutionary pressure has led to protection of the electron spins from irreversible loss of coherence so that the underlying quantum dynamics can survive in a noisy biological environment.[182] Kattnig *et al.*[142] (2016) addressed this question for a structurally characterized model cryptochrome with the aim of assessing the effects of spin relaxation on the performance of the protein as a compass sensor, also being expected to share many properties with the putative avian receptor protein. Both flavin–tryptophan and flavin $-Z^{\cdot}$ radical pairs are studied ($Z^{\cdot}$ is a radical with no hyperfine interactions). Relaxation is considered to arise from modulation of hyperfine interactions by liberating motions of the radicals and fluctuations in certain dihedral angles.

The potential involvement of coherent spin dynamics in the "warm, wet and noisy" environment of a living cell has led to the inclusion of avian magnetic sensing in the currently fascinating, as well established field of quantum biology[145, 146, 148, 149] 137–141. The biophysical mechanism by which all of migratory songbirds extract directional information from the Earth's magnetic field is something of a mystery.[150] Very recently developed, most promising hypothesis involves light-induced, magnetically sensitive chemical reactions in the birds' retinas.[143] By now, it has already been established that photo-excitation of the Flavin adenine dinucleotide (FAD) chromophore in cryptochrome proteins is responsible for triggering a sequence of intra-protein electron transfers along a chain of three/four tryptophan (Trp) residues, producing a FAD–Trp radical pair in a highly non-equilibrium (initially singlet) electron spin state.[143]

Electron-nuclear hyperfine interactions within the radicals, break the symmetry of the two electron spins and drive coherent interconversion of the singlet and triplet states at MHz frequencies. The corresponding oscillations being sensitive to the strength and the direction of an external magnetic field via the electron Zeeman interaction, happen in the anisotropic components of the hyperfine and dipolar interactions. If the singlet and triplet states of the radical pair react to give distinct products, the direction of the geomagnetic field vector with respect to the bird's head direction could be encoded in the quantum yield of one of the products, acting as a signaling state.[119] However, it should be noted that even though the radical pair (RP) mechanism is an established phenomenon supported by hundreds of laboratory-based studies of organic radical reactions with a well developed theoretical basis,[154] the problem, i.e., whether it provides birds with a magnetic compass bearing, is less clear.

Blue-light irradiation of purified cryptochromes produces magnetically sensitive FAD–Trp radical pairs, appearing fit for this purpose as magnetoreceptors[151,172] and hence a cryptochrome-based radical pair sensor would be compatible. Importantly, several corresponding observations suggest the avian compass to be light-dependent, detecting the inclination rather than the polarity of the Earth's field which can be disrupted by radio frequency electromagnetic fields. Hore *et al.*[66] opined that spin dynamics of radical pairs are fundamentally quantum mechanical. Unless the electron spins are strongly exchange-coupled, the initial singlet state, being a coherent superposition of the spin eigenstates of the pair, becomes oscillatory, time-dependent at frequencies, corresponding to eigenvalue differences.

Now, as already mentioned earlier, the crucial requirement for a magnetic field as weak as that of Earth's (50 mT) to affect the spin dynamics is that the coherence must persist for at least one period ($\approx$700 ns) of the electron Larmor frequency ($\approx$1.4 MHz). But, this condition is not particularly restrictive, meaning that the electron spin relaxation times of radicals in cryptochromes could be as long as $\approx$1 ms.[142,143,182] Including that, field-induced changes in the instantaneous probability, clarifying whether the radical pair is singlet or triplet, determine the probability of the pair, reacting along spin-selective pathways. This is translated then into reaction product yields, depending on the intensity and direction of the magnetic field. Importantly, as the electron spins are not at thermal equilibrium, it is irrelevant whether all the magnetic interactions are within the orders of magnitude weaker than kBT.

Radical pair magnetoreception has proved itself to be popular amongst those, interested, especially in quantum information and quantum computation, thus prompting speculations about magnetic sensing devices inspired by the quantum physics of migratory birds. But, one vital as well as crucial question always remains, i.e., *how far "quantum" is this mechanism?* Because, even though, the photochemical formation of the radical pair in a non-equilibrium electronic spin state is undeniably important, if the electron spins were always at thermal equilibrium (25% singlet, 75% triplet), weak Zeeman interactions *would never be detectable* in the reaction product yields. Then the same old question survives, i.e., whether one can *only quantitatively*' account for the effects of coherent singlet-triplet interconversion using quantum mechanics. Importantly, it should be mentioned in this context, in the 1970s, Schulten and Wolynes (SW) proposed a semiclassical approximation in which the electron spin in each radical precesses around a hyperfine-weighted sum of classical nuclear spin vectors, assumed to be fixed in space.[172]

In an extraordinarily and far-sighted proposal, it was Klaus Schulten who suggested that hyperfine interactions between electron and nuclear spins combined with Zeeman interactions with the geomagnetic field in molecules generating photoinduced radical pairs could form the basis of a chemical magnetic compass, with the help of a biomagnetic sensory mechanism, based on magnetic field modulated coherent electron spin motion. This approach has been proved to be in excellent agreement with exact quantum simulations for a sufficiently large number, N, of nuclear spins. For the radical pair hypothesis to be relevant for avian magnetoreception, one key requirement is that molecules having the needed biophysical characteristics, exist in the eye of migratory birds and the only vertebrate proteins, known to form radical pairs upon photoexcitation are the cryptochromes.

5.4.17 *Quantumness of Radical Pair Mechanism in Magnetoreception*

Bradlaugh *et al.*[79] discussed in his work in detail about geomagnetic information via a quantum effect on a light-initiated radical pair reaction, i.e., the role played by the flavoprotein CRYPTOCHROME (CRY), now generally, believed to be a magnetosensor. By now, we are already aware that the radical pair mechanism of magnetoreception has been the emerging field of Quantum Biology, based on the strength of the absolute requirement that the radical pair must be in a coherent superposition of the quantum states of the two electron spins. In fact, very recently, the concept that the initial electronic singlet state of the radical pair is quantum mechanically entangled [although the involvement of the entanglement, believed to be a satisfactorily valid approach, earlier]– has been conferred as *having no advantage* in terms of the general operation of the compass,[159] nor is it essential for the existence of the spike. Hishcock *et al.*[26] showed that the spin dynamics of long-lived radical pairs in weak magnetic fields can be described by a semiclassical approximation that becomes increasingly accurate as the number of nuclear spins is increased.[153, 154] If the behavior of a realistic radical pair magnetoreceptor can be satisfactorily modeled in terms of classical rather than quantum oscillations, then arguably it does not belong under the quantum biological umbrella.

However, the spike discussed here is undeniably a quantum effect, arising from the mixing of states associated with avoided energy-level crossings, and is not captured by the semiclassical theory. In this sense, radical pair magnetoreception may be more of a quantum phenomenon than hitherto realized. Hore and his group,[23] in 2016, presented an extension to the SW approach[172] in which each individual nuclear spin is allowed to have a precession around the electron spin to which it is coupled.[153] Their method approaches quantitative agreement by applying quantum mechanics for large numbers of coupled nuclear spins, and, unlike the SW theory, accurately captures the effects of weak external magnetic fields on long-lived radical pairs with isotropic hyperfine couplings. Although, it has been tested against exact quantum mechanical calculations, for systems with anisotropic hyperfine interactions,[158] but not been tested for systems in which both anisotropic interactions and electron spin coupling are involved.

Comparison of quantum and (semi)classical calculations is a standard way of assessing "quantumness" in many of other contexts.[75] If classical motions of classical spin vectors could adequately describe the operation of a radical pair compass sensor then avian magnetoreception would not be a quantum biological phenomenon. In the following report, we have followed Hore & his group's most recent work exclusively though in short form, with the aim of evaluating the accuracy with which three semiclassical methods as mentioned below, reproduce quantum mechanical calculations of the reaction product yields of cryptochrome-based radical pairs.

5.4.18 *Semi-classical Methods: Quantum Mechanical Calculations of the Reaction Product Yields of Cryptochrome-Based Radical Pairs*

Mauritzen, famously commented that '*The biophysical mechanism by which migratory songbirds extract directional information from the Earth's magnetic field is something of a mystery*[150]'. We hereby, will try to give an outline of the theory, as proposed and applied by Hore's group as well, in describing the radical pair (RP) spin dynamics. One of the most popular theories that incoming photons excite light-sensitive proteins called cryptochromes in the retina of a bird's eyes, is based on the results, i.e., in an electron being transferred between two molecules within the protein.

Up till recently, such most promising hypothesis is based on the involvement of light-induced, magnetically sensitive chemical reactions, occurring in the birds' retinas.[119] Photo-excitation of the flavin adenine dinucleotide (FAD) chromophore in cryptochrome proteins is thought to trigger a sequence of intra-protein electron transfers along a chain of three or four tryptophan (Trp) residues which produce a FAD–Trp radical pair in a highly non-equilibrium (initially singlet) electron spin state.[151] Electron-nuclear hyperfine interactions within the radicals break the symmetry of the two electron spins and drive coherent interconversion of the singlet and triplet states at MHz frequencies. These oscillations are sensitive to the strength and the direction of an external magnetic field via the electron Zeeman interaction and the anisotropic components of the hyperfine and dipolar interactions. If the singlet and triplet states of the radical pair react to give distinct products, the direction of the geomagnetic field vector with respect to the bird's head direction could be encoded in the quantum yield of one of the products acting as a signaling state.

Importantly, radical pair recombination reactions are known to be sensitive to extremely weak magnetic fields and can therefore be said to function as molecular magnetoreceptors. The recently developed classic example is a carotenoid-porphyrin-fullerene (C•+PF•−) radical pair that has been shown to provide a "proof-of-principle" for the operation of a chemical compass.[21] Previous simulations of this radical pair have employed semiclassical approximations which are routinely applicable to its 47 coupled electronic and nuclear spins. However, calculating the exact quantum mechanical spin dynamics presents a significant challenge and has not been possible until now. Here, Lachlan P.Lindoy *et al.*[149] applied a recently developed method to perform numerically converged simulations of the C•+PF•− quantum mechanical spin dynamics, including all coupled spins. A comparison of these quantum mechanical simulations with various semi-classical approximations reveals that, while it is not perfect, the best semi-classical approximation does capture essentially all of the relevant physics in this problem.

In recent times, the most likely mechanisms of inner dynamics of the magnetic compass sense in migratory songbirds relies on the coherent spin dynamics of pairs of photochemically formed radicals in the retina. Blue-light irradiation of purified cryptochromes produces magnetically sensitive FAD–Trp radical pairs which appear to be fit for purpose as magnetoreceptors.[110,143] A cryptochrome-based radical pair sensor would be compatible with the observations that the avian compass is light-dependent[24e,26] detects the inclination, rather than the polarity of the Earth's field[24(b,c)] and can be disrupted by radio frequency electromagnetic fields.[36,37a] Thus, spin-conserving electron transfer reactions are supposed to result in radical pairs whose near-degenerate electronic singlet and triplet states inter-convert coherently as a result of hyperfine exchange, and dipolar couplings and, crucially- for a compass sensor, i.e., Zeeman interactions with the geomagnetic field.

Thus the yields of the reaction products can be influenced, following this process by magnetic interactions with value, by a million times smaller than kBT. But at this point, the question arises that whether one can account for the coherent spin dynamics by only using quantum mechanics. According to Hore and his group's[135] (2020) opinion, it is only the semiclassical approximations to the spin dynamics of radical pairs which, according to their proposal, can provide a satisfactory description of the anisotropic product yields. However, when there is no electron spin–spin coupling, a situation, quite unlikely to be consistent with a magnetic sensing function. What they concluded is that although these methods perform reasonably well for shorter-lived radical pairs with stronger electron-spin coupling, the 'accurate' simulation of anisotropic magnetic field effects relevant to magnetoreception 'seems' to require '*full quantum mechanical calculations*'.

5.4.19 *Spin Dynamics Related to Radical Pairs; Fay & Hore's Approach (2020): Spin Hamiltonian and Dynamical Equations*

The spin dynamics, related to the radical pairs are fundamentally quantum mechanical. Unless the electron spins are strongly exchange-coupled, the initial singlet state is a coherent superposition of the spin eigenstates of the pair which, then result in oscillatory time-dependence at frequencies corresponding to eigenvalue differences.

The main requirement for a magnetic field as weak as the Earth's ($\approx$50 mT) to affect the spin dynamics is that the coherence should persist for at least one period ($\approx$700 ns) of the electron Larmor frequency ($\approx$1.4 MHz). This condition is not particularly restrictive: electron spin relaxation times of radicals in cryptochromes could be as long as ≈1 ms[136(a&b)].

As a next, field induced changes in the instantaneous probability whether the radical pair is singlet or triplet, determine the probability of the pair which reacts along spin-selective pathways, translating into reaction product yields that depend on the intensity and direction of the magnetic field. It is quite important to be mentioned here that as the electron spins are not at thermal equilibrium, it is irrelevant whether all the magnetic interactions are of the orders of magnitude weaker than kBT. Again, in the present case of the two resulting free-radical molecules, each having an unpaired electron but produced simultaneously, the spins of the two electrons are correlated, forming a coherent quantum state, get affected by weak external magnetic fields, such as the geomagnetic field. This interaction affects the chemical reactivity of the free-radical molecules, including the signals ultimately that they send to the brain. But, even if this sounds plausible in theory, no one has explained exactly how this process allows the Earth's magnetic field to be measured to within 5° precision.

To be precise, in fact, the potential involvement of the spin dynamics and that also in association with the "warm, wet and noisy" environment of living cells have led to the inclusion of avian magnetic sensing, in the arena of "quantum biology".[137–141] Very recently, theory related to radical pair magnetoreception has turned into an interested arena in quantum information and quantum computation,[6,40,81,120,134] as a result of which, the development has prompted speculations about magnetic sensing devices inspired by the quantum physics of migratory birds. Comparison of quantum and (semi)classical calculations is a standard way of assessing "quantumness", also, in many other contexts.[152] If classical motions of classical spin vectors could adequately describe the operation of a radical pair compass sensor, then avian magnetoreception would not be a quantum biological phenomenon and this is the vital factor of necessity and importance. This very fact led Hore and his group[135] in evaluating the accuracy so far as the three semiclassical methods, as followed here, are concerned in reproducing quantum mechanical calculations of the reaction product yields of cryptochrome-based radical pairs.

The comparison of quantum and (semi)classical calculations is to be considered a standard way of assessing the so called "quantumness" in other contexts[135] if classical motions of spin vectors could be shown adequately describing the operation of a radical pair compass sensor. In that case, avian magnetoreception would not be considered as a quantum biological phenomenon. Fay *et al.*[153] evaluated in their pioneering works which have proved accurately how three semiclassical methods reproduce quantum mechanical calculations of the reaction product yields of cryptochrome-based radical pair.

5.4.20 *The Radical Pair Spin Dynamics: Spin Hamiltonian and Dynamical Equations*

In the following section, we try to describe the quantum master equation as developed by Fay *et al.* (2020)[135] together with the effective spin Hamiltonian under which the spin density operator is considered to be evolved. This has been followed by describing here the methods, then employed, performing the exact quantum mechanical and approximate semiclassical calculations, in order to avail the product quantum yields of the radical pair reactions, followed by calculating their dependence on the direction of the applied magnetic field. In order to develop spin Hamiltonian and corresponding dynamical equations, the possible contribution, caused by the presence of the anisotropic magnetic field effects on the radical pair dynamics, has also been taken into account. To do this, the following quantum master equation for the spin density operator of the radical pair,[148,80,153,154] i.e., $\hat{\rho}(t)$ can be expressed as,

$$\frac{d}{dt}\hat{\rho}(t) = -\frac{i}{\hbar}[\hat{\mathbf{H}}, \hat{\rho}(t)] - \left\{\frac{k_s}{2}\hat{P}_s + \frac{k_T}{2}\hat{P}_T, \hat{\rho}(t)\right\} \tag{5.22}$$

In the above expression, it is important to note that the notation $[.,.]$ represent commutator, and .,. the corresponding anticommutator whereas k_s and k_t are the first order recombination rate constants for the singlet and triplet pairs[135] respectively. Here, $\hat{P}_s$ and $\hat{P}$ are the projection operators onto the singlet and triplet electronic subspaces, the $\hat{H}$ being the effective spin Hamiltonian for the radical pair. This can be expressed as

$$\begin{aligned}\hat{\mathbf{H}} = {} & g_1\mu_B\mathbf{B}\cdot\hat{\mathbf{S}}_1 + g_2\mu_B\mathbf{B}\cdot\hat{\mathbf{S}}_2 \\ & - J\,\hat{\mathbf{S}}_1\cdot\hat{\mathbf{S}}_2 + \hat{\mathbf{S}}_1\cdot\mathbf{D}\cdot\hat{\mathbf{S}}_2 \\ & + \sum_{k=1}^{N_1}\hat{\mathbf{I}}_{1,k}\cdot\mathbf{A}_{1,k}\cdot\hat{\mathbf{S}}_1 \\ & + \sum_{k=1}^{N_2}\hat{\mathbf{I}}_{2,k}\cdot\mathbf{A}_{2,k}\cdot\hat{\mathbf{S}}_2\end{aligned} \tag{5.23}$$

The first two terms described above, are the Zeeman interactions of the electron spins, i.e., $\hat{\mathbf{S}}_i$, with the external magnetic field, $\mathbf{B}$, the third term being the exchange interaction of the electron spins, where J is the coupling constant, and the fourth term is the dipolar interaction of the two electron spins with dipolar coupling tensor $\mathbf{D}$. Here, μ_B depicts the Bohr magneton and g_i is the g-value of radical i. Throughout their works, Fay and his group[135] used unit-less spin operators so that, the last two terms in equation (5.2) are the hyperfine interactions with the nuclear spins, $\hat{\mathbf{I}}_{i,k}$, in each radical, where $\mathbf{A}_{i,k}$ is the hyperfine coupling tensor for nuclear spin k in radical i.

As an initial case, the radical pair, here, has been considered, to be in a singlet state, having the following initial condition, i.e.,

$$\hat{\rho}(0) = \frac{1}{\mathbf{Z}}\hat{P}_s \tag{5.24}$$

Z being the dimensionality of the full nuclear spin Hilbert space. Next important part is to calculate the quantum yield of the product, formed by the spin-selective recombination of the singlet state, i.e., Φ_s, which can be written as

$$\Phi_S = k_s \int_0^{\infty} P_S(t)dt \tag{5.25}$$

Here, $P_S(t)$ is for the singlet probability at time t,

$$p_S(t) = \mathbf{Tr}[\hat{P}_S\hat{\rho}(t)] \tag{5.26}$$

5.4.21 *The Effects of Semiclassical Approximations to the Spin Dynamics of Radical Pairs for the Model System of Cryptochrome Adopted*

Fay *et al.* (2020)[141] adopted two models of radical pairs, based on the photochemistry of cryptochromes. Among them,

- One is $[FAD^{\dot{-}}TrpH^{\dot{+}}]$ which is magnetically sensitive state of the protein formed by electron transfer from the Trp-triad or tetrad to the photo-excited flavin chromophore
- The other is $[FAD^{\dot{-}}Z^{\cdot}]$, a hypothetical magnetic sensing state of cryptochrome,[36,155] in which $Z^{\cdot}$ is a radical with no hyperfine-coupled nuclear spins.

These have been named as FAD-TrP and FAD-Z respectively.

As presently, the lifetimes, spin-coupling tensors and other properties of the radicals have not yet been known precisely in vivo. Fay *et al.* considered simplified models of these species, following only the basic interests, i.e., in order to find the prime ability of semi-classical approximations, with the aim of capturing the anisotropy of the singlet yield, Φ_s. For this, spin relaxation which generally diminishes the role of quantum mechanical coherence in the spin dynamics, have been ignored. However, it is quite important to state that as such, these models have been likely to pose the largest challenge for semiclassical methods, and, been under great thrust of further research.

To start with, we first consider models of FAD–Trp and FAD–Z in the absence of electronspin coupling, with the number of hyperfine-coupled nuclear spins in each radical truncated at 7 or 11. The hyperfine coupling tensors were taken from reference.[143] Together with this we have extended our consideration in describing models with non-zero electron-spin coupling. The dipolar interaction parameters,[164] ($D = -0.38mT, E = 0$) were calculated for the FAD–TrpC radical pair in Drosophila cryptochrome, and the exchange interaction was taken to be, $J = 0.224mT$. The resulting total coupling tensor, $\mathbf{C} = \mathbf{D} - 2J\mathbf{1}$, where, the value of $\mathbf{C}$ has been taken from

$$C = \begin{pmatrix} -0.382276 & 0.292979 & -0.146796 \\ 0.292979 & -0.652196 & 0.229243 \\ -0.146796 & 0.229243 & -0.309528 \end{pmatrix} mT \tag{5.27}$$

For the comparison, the same C have been used for the hypothetical FAD–Z pair for which J and D are unknown.

Birds seem to use a light dependent radical-pair based magnetic compass for the navigation and orientation, although, the basic sensory mechanisms underlying magnetoreception remain still elusive. Quite importantly, among many of the models, only Fay *et al.* (2020) considered symmetric recombination, where $k_S = k_T = k = 10^6\ \mathrm{s}^{-1}$, together with the g-values taken for both the electron spins to be the free-electron g-value, i.e., $g_1 = g_2 = g_e$. For each model, they have calculated the singlet quantum yield as a function of the angle, θ, between the applied magnetic field and the normal to the plane of the tricyclic aromatic ring system of the FAD radical.

$$\mathbf{B}(\theta) = B(\sin\theta, 0, \cos\theta)^T \tag{5.28}$$

$\approx$ with $B = 1$ mT. This field is approximately 20 times stronger than the Earth's magnetic field ($\approx$50 mT). According to Fay *et al.* the reason they have chosen so, is, simply because of the fact that, it produces larger singlet yield anisotropies which are faster to converge with respect to the number of Monte-Carlo samples both in the semiclassical and quantum mechanical calculations for the FAD–Trp radical pair with electron-spin coupling. However, they found this field strength to be still sufficiently small and that all spin–spin interactions to be highly non-perturbative in nature.

5.5 Application of Quantum Dynamics of FAD-Z in the Theory of Navigation of Bird

5.5.1 *Application of FAD-Z*

Taking the case of FAD-Z, equations can be evaluated by doing the numerical diagonalization of the Hamiltonian, so that the singlet quantum yield can be obtained as,

$$\Phi_s = \frac{k^2}{Z}\sum_{n,m}\frac{|\langle n|\hat{P}_s|m\rangle|^2}{[(\epsilon_n - \epsilon_m)/\hbar]^2 + k^2} \tag{5.29}$$

In the above expression, $|n\rangle$ represents an eigenstate of $\hat{H}$ having the eigen value ϵ_n[17].

Now, in the absence of electron spin coupling in FAD-Trp radical pair, the Hamiltonian is separable into two lower-dimensional single-radical terms, i.e.,

$$\hat{H} = \hat{H}_1 + \hat{H}_2$$

Hence, following such condition, the single-radical spin correlation tensor $R^{(i)}_{\alpha,\beta}(t)$ is possible to evaluate as follows:

$$R^{(i)}_{\alpha,\beta}(t) = \frac{1}{Z_i \mathbf{Tr}_i\left[\hat{S}_{i,\alpha}e^{+i\hat{H}_{it/\hbar}}\hat{S}_{i,\beta}e^{-i\hat{H}_{it/\hbar}}\right]} \tag{5.30}$$

Z_i, here, is the dimensionality of the nuclear-spin Hilbert space for radical i and $Tr_i[\ldots]$ denotes the trace over the full spin Hilbert space of radical i. Now the single probability can be calculated using the above singlet probability by having[153]

$$P_{S(t)} = \left(\frac{1}{4} + \sum_{\alpha,\beta} R^{(1)}_{\alpha,\beta}(t) R^{(2)}_{\alpha,\beta}(t)\right) e^{-kt} \tag{5.31}$$

As a next, it is integrated numerically for calculating Φ_S.

For the largest model system considered, i.e., FAD-Trp with electron-spin coupling and 7 nuclear spins in each radical, we employ a coherent state sampling method to evaluate $P_s(t)$. Now, it is possible to express $P_s(t)$, using the resolution of the identity operator, in terms of coherent spin states as,

$$P_S(t) = \frac{1}{(4\pi)^{N_1+N_2}} \int d\mathbf{\Omega}, e^{-kt} \langle \mathbf{S}, \Omega; t|\hat{P}_S|\mathbf{S}, \Omega; t\rangle \tag{5.32}$$

In the above equation,

$$|S, \mathbf{\Omega}; t\rangle = e^{-i\hat{H}t/\hbar}|S, \mathbf{\Omega}\rangle$$

and

$$|S, \mathbf{\Omega}\rangle = |S\rangle \otimes |\Omega_{1,1}\rangle \otimes \cdots \otimes |\Omega_{2,N_2}\rangle.$$

Here, $|\Omega_{i,k}\rangle$ is a coherent spin state for nucleus k in radical i, specified by a pair of angles $\Omega_{i,k} = (\theta_{i,k}, \phi_{i,k})$ sampled from the surface of a sphere. The time-evolved $|S, \mathbf{\Omega}; t\rangle$ state can be evaluated using "Krylov subspace" methods. Importantly, the crucial efficiency of such method comes from Monte-Carlo sampling of the integral, dramatically reducing to the computational cost for large coupled spin systems.

5.6 Semiclassical Dynamics for Calculating Singlet Quantum Yields

As a next, here our conjecture is to give a glimpse of three semiclassical methods applied for calculating the quantum yields, especially for singlet quantum.

5.6.1 *Schulten–Wolynes (SW) Theory*

In the present case, following the Schulten–Wolynes (SW) approximation, the quantum mechanical spin operators $\hat{\mathbf{I}}_{i,k}$ have been replaced with static classical spin vectors $\mathbf{I}_{i,k}$ of length $\sqrt{\mathbf{I}_{i,k}(\mathbf{I}_{ik}+1)}$, also taking trace over the nuclear spins which has been replaced with an integral over all orientations of the nuclear spin vectors, i.e.,

$$\mathbf{Tr}_{i,k} \leftarrow (4\pi)^{-1}(2\mathbf{I}_{i,k}+1) \int d\Omega_{i,k}$$

This approximation gives the result to the electron spin density operator as,

$$\hat{\sigma}(t) = Tr_N[(\hat{\rho})(t)]$$

which can be expressed as

$$\hat{\sigma}_{SW}(t) = \frac{1}{(4\pi)^{(N_1+N_2)}} \int d\mathbf{\Omega}\hat{\sigma}_{SW}(t, \mathbf{\Omega}) \tag{5.33}$$

In the above equation, $\hat{\sigma}_{SW}(\mathbf{\Omega})$ is the electron spin density operator evolving in the presence of the static nuclear spin vectors for a particular orientation $\mathbf{\Omega}$,

$$\frac{d}{dt}\hat{\sigma}_{SW}(t, \Omega) = -\frac{i}{\hbar}[\hat{H}_{SW}(\Omega), \hat{\sigma}_{SW}(t, \Omega)] - \left\{\frac{k-s}{2}\hat{P}_s + \frac{k_T}{2}\hat{P}_T, \hat{\sigma}_{SW}(t, \Omega)\right\} \tag{5.34}$$

In the above equation, $\hat{H}_{SW}(\mathbf{\Omega})$ is the spin Hamiltonian $\hat{H}$ with all nuclear spin operators $\hat{\mathbf{I}}_{i,k}$ replaced with static vectors $\hat{\mathbf{I}}_{i,k}$ in a given orientation $\Omega_{i,k}$. Thus the singlet probability $P_s^{SW}(t)$ can be achieved with the help of Schulten-Wolynes approximation to the singlet probability $P_S^{SW}(t)$. However, *this approximation gives the exact value* only in the absence of hyperfine-coupled nuclei in either radical together with the typical case, i.e., within the limit of very large large number of hyperfine-coupled nuclei.

5.6.2 *Improved Semi-Classical Dynamics*

Three different semiclassical methods, as stated by Fay *et al.*, have been described in the following section where, the approach, as employed by Schulten–Wolynes (SW) theory, starts by adapting SW approximation, i.e., by replacing mechanical spin operators $\hat{\mathbf{I}}_{i,k}$ with static classical spin vectors of length $\sqrt{I_{i,k}(I_{i,k}+1)}$, and the trace over the nuclear spins is replaced with an integral over all orientations of the nuclear spin vectors,

$$Tr_{i,k} \rightarrow (4\pi)^{-1}(2I_{i,k}+1)\int d\Omega_{i,k}$$

Then, the resulting approximation to the electron spin density operator, $\hat{\sigma}(t) = Tr_N[\hat{\rho}(t)]$, is given by,

$$\hat{\sigma}_{SW}(t) = \frac{1}{(4\pi)^{N_1+N_2}} \int d\mathbf{\Omega}\hat{\sigma}_{SW}(t, \mathbf{\Omega}) \tag{5.35}$$

Here, $\hat{\sigma}_{SW}(t, \mathbf{\Omega})$ is the electron spin density operator that evolves in the presence of the static nuclear spin vectors, in given orientation, $\mathbf{\Omega}$, so that,

$$\frac{d}{dt}\hat{\sigma}_{SW}(t,\Omega) = -\frac{i}{\hbar}[\hat{H}_{SW}(\Omega), \hat{\sigma}_{SW}(t,\Omega)] - \frac{k_s}{2}\hat{P}_s + \frac{k_T}{2}\hat{P}_T, \quad \hat{\sigma}_{SW}(t,\Omega) \tag{5.36}$$

$\hat{H}_{SW}(\mathbf{\Omega})$ depicts the spin Hamiltonian $\hat{H}$ having all nuclear spin operators $\hat{\mathbf{I}}_{i,k}$ replaced with static vectors $\mathbf{I}_{i,k}$ in a given orientation $\Omega_{i,k}$. This consideration gives the Schulten-Wolynes (SW) approximation to the singlet state probability as $P_S^{SW}(t)$. In absence of any hyperfine-coupled nuclei in either radical, and also in the limit of a very large number of hyperfine-coupled nuclei, this approximation is exact, provided, all of the hyperfine coupling nuclei in either radical and, also in the limit of a very large number of hyperfine coupled nuclei decaying at least as $\frac{1}{\sqrt{N}}$.

5.6.3 *Improved Semiclassical Theory 1 (SC1)*

Among all the three approaches, in the following alternative approach, the singlet projection operator can be considered as

$$\hat{P}_S = \frac{1}{4} - \hat{\mathbf{S}}_1 \cdot \hat{\mathbf{S}}_2 \tag{5.37}$$

As a next, the singlet probability can be written in terms of singlet $\hat{P}_S$ autocorrelation function, as

$$P_s(t) = \frac{1}{Z}Tr\left[\hat{P}_s(t)\hat{P}_s(0)\right]e^{-kt} \tag{5.38}$$

At this stage, the semiclassical approximation to this, in terms of one-spin variable (the model, now, onwards, called as SC1), can be obtained by replacing all operators with classical vectors, i.e.,

$$\hat{\mathbf{I}}_{i,k}(t) \to \mathbf{I}_{i,k}(t); \quad \hat{\mathbf{S}}_i(t) \to \mathbf{S}_i(t)$$

having length $\sqrt{I_{i,k}(I_{i,k}+1)}$ and $\sqrt{3}/2$. Also, it is needed to replace

$$\hat{P}_S(t) \to P_S(t, \mathbf{\Omega}) = 1/4 - \mathbf{S}_1(t) \cdot \mathbf{S}_2(t)$$

for building the criteria of the model to be represented.

Finally, the trace is replaced with an integral over the orientations of all the spin vectors which evolve following the Heisenberg equations of motion, and *quite interestingly, again replacing all quantum mechanical operators with classical vectors.* Here, the initial orientation of each vector is sampled independently from the surface of a sphere, so that the SCI approximation to the singlet probability can finally be expressed as

$$P_S^{SC1}(t) = \left(\frac{1}{2\pi}\right)^2 \left(\frac{1}{4\pi}\right)^{N_1+N_2} \int d\mathbf{\Omega} e^{-kt} P_S(t, \mathbf{\Omega}) P_S(0, \mathbf{\Omega}) \tag{5.39}$$

Ω, in the above expression, includes also the orientations of the electron spin vectors. Importantly, the semiclassical approximation can also be derived as the classical limit of the above correlation function from the Weyl-Wigner-Moyal formalism for spin. Another necessary factor to be mentioned here is that the above approximation will be exact only in the absence of any hyperfine-coupled nuclei or, of any electron-spin coupling. But, in the presence of any other kind of electron-spin coupling, it is no longer exact. Importantly, this particular observation motivated the above mentioned group to introduce the semiclassical approximation involving two-electron spin variables, naming it, as 2(SC2).

5.6.4 *Semi Classical Approximation 2(SC2)*

However, as the SC1 approximation has been found not exact in the presence of electron-spin coupling, especially happening, in the absence of any hyperfine-coupled nuclei in case of either radical, in order to rectify that criteria, two electron spin variables have been introduced by Hore's group in SC2 approximation. As a result, the nuclear spin variables has been considered as coupled to the electron spin variables as before,[166] i.e.,

$$\frac{d}{dt}\mathbf{I}_{i,k}(t) = \left(\mathbf{A}_{i,k} \cdot \mathbf{S}_{i,k}(t)\right) \times \mathbf{I}_{i,k}(t) \tag{5.40}$$

In the above equation, the initial values of $\mathbf{S}_i(t)$ and $\mathbf{I}_{i,k}(t)$ are sampled as in the SC1 approximation. Interestingly, SC1 and SC2 approximations agree exactly when the electron-spin coupling tensor is zero, i.e., $\mathbf{C} = \mathbf{0}$, but importantly, *predict different dynamics in the presence of electron spin coupling.* In such a case, the SC2 approximation becomes *exact* in the absence of hyperfine-coupled nuclei when $\mathbf{C} \neq \mathbf{0}$, but predict *different* dynamics in the presence of electron-spin coupling, the SC2 approximation being exact in the absence of hyperfine-coupled nuclei when $\mathbf{C} \neq 0$.

5.6.5 *Comparison Among SC1, SC2 and SW Models and Conclusion*

From detail comparison among the computational results, obtained from the above methods, the following observations have been made by Fay *et al.* and arrived finally to the following observations: First important point to be noticed here is that when the two radicals do not interact, the SC1 and SC2 semiclassical methods agree well with quantum mechanics for both the cryptochrome-based radical pairs. Therefore, the question, now, to be posed becomes: could a radical pair with negligible electron-spin coupling feasibly form the basis of a geomagnetic compass sensor? If the answer is yes, 'then radical pair magnetoreception arguably has **no place** in quantum biology'. Both the exchange and dipolar interactions depend on the distance, R, between the radicals, the former approximately exponentially ($e^{-\beta R}$), the latter being as R^{-3}. For example, in Drosophila cryptochrome, R for FAD–Trp is either 1.91 nm or 2.21 nm, depending on whether the third or the fourth tryptophan of the electron transfer chain is involved.

In the above approach, followed by Fay *et al.*[135] (2019), it has been expected that the semiclassical approaches would adequately reproduce the quantum spin dynamics only when D and J are smaller than the strength of the Earth's magnetic field (i.e. $(50 \leq mT)$ which, for the dipolar coupling, requires $RT \geq 3.82$ nm.[167] According to the views of Fay *et al.*'s group, such a large distance between spin-correlated radicals in cryptochromes could be achieved only if the electron transfer chain could be extended in order to include a binding-partner protein. However, the rate of back electron transfer from the singlet state of the radical pair varies approximately as $e^{-\beta R}$, being almost certainly slower than 10^5 s^{-1} for RT 2 nm.[167] Not only that as spin relaxation in the radicals is unlikely to be much slower than 1 μs,[136b] all spin coherence in a radical pair with $4R \geq 2$ nm would be lost before significant recombination could take place and no magnetic field effects would arise in such situation. It therefore seems highly improbable for the magnetic sensitivity to be significantly compatible with sufficiently weak electron-spin coupling. Importantly, as commented by Fay *et al.* the only way that this restriction could be circumvented is, if really, Kattnig's paramagnetic scavenging proposal turns out to be right.[176]

Next important point to be discussed following such approach, is that the semiclassical approximations may provide unsatisfactory descriptions of the spin dynamics only in three other situations, for example,

- when the two spin-selective rate constants, kS and kT, differ significantly;
- when pronounced avoided level-crossing effects are present in FS; and
- when the radical partnering with FADc, contains only a few (e.g. 2 or 3) hyperfine-coupled nuclear spins.

5.7 Directional (Heading) Precision: Presence of the Quantum Needle of the Avian Magnetic Compass in the Eye of Migratory Bird

That, the long-lived spin coherence in proteins, found in the eyes of migratory birds could explain how the creatures are able to navigate along the Earth's magnetic field with extraordinary precision has been one of the pathfinding outcome of researchers in the UK and Germany who have created a new realistic model of cryptochrome proteins, based on advanced simulations of nuclear and electron spins. The team also has provided an explanation for how the avian magnetic compass has been optimized by evolution.

It has been an established fact that Migratory birds have a light-dependent magnetic compass, the mechanism of which is thought to involve radical pairs formed photochemically in cryptochrome proteins in the retina. Before the successful contribution regarding the theoretical descriptions of this compass, developed by Hitchcock *et al.*, thus far, it has been unable to account for the high precision with which birds are able to detect the direction of the Earth's magnetic field. After coherent spin dynamics simulations have been employed in exploring the "behavior" of realistic models of cryptochrome-based radical pairs it has been established that when the spin coherence persists for longer than a few microseconds, the output of the sensor contains a sharp feature, referred to as a **"spike"**. This spike arises from avoided crossings of the quantum mechanical spin energy-levels of radicals formed in cryptochromes. Such a feature could deliver a "heading precision", sufficient to explain the navigational behavior of migratory birds in the wild. Firstly, their results offered

- new insights into radical pair magnetoreception,
- suggest ways in which the performance of the compass could have been optimized by evolution,
- may provide the beginnings of an explanation for the magnetic disorientation of migratory birds exposed to anthropogenic electromagnetic noise, and

- suggest that radical pair magnetoreception may be more of a quantum biology phenomenon than previously realized.

By now, it has already been a quite established fact that Migratory birds have a light-dependent magnetic compass.[25,65,122,166] The primary sensory receptors are located in the eyes,[167] and directional information is processed bilaterally in a small part of the fore-brain, accessed via the thalamofugal visual pathway. The evidence, thus far obtained, currently points to a chemical sensing mechanism based on photo-induced radical pairs in cryptochrome flavoproteins, present in the retina. In the Earth's magnetic field,[168] anisotropic magnetic interactions within the radicals are thought to be responsible for this, giving rise to intracellular levels of a cryptochrome signaling state that depend on the orientation of the bird's head.

Supporting this proposal, the photochemistry of isolated cryptochromes in vitro, has been found to respond to applied magnetic fields in a manner such that it becomes quantitatively consistent with the radical pair mechanism. Aspects of such radical pair hypothesis have also been explored in a number of theoretical studies, the majority of which have concentrated on the magnitude of the anisotropic magnetic field effect even though, very little attention has been devoted to this aspect of the problem. This has been addressed in an extensive way, by the group of Hithcock *et al.*[26] which however, addressed here, in quite a short form, taking the precision of the compass bearing, available from a radical pair sensor, as detailed by the pioneering works of Solov'yov *et al.*

To migrate successfully over large distances, it is not sufficient, simply to distinguish north from south (or pole-ward from equator ward). For example, a bar-tailed godwit (Limosa lapponica baueri) was tracked by satellite flying from Alaska to New Zealand in a single 11,000-km nonstop flight across the Pacific Ocean[177] in which situation, a directional error of more than a few degrees, could have been definitely fatal. As the magnetic compass seems to be the dominant source of directional information,[130] and the only compass available at night under an overcast (but not completely dark) sky, migratory birds must be able to determine their flight direction with high precision using their magnetic compass. Extensive studies have shown that migratory songbirds can detect the axis of the magnetic field lines with an accuracy better than 5°.[172,173]

Thus, any kind of plausible magnetoreception hypothesis must be able to explain how such a directional precision can be achieved. Previous simulations of radical pair reactions show only a weak dependence on the direction of the geomagnetic field and therefore cannot straightforwardly account for the magnetic orientation of birds in thc wild. Theoretical treatments of radical pair-based magnetoreception typically involve simulations of the quantum spin dynamics of short-lived radicals in Earth strength ($\approx$50 μT) magnetic fields. The general aim is to determine how the yield of a reaction product depends on the orientation of the reactants with respect to the magnetic field axis. A crucial element in all such calculations is the presence of nuclear spins whose hyperfine interactions are the source of the magnetic anisotropy.[178] Most studies have focused on the use of idealized spin systems comprising the two electron spins, one on each radical, augmented by one or two nuclear spins. Only a handful has attempted to deal with realistic, multinuclear radical pairs.

The other critical ingredient in such simulations is the lifetime of the electron spin coherence: if the spins dephase completely,- before the radicals have a chance to react, there can be no effect of an external magnetic field.[134] Several studies have assumed, explicitly or implicitly, that the spin coherence persists for about a microsecond, i.e., the reciprocal of the electron Larmor frequency (1.4 MHz) in a $50 - \mu T$ field. Either, because the spin system was grossly oversimplified, or because of such restriction on the spin coherence time, previous theoretical treatments have generally predicted the reaction yield to be a gently varying (often approximately sinusoidal) function of the orientation of the radical pair in the geomagnetic field. Although capable of delivering information on the direction of the field, such a compass would not provide a precise heading. A more sharply peaked dependence on the field direction would be needed to achieve a compass bearing with an error of 5° or less.

The group of Hore[118] explored the behavior of cryptochrome-inspired radical pairs with multinuclear spin systems and long-lived ($\geq$1 μs) spin coherence, as a result of which they reached to the conclusion that a cryptochrome-based radical pair compass have evolved with a heading precision, sufficient to explain the navigational behavior of migratory birds both in the laboratory and in the wild as well. They have demonstrated that a radical pair magnetoreceptor may be capable of much higher angular precision than previously thought possible. More specifically, a version of such kind of radical pair model has been presented by this group where it could potentially explain the magnetic compass precision, observed for night-migratory songbirds.[172,173] The special feature making this feasible, referred to as a **'spike'**, emerges naturally for cryptochrome-based radical pairs having the lifetime of the spin coherence exceeding 1 μs. Product yields of radical pair reactions have been calculated by solving a Liouville equation containing

- the internal magnetic (hyperfine)interactions of the electron spin with the nuclear spins in each radical,
- the magnetic (Zeeman) interactions of the two electron spins with the external magnetic field, and
- appropriate spin-selective reactions of the singlet and triplet states of the radical

As a starting point, we modeled [FAD•− TrpH•+], the radical pair, responsible for the magnetic sensitivity of isolated cryptochrome molecules in vitro.[151] It consists of the radical anion of the noncovalently bound flavin adenine dinucleotide (FAD) cofactor and the radical cation of the terminal residue of the "tryptophan (Trp) triad" electron transfer chain, within the protein. The initial state of the spin system was a pure singlet. Two approximations were introduced to make simulations of the 16-spin system computationally tractable with the following conditions:

- exchange and dipolar interactions between the radicals were assumed to be negligible, and
- the singlet and triplet states were assumed to react to form distinct products with identical first order rate constants, k.

In such case, the lifetime of the radical pair, τ, is defined as the reciprocal of k. As a measure of the available directional information, Hithcock *et al.* calculated Φ_s, i.e., the fractional yield product formed from the singlet state of the radical pair.

The variation of Φ_s for [FAD•−TrpH•+] has been shown as the direction of a $50 - \mu T$ magnetic field, specified by the angle θ, is rotated in the z-x plane of the flavin. When $\Theta = 0$, the field is parallel to the flavin-z axis. With a lifetime $\tau = 1$ μs, Φ_s exhibits a shallow minimum around $\Theta = 90°$ and maxima near 0° and 180°, as found previously.[163] Shorter lifetimes gives even weaker angular variation. As the lifetime is prolonged from 1 μs toward 100 μs, the dependence of Φ_s on Θ becomes increasingly structured, and a prominent spike emerges, strengthens, and narrows.

Centered accurately at $\Theta = 90°$, this feature occurs when the magnetic field is in the plane of the flavin ring system (parallel to the x-axis). As τ is increased beyond 100 μs, the only change is that the spike grows (by roughly a factor of 3 as $\tau \to \infty$). The anisotropy of Φ_s can be seen more clearly from polar plots of the same data (Fig. 1C in the literature) after subtraction of the isotropic components. As the lifetime is prolonged, the anisotropy grows, and Φ_s depends more strongly on Θ. As expected from time reversal symmetry, Φ_s is invariant to 'inversion of the direction of the magnetic field', — a property shared by the avian magnetic compass.[24b] Very similar behavior was found when the magnetic field was rotated in the molecular ZY plane. In fact, Φ_s has roughly axial symmetry around the molecular z axis, apart from a tilted feature arising predominantly from the indole nitrogen of TrpH•+, for $\tau = 10\mu$sec. The spikes, observed in Fig. 1B and 1C at $\theta = 90°$ are, in fact, cross-sections through the thin equatorial disk, produced when the magnetic field axis is close to the x-y plane of the flavin. They are the representations of a bird's perception of the directional information which is delivered by an array of cryptochrome, containing magnetoreceptor cells, distributed around the retina, for example, in the present case, for a bird in the northern hemisphere looking horizontally toward magnetic north in a $50 - \mu T$ magnetic field with a 66° inclination. As the lifetime τ is prolonged, and the spike becomes stronger, the spot indicating the axis of the geomagnetic field lines becomes more intense and less diffuse.

5.7.1 *Origin and Conditions for the Formation of the Spike in* Φ_s

Finally, a degree of rotational disorder among the magnetoreceptor cells[168] can be modelled by averaging the polar plot in Φ_s over a 360° rotation around a chosen axis. If this axis is in the x-y plane of the flavin, the thin blue equatorial disk into the needle-shaped object ($\tau = 10\ \mu$ sec), Fig. 1F appears to be ideal for determining a precise compass bearing. It is important to mention that radical pair sensor is an 'inclination' compass rather than a 'polarity' compass.

So far as the origin of the spike in Φ_s is concerned, Hishcock *et al.* noted an approximate axial symmetry of Φ_s for $\tau = 1\mu$sec which principally, have been attributed to the two nitrogens, N5 and N10, in the central ring of the FAD•− radical.[89,178] N5 and N10 are known as the only nuclei in [FAD•−TrpH•+] with hyperfine tensors which, like Φ_s, posses approximately axial symmetry around the flavin-z axis and hence, therefore, seems probable that they also play a role in creating the spike that arises when $\tau > 5$ μs. Importantly, this prediction, made by Hishcock *et al.*, has been confirmed in detail, showing Φ_s, for a very slightly modified version of $[FAD^{\dot{-}}TrpH^{\dot{+}}]$. The z components of the hyperfine interactions of N5 and N10 in flavin radicals are found to be large. But, anyway, the x and y components have small but nonzero absolute values. The thin equatorial disk, as elaborated in Fig. 1D, can unambiguously be attributed to the avoided crossings of the quantum mechanical energy levels of the radical pair spin Hamiltonian, as a function of the magnetic field direction, predicting that the line shape of a crosssection through the disk (i.e., the spike) will be an upside-down Lorentzian. Quite interestingly, when A_{xx} and A_{yy} for both nitrogens are set to zero, *the avoided crossings become level crossings and the spike vanishes.*

With the aim of obtaining further insight into the origin of the spike, Hithcock *et al.* performed simulations, especially, for three radical pairs related to [FAD•− TrpH•+]. From the simulations of flavin-containing radial pairs, it appears that radical partners FAD•− should have at least one nucleus with an anisotropic hyperfine interaction which condition is amply fulfilled by TrpH•+, in which the indole nitrogen and the aromatic hydrogens — all interact anisotropically with the electron spin.[163, 177] However, we can infer from these exploratory studies that the spike in Φ_s is not excessively sensitive to reasonably rapid, relatively low amplitude motions, likely to occur for the radicals in their binding sites in cryptochrome. A quite crucially important message, we take from these calculations, is that radical motions on timescales faster than about 1 ns could allow the spikiness of Φ_s to survive.

Interestingly, so far as the precision of the Compass Bearing is concerned, it is already an established fact that the directional information available from the knowledge of Φ_s, will inevitably be degraded by 'stochastic noise' in the detection system. Also, for a given noise level, a sharper and stronger spike will deliver a more precise compass bearing. Hithcock *et al.* tried to quantify the effects of such detector noise on the signals in Fig. 5.1 out of which, a precision of 1° is found to be needed in order to avail precision for $\tau = 1$ μsec, the noise level would have to be $\approx$40 times smaller than when $\tau = 1$ μsec, i.e., if such an improvement in signal-to-noise had to be achieved by averaging repeated measurements, it would take $\approx 40^2 = 1{,}600$ times longer for a radical pair with a lifetime of 1 μsec than for one, with 10 μsec.

The third, rather crucially major condition for the emergence of the spike is that the spin coherence times of the radicals should be longer than 1 μs, meaning that the libration of the radicals within their binding pockets must be of relatively low amplitude but not too sluggish. As such, motions are determined by the interactions of the radicals with the associated protein environment. This is an property that could have been optimized by evolution. The spike emerges only when the coherence time exceeds 1 μs, its presence could explain why slow relaxation might have evolved. Moreover, it may now become possible to understand how radio frequency fields, in particular broadband anthropogenic electromagnetic noise (sometimes called electrosmog)[96], interferes with the operation of the avian compass: not because all anisotropy is destroyed, but because the spike is attenuated. It remains to be followed further, whether the spin relaxation can be slow enough to explain the reported effects.[102]

As per the finding of Hishcock *et al.*, fundamental requirement for the occurrence of a pronounced spike in the reaction yield (Φ_s) is one of the radicals, i.e., $FAD^{\bullet -}$ or at least something closely resembling it. Particularly, the two nitrogen nuclei (N5 and N10) in the central ring of the tricyclic flavin ring system, appear to have almost ideal magnetic hyperfine interactions.[178] The width and height of the spike can be tuned by adjusting the transverse components (A_{xx} and A_{yy}, implying that random mutations in the sequence of the protein in the neighborhood of the FAD could have provided evolution with the scope in optimizing the compass precision.

However, as emphasized by Hithcock *et al.*, the precise values of the hyperfine parameters required to produce a substantial spike in Φ_s, are not crucial and have neither been guessed nor carefully chosen; but obtained directly from independent molecular orbital calculations. Finally, they also concluded that the **prerequisite for spiky behavior is the radical that partners the FAD•−, must having at least one appreciably anisotropic hyperfine interaction**. This condition is certainly satisfied by the TrpH•+ radical, formed by photo-induced electron transfer along the Trp-triad in cryptochrome. It is also consistent with the oxidized form of ascorbic acid (Asc•−), a radical that has been tentatively suggested as an alternative to TrpH•+, on the basis that [FAD•− Asc•−] is expected to show much larger magnetic field effects than [FAD•− TrpH•+], by virtue of the small hyperfine interactions in Asc•−.[178] However, a spike would not be expected for a [FAD•− Z•] radical pair, in which Z• is a radical, as it is completely devoid of hyperfine interactions, such as superoxide, $O_2^{\bullet -}$. However, very recent experiments,[174] designed to replicate the earlier study under much more stringently controlled conditions, failed to find specific effects at the Larmor frequency. In contrast, very weak broadband fields were found to disrupt the birds' magnetic compass orientation capabilities.[96,174] These new findings are consistent with radical pairs that have significant hyperfine interactions in both radicals, e.g., [FAD•− TrpH•+] and [FAD•− Asc•−].

The **third** major condition for the emergence of the spike is that the spin coherence times of the radicals should be longer than 1 μs. This means that the librations of the radicals within their binding pockets must be of relatively low amplitude but not too sluggish. Such motions being determined by the interactions of the radicals with the protein environment, could have been optimized by evolution. That spin relaxation much slower than 1 μs, has been invoked before, to explain the apparent sensitivity of birds to weak (nano tesla) monochromatic radiofrequency fields.[44,180] The problem, proposed here is that if there is no possibility of a spike, a coherence time of 1–2 μs is sufficient to achieve the optimum compass performance without having any evolutionary pressure in prolonging relaxation times beyond this point.[182,183] As the spike emerges only when the coherence time exceeds 1 μs, its presence could explain why 'slow relaxation' might have evolved. Moreover, it may now become possible to understand how radiofrequency fields, in particular, broadband anthropogenic electromagnetic noise (sometimes called electrosmog), interferes with the operation of the avian compass: **not because all anisotropy is destroyed, but because the spike is attenuated.**

5.7.2 *Exchange and Dipolar Interactions: Effects on Spike Formation*

Preliminary simulations using the toy radical pair model suggest that the spike can be affected by the exchange and dipolar interactions of the electron spins in the two radicals. These effects, probably less pronounced for radicals with many nuclear spins, are quite difficult to verify because the Liouville-space calculations required when electron-electron interactions are included which makes the calculations prohibitively slow for realistic spin systems. The effects of spin-spin interactions may also be reduced by the partial cancellation of the exchange and dipolar contributions predicted for a radical pair with a separation close to that of $F\dot{A}D^{-}$ and $TrpH^{\dot{+}}$ in cryptochrome.[184] Finally and quite importantly, if in a migratory bird cryptochrome, as in Xenopus laevis[93,185,186] photolyase, there are four

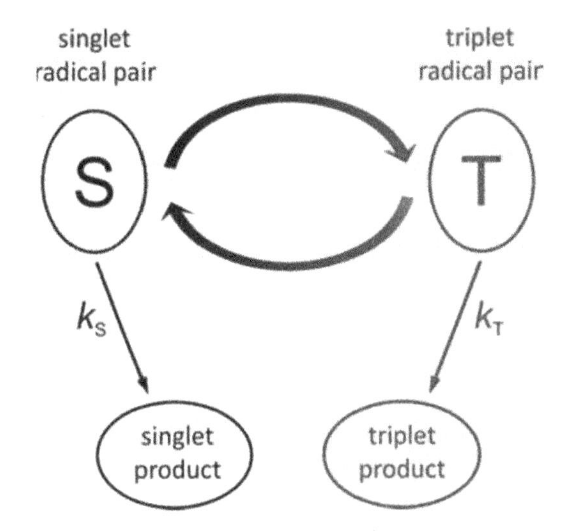

Figure 5.9. 'Singlet' (S) and 'triplet' (T) refer to the states of the two electron spins, respectively, one in each radical. The curved arrows, as depicted in the figure above, is meant for the coherent spin dynamics arising from the combined effects of Zeeman and hyperfine interactions. Here, the straight arrows describes in the above figure the spin-selective reaction steps.

instead of three Trp- residues involved in photoreduction of the FAD, then the magnetically sensitive FAD-Trp radical pair could have a larger distance between the radical centers and consequently smaller spin-spin interactions.

5.7.3 *Calculation of Φ_s in the Absence of Molecular Motion*

Magnetic field effects were modeled by means of the following reaction scheme:

In the above figure, 'Singlet' (S) and 'triplet' (T) refer to the states of the two electron spins, respectively, one in each radical. The curved arrows, as depicted in the figure above, is meant for the coherent spin dynamics arising from the combined effects of Zeeman and hyperfine interactions. Here, the straight arrows describes in the above figure the spin-selective reaction steps. The radical pair is created in a singlet state by spin-conserving electron transfer. Singlet and triplet radical pairs here have been shown as undergone by separate spin-conserving but having reverse electron transfer reactions, in order to form distinct singlet and triplet products respectively.

Abstract examples of such reactions can be stated as follows:

$$^{s}[A^{\dot{+}}B^{\dot{-}}] \longrightarrow {}^{s}A + {}^{s}B$$

and

$$^{T}[A^{\dot{+}}B^{\dot{-}}] \longrightarrow {}^{T}A + {}^{s}B$$

where, ${}^{s}A$ and ${}^{s}B$ are diamagnetic (closed shell, singlet state) molecules with no unpaired electron spins and ${}^{T}A$ a paramagnetic molecular triplet. In the model, the rate constants of these two reactions are identical: $k_S = k_T = k = 1/\tau$. We calculate Φ_s, the fractional yield of the singlet product, once all radical pairs have reacted. Φ_s is related to the yield of the triplet product by $\Phi_T = 1 - \Phi_s$.

The singlet yield Φ_s has been calculated as as,

$$\Phi_s = k \int_0^{\infty} p_s(t) e^{-kt} dt \tag{5.41}$$

in which $k = k_s = k_T$ represents the recombination rate constant and $p_s(t)$ the fraction of radical pairs in the singlet state at time t:

$$p_s(t) = \frac{1}{4} + \sum_{p=x,y,z} \sum_{q=x,y,z} R_{pq}^{A}(t) R_{pq}^{(B)}(t)$$

having the value for R, as,

$$R_{pq}^{(m)}(t) = \frac{1}{Z_m} \mathrm{Tr}\left[\hat{S}_{mp} e^{-i\hat{H}_m t} \hat{S}_{mq} e^{i\hat{H}_m t}\right]$$

and

$$Z_m = \prod_{j=1}^{N_m} (2l_{mj} + 1)$$

Here, $\hat{S}_{mp} (p = x, y, z)$ are the electron spin operators and $\hat{H}_m$ is te spin Hamiltonian of radical m $(m = A, B)$. For the present case, I_{mj} is the spin quantum number of nucleus j in radical m and N_m is the number of nuclei in radical m.

It is important to mention here that in the above case, i.e., in the case of two radicals: spin Hamiltonian for each radical contains terms for the Zeeman interaction (Z) of the electron spin with the applied magnetic field together with the various hyperfine interactions (HFI), i.e.,

$$\hat{H}_m = \hat{H}_{m,z}(\Theta, \phi) + \hat{H}_{m,HFI}$$

where, we have,

$$\hat{H}_{m,z}(\Theta,\phi) = \gamma_e B_0 \left[\hat{S}_{m,x}\sin\Theta\cos\phi + \hat{S}_{m,y}\sin\Theta\sin\phi + \hat{S}_{m,z}\cos\Theta\right]$$

and

$$\hat{H}_{m,HFI} = \sum_{j=1}^{N_m}\left[\alpha_{mj}^{iso}\cdot\hat{\mathbf{S}}_m\cdot\hat{\mathbf{I}}_{mj} + \hat{\mathbf{S}}_m\cdot\mathbf{T}_{mj}\cdot\hat{\mathbf{I}}_{mj}\right]$$

In the above equation B_0 is the strength of the external magnetic field, Θ and ϕ define its direction with respect to the radical pair.

5.7.4 *Experimental Evidence for the Presence of Spike, Responsible for the Precision of the Avian Magnetic Compass*

In order to establish the phenomenon that determine whether a spike is really responsible for the precision of the avian magnetic compass, although direct detection might be quite challenging, however it should be possible to discover that the proper conditions could exist in a cryptochrome and that would be compatible with the existence of a spike. Hence it must be established which of the four known avian cryptochromes[187] plays a role in compass magnetoreception. Then, its structure will be known making it possible to determine more about the librational motions of the radicals and the spin relaxation they produce. It seems quite probable that the magnetic and dynamic properties of a cryptochrome that has evolved as a compass sensor, would differ significantly from those of cryptochromes that do not have a magnetic sensing function. Not only that, it also appears likely that the properties of such a protein in vivo, will differ from those of the isolated protein in vitro, for example, as a result of binding to signaling partners or attachment to whatever intracellular structures are responsible for alignment and/or immobilization of the protein.[136] Another approach would be to extend the behavioral experiments in which broadband sub-nanotesla electromagnetic noise was found to prevent European robins from using their magnetic compass.

The next important and crucial problem, as well, arises, whether it is possible to determine, i.e., whether a spike is really responsible for the precision of the avian magnetic compass? Although direct detection might be challenging, it should be possible to discover whether proper and suitable conditions could exist in a cryptochrome such that it would be compatible with the existence of a spike. Once it has been established which of the four known avian cryptochromes[187] plays a role in compass magnetoreception, and its structure is known, it will be possible to determine more about the librational motions of the radicals and the spin relaxation they produce. It seems probable that the magnetic and dynamic properties of a cryptochrome that has evolved as a compass sensor, would differ significantly from those of cryptochromes that do not have a magnetic sensing function. It also appears likely that the properties of such a protein in vivo will differ from those of the isolated protein in vitro, for example, as a result of binding to signaling partners or attachment to whatever intracellular structures are responsible for alignment and/or immobilization of the protein.[136]

5.7.5 *Visual Modulation Patterns in the Eye of Migratory Birds, When Flying*

"Visual modulation patterns", first calculated by Ritz *et al.* are crude representations of a bird's perception of the compass information delivered by an array of cryptochrome- containing magnetoreceptor cells distributed around its retina. The proteins are assumed to be identically oriented in every cell and each cell to be identically oriented with respect to the local retina normal. Cells at different locations in the retina have different orientations with respect to the geomagnetic field and therefore deliver different directional information. The assumption is that either by comparing the signals from different parts of the retina or by performing head scans, or both, the bird would obtain information sufficient to orient itself. As noted by Ritz *et al.* different arrangements of the cryptochromes, e.g. a perpendicular orientation of the proteins in neighbouring cells, can result in different sensitivity to the geomagnetic field. However, the clarity of the sensory information for these alternative arrangements would be similar to that of the simulations used here for illustrative purposes. The visual modulation patterns for $[FAD^{\dot{-}}\ TrpH^{\dot{+}}]$ in Fig. 1E were calculated as described by Lau *et al.* The cells were assumed to have no preferred orientation with respect to rotation around the local retina normal. This ordering is not unlike that of the rod and one visual receptor cells in the retina (one of the proposed locations for cryptochrome magnetoreceptors. The signals were therefore integrated over the angle η. The z-axes of the FAD molecules within each cell were assumed to be perpendicular to this symmetry axis so that the η- averaging of ϕ_s is around an axis in the xy-plane of the flavin ring system. Fig. S3 summarizes the geometry of the model and the various rotations. In the visual modulation patterns, θ varies from zero at the centre of the pattern to 90° at the edge. ϕ increases anticlockwise from zero at the bottom

5.7.6 *Angular Precision of Radical Pair Compass Magnetoreceptors*

Although it has been known for almost more than 50 years that small night-migratory songbirds have a light-dependent magnetic compass sense to help them navigate thousands of kilometers every year, still the biophysical mechanism remains largely unknown. The leading hypothesis is that the Earth's magnetic field can alter the course of photochemical reactions in the birds' eyes even though the energies involved are a million times smaller than the thermal energy, kBT.

These night-Migratory songbirds have a light-dependent magnetic compass sense helping them navigate the thousands of kilometers that separate their breeding and wintering grounds. How this compass works is something of a mystery.[191] Currently, the leading hypothesis for the primary detection event involves the photochemical production of transient radical pairs in cryptochrome proteins contained in the birds' retinas. The spin dynamics of these paramagnetic reaction intermediates could provide a mechanism by which the direction of the Earth's magnetic field (~50mT) is encoded in the quantum yield of a signaling state of the protein. As it is an well known fact that light is a crucial component of this hypothesis. Blue **light** is required to excite the flavin chromophore and trigger the intraprotein electron transfers which is responsible for generating the radical pairs. But, a crucially important question arises at this juncture, placing a fundamental question, i.e., is this mechanism is viable, given the light condition, quite dimmed in nature, experienced by nocturnal migrants? This issue has recently been addressed by Hitchcock *et al.*, who used an information theory approach to obtain a strict lower bound on the angular precision about which we are aware that a bird could orient itself using a cryptochrome-based radical pair sensor. The method, applied by this group, has the extra advantage by which it avoids having to make guesses about the nature and efficiency of signal transduction and post-processing in vivo, instead calculating the best-case precision. They arrived at the conclusion which concluded that the average photon flux on a **cloudless and moonless night** (~3×10^4 lux (11-ref same as above)) might not be sufficient for at least some of the currently considered radical pair models.

5.7.7 *Application of Quantum Biology in Radical Pair Mechanism and the Process of Magnetoreception*

Radical pair recombination reactions are known to be sensitive to extremely weak magnetic fields and can therefore be said to function as molecular magnetoreceptors. The recently developed classic example is a carotenoid-porphyrin-fullerene (C•+PF•−) radical pair that has been shown to provide a "proof-of-principle" for the operation of a chemical compass.[24] Previous simulations of this radical pair have employed semiclassical approximations, which are routinely applicable to its 47 coupled electronic and nuclear spins. However, calculating the exact quantum mechanical spin dynamics, presents a significant challenge and has not been possible until now. Recently, Lachlan P. Lindoy *et al.*[200] applied a recently developed method to perform numerically converged simulations of the C•+PF•− quantum mechanical spin dynamics, including all coupled spins. A comparison of these quantum mechanical simulations with various semi-classical approximations reveals that, while it is not perfect, the best semi-classical approximation does capture essentially all of the relevant physics in this problem.

The radical pair of magnetoreception has found a place in the emerging field of Quantum Biology[147,189,190] on the strength of the absolute requirement that the radical pair must be in a coherent superposition of the quantum states of the two electron spins. Though, recently, it has been showed that the spin dynamics, i.e., the initial electronic singlet state of the radical pair is quantum mechanically entangled, although the entanglement, as such, confers no advantage in terms of the general operation of the compass, nor is it essential for the existence of the spike. Recently it has been showed by Hishcock *et al.* that the spin dynamics of long-lived radical pairs in weak magnetic fields can effectively be described by a semiclassical approximation whose accuracy increases accurately with increase in the number of nuclear spins. Instead of putting many more nuclear spins, considering realistic couplings to the electron spins, specifically, — assuming the spin coherence of the paired radicals to be much longer lived than in previous models, they observed a marked "spike" in the yield of the signaling state whenever the spin coherence times in its model, were set to be longer than a few microseconds, produced by spin-selective reactions of the radical pairs. As the spin lifetime is prolonged from 1 μsec toward 100 μsec, the spike 'emerges, strengthens and narrows'.

If the behavior of a realistic radical pair magnetoreceptor can be satisfactorily modeled in terms of classical rather than quantum oscillations, then arguably it does not belong under the quantum biological umbrella. However, the spike discussed here is undeniably a quantum effect, arising from the mixing of states associated with avoided energy-level crossings, and is not captured by the semiclassical theory. In this sense, radical pair magnetoreception may be more of a quantum phenomenon than hitherto realized. Such a crucial feature could provide birds with directional information in explaining their navigational behavior with sufficient precision. Hore *et al.* pointed to a crucial observation, i.e., while the exact chemical signal produced in birds, is unknown, it is almost certainly a form of the cryptochrome protein, having a different shape in turn which allows the protein to have different interactions with other proteins, starting the 'neural signaling process'. The long-lived spin coherences provide enough time "for the interconversion of the singlet and triplet states of the radical pairs, and therefore the yield of the signaling state, to respond sensitively to the direction of the geomagnetic field" whereas, as stated before, radical pairs, not going forward to the signaling state, revert to the original protein structure.

As has already been explained, the process in magnetoreception in radical pair mechanism has found a place in the emerging field of Quantum Biology on the strength of the absolute requirement that the radical pair must be in a coherent superposition of the quantum states of the two electron spins. In fact, the initial electronic singlet state of the radical pair is quantum mechanically entangled [although the entanglement, as such, confers no advantage in terms of the general operation of the compass, nor is it essential for the existence of the spike]. Thus the group led by Hore showed in their works that the spin dynamics of long-lived radical pairs in weak magnetic fields can be described by a semiclassical approximation that becomes increasingly accurate as the number of nuclear spins is increased. If the behavior of a realistic radical pair magnetoreceptor can be satisfactorily modeled in terms of classical rather than quantum oscillations, then arguably it does not belong under the quantum biological umbrella. However, the spike discussed here is undeniably a quantum effect, arising from the mixing of states associated with avoided energy-level crossings whch is not possible to capture by the semiclassical theory. In this sense, radical pair magnetoreception may be more of a quantum phenomenon than hitherto realized.

Finally–to reveal the inner working of the avian compass, another crucially important and significant step in this direction, a novel aspect, pointed out and explained by Hore and his group, is that, though coherence phenomena are not expected to be long-lived in most biological systems, birds could have evolved a molecular system that do not rapidly lose coherence, i.e., the source of the spin relaxation that would destroy the coherence, is the stochastic fluctuations of the radicals in their binding sites in the protein so that, at this juncture, if mutations in the proteins make those motions both fast and relatively of low amplitude, then the loss of coherence should be inefficient, thus allowing the spin-correlated state to persist, perhaps, for "5–10 μm", which would be sufficient to produce the spikes in the yield of the signaling state.

By now, we are already aware of the two main theories of avian magnetorecption — (1) one theory holds that birds navigate using particles of magnetite, located in their head; (2) the other holds that the birds sense the magnetic field as a the result of photo excitation in the eye whose de-excitation product is affected by the magnetic field. The excitation, which occurs inside the cryptochrome protein in the presence of blue light, produces a radial pair whose singlet and triplet states are essentially degenerate. In the presence of a magnetic field, more radical pairs are pushed into the triplet state, leading to an increased production of triplet de-excitation products. The radical pair thus acts as a quantum chemical compass in the eye — implying that birds can "see" magnetic fields. Interestingly, very recently, in the study,[192] "Cellular autofluorescence is magnetic field sensitive", Noboru Ikeya and Jonathan Woodward found that fluoresence from crytochromes in living cells is affected by an applied magnetic field, demonstrating that the de-excitation pathway of the radical pair is indeed sensitive to a magnetic field.

5.7.8 *Theory of Magnetoreception & Navigation of Bird*

Two decades after Yeagley's[200, 201] original publication it was an extraordinarily striking finding of Keetan[199] that under condition of heavy overcast, pegions have been found not only be well oriented but also, not affected in their orientations, even by clock-shifts which ultimately led Keetan to reconsider the possibility that might be the geomagnetic field providing the directional information to the navigating birds- importantly -in the 'absence of the sun'. Keetan also observed that magnets attached to the experienced home pigeons (to the beak) resulted in disorientation whereas no such disorientation observed when there was similar releases done during the same period of time but the sun 'being visible'. Interestingly, many of these birds somehow managed in coming back to their destinations, sometimes, just as rapidly as the control birds did without magnets which fact appears that whatever disturbance the magnets produce, the birds are able to compensate for it, relatively quickly–correcting their orientations, even though the magnets remain attached to their bodies. In fact, this way, existence of large number of magnetic-particles — probably magnetite and -likely biogenic in animal tissue — has been found to posses an important implications for the study of magnetic field sensitivity in birds.

Richard Holland *et al.*,[192] in their pioneering approach, used the technique of the classic "Kalmijn-Blakemore" pulse re-magnetization technique in their experiment, whereby the polarity of cellular magnetite is reversed. The results demonstrate that the big brown bat Eptesicus fuscus uses single domain magnetite in order to detect the Earths magnetic field and the corresponding response indicates a polarity based receptor. They established the fact that the polarity detection is a prerequisite for the use of magnetite as a compass and hence suggested that big brown bats use magnetite to detect the magnetic field as a compass. Their results also indicated the possible roles of the sensory cells in bats which contain freely rotating magnetite particles, which appears not to be the case in birds. They, thus, suggested the crucial importance of the ultra-structure of the magnetite containing magnetoreceptors, in understanding the role and mechanism as well, played by the magnetoreception in animals too. Holland also did put a strong suggestion which states that birds are capable of true navigation, together with their ability to return to a known goal from a place they have never visited before. This is demonstrated most spectacularly during the vast migratory journeys made by these animals year after year, often between continents and occasionally global in nature.

Importantly, it has been observed that hypothesized mechanism of magnetite-based **magnetic field perception** in animals depend upon the alignment of arrays of single-domain magnetic grains in the ambient magnetic field. The detection of the average direction of alignment by the nervous system would provide the basis for a compass sense, while detection of the variance of the alignment about the average direction would provide the basis for intensity perception. That is why it has been found as possible for European starling, flying in a wind tunnel being maintained constant to within 0.1–0.4°C for extended periods of time or so. This cause may also play an important role in limiting the sensitivity of a hypothetical geomagnetic field intensity detector. As mentioned already, interestingly, it has been observed that in poikilothermic animals, for example, insects, amphibians, reptiles, and some fish,- body temperature can fluctuate over a wide range of degrees which is an well known factor, being responsible in limiting the ability of such animals to exhibit detailed positional information from the geomagnetic field. However, like birds and animals, certain fishes (for example, TUNA) and insects (honey bees, for example) also posses various means of regulating body temperature to keep it relatively constant under certain conditions. Added to this, another interesting and important information states about the presence of the magnetic remanence in the head region of all turtles which can also be found among their hatch-lings.

Based on all of such information, conclusively it can be stated now about the apparent occurrence of quite fine-grained, permanently magnetic material in several kinds of tissues presents the possibility of playing several roles. Not only that these kind of particles also play the significant role, presumably utilizing one or more of the magnetite's unique properties of ferromagnetism, i.e., high density, hardness and conductivity. Electron microprobe analysis have already revealed that the extracted particles were rich in iron and contained no measurable titanium or chromium as would be expected for magnetite isolated from rock. With respect to the possible utilization of biogenic magnetite by birds and other animals as a sensory basis for geomagnetic field detection, it is tempting to say now, as Dirac wrote a half century ago in another context involving electromagnetism, that "**under these circumstances one would be surprised if Nature had made no use of it**".

Though great progress achieved in understanding the mystery of how birds navigate using magnetic cues, it still remains something of a mystery as to how they sense the magnetic field. However, great progress has been achieved in conclusive suggestions that birds sense magnetic values through a light-sensitive molecule called cryptochrome, or through sensory cells containing magnetic iron oxide particles — but definitive concrete evidence for either of these has not yet been provided. Also, behavioural evidence continues to underscore how the Earth's magnetic field is crucial in helping some birds make their epic journeys to breed each year — providing a global positioning system that might just provide birds with a complete navigational map of the world. At present, though the role of magnetic cues for compass orientation has been confirmed in numerous animals, the mechanism of detection is still being debated. Two hypotheses have been proposed, one based on a light dependent mechanism, apparently used by birds and another based on a "compass organelle" containing the iron oxide particles magnetite (Fe3O4). For example, bats have recently been shown to use magnetic cues for compass orientation but the method by which they detect the Earth's magnetic field still remains quite puzzlingly, not yet settled in a concrete fashion.

To date, four different kind of cryptochromes have been detected in the retina of several bird species. These posses one key requirement, i.e., the molecules having the needed biophysical characteristics exist in the eye of migratory birds. Presently, the only vertebrate proteins, known to form radical pairs upon photoexitation are the cryptochromes. These are flavo-proteins, sharing moderate amino acid, similar to photolyases but, importantly, do not show photolyase activity. In plants and various animal species, they are involved in blue-light-dependent pathways and in the circadian clock. Mammalian cryptochromes involved in the circadian rhythm, are mainly localized in cell nuclei, whereas magnetoreceptive cryptochromes should be located in the cytosol and be associated with membranes and/or the cytoskeleton. Up till now (i.e., 2017), four different cryptochromes have been found in the retina of several bird species.

In vertebrates, cryptochromes have been found to be the only class of proteins that form radical pairs upon photo-excitation. Therefore, currently, they are the only candidate proteins for light-dependent magnetoreception. Cryptochrome 4 (Cry4) is particularly interesting because it has only been found in vertebrates that use a magnetic compass. However, its structure and localization within the retina has remained unknown. Anja Gunthur *et al.* in their work, sequenced night-migratory European robin (Erithacus rubecula) Cry4 from the retina and predicted the currently unresolved structure of the erCry4 protein. According to their finding it has been concluded that erCry4 should bind Flavin and also that Cry1a, Cry1b, and Cry2 mRNA display robust circadian oscillation patterns. A quite interesting and important observation of this group claim that, Cry4 shows only a weak circadian oscillation and being compared with the relative mRNA expression levels of the cryptochromes during the spring and autumn migratory seasons relative to the non-migratory seasons,- in European robins and domestic chickens (Gallus gallus), the Cry4 mRNA expression level in European robin retina is significantly higher during the migratory season compared to that in non-migratory seasons but this has not been noted in case of chicken retinas. Not only that Cry4 protein is specifically expressed in the outer segments of the double cones and long wave length single cones in European robins and chickens. A localization of Cry4 in double cones seems to be ideal for light-dependent magnetoreception. Considering all of the data presented here, especially including its localization within the European robin retina, the Anju Gunthur and Mouritson *et al.* arrived at the conclusion that it is highly probable that the binding of Flavin, and its increased expression during the migratory season in the migratory bird but not in chicken, Cry4 could be the magnetoreceptive protein. Thus, there exists still two classes of main theories of avian magnetorecption, i.e., —

- One theory holds that birds navigate using particles of magnetite located in their head,
- The other holds that the birds sense the magnetic field as a the result of photo excitation in the eye whose de-excitation product is affected by the magnetic field. The excitation, which occurs inside the cryptochrome protein in the presence of blue light, produces a radial pair whose singlet and triplet states are essentially degenerate. In the presence of a magnetic field, more radical pairs are pushed into the triplet state, leading to an increased production of triplet de-excitation products. The radical pair thus acts as a quantum chemical compass in the eye — implying that birds can "see" magnetic fields.

Another typical characteristics of the bird aviation mechanism has been noted and studied in detail by D.E.Manopoulos *et al.* in case of bird navigation, pointing to the dependence on strength of magnetic field in radical pair recombination reactions which are known to be quite sensitive in the case of application of magnetic fields. The application of a weak magnetic field reduces the singlet yield of a singlet-born radical pair, whereas the application of a strong magnetic field increases the singlet yield. The high field effect arises from energy conservation: when the magnetic field is stronger than the sum of the hyperfine fields in the two radicals, $S \rightarrow T_{\pm}$ transitions become energetically forbidden, thereby reducing the number of pathways for singlet to triplet interconversion. In fact, here, the low field effect arises from symmetry breaking: the application of a weak magnetic field lifts degeneracies among the zero field eigenstates and increases the number of pathways for singlet to triplet interconversion. However, the details of this effect are more subtle which Manolopoulos *et al.* presented with a complete analysis of the 'low field effect' in a radical pair containing a single proton and in a radical pair in which one of the radicals contains a large number of hyperfine-coupled nuclear spins. They arrived at the new findings, i.e., the new transitions that occur when the field is switched on are between S and T_0 in both cases, and not between S and $T_{\pm}$ as has previously been claimed. They illustrated their result by using it in conjunction with semiclassical spin dynamics simulations which accounts for the observation of a biphasic-triphasic-biphasic transition with increasing magnetic field strength This has been described as the magnetic field effect on the time-dependent survival probability of a photo-excited carotenoid-porphyrin-fullerene radical pair.

5.7.9 *Molecular Basis of Cryptochrome Dependent Magnetosensitivity*

Bradlaugh *et al.*[190] discussed in his work in detail of the flavoprotein CRYPTOCHROME (CRY), now generally, believed to be a magnetosensor, providing geomagnetic information via a quantum effect on a light-initiated radical pair reaction.

The present situation is such that even after the tremendous efforts having considerable physical and behavioral data to support this view, the precise molecular basis of animal magneto-sensitivity remains frustratingly unknown. A key reason for this is the difficulty in combining molecular and behavioral biological experiments with the sciences, i.e., of magnetics and spin chemistry. By now, we are already aware about certain capabilities of bird related to flying, i.e., birds can use the geomagnetic field for compass orientation. Behavioral experiments, mostly with migrating passerines, revealed three characteristics of the avian magnetic compass: (1) it works spontaneously only in a narrow functional window around the intensity of the ambient magnetic field, but can adapt to other intensities, (2) it is an "inclination compass", not based on the polarity of the magnetic field, but the axial course of the field lines, and (3) it requires short-wavelength light from UV to 565 nm Green. The Radical Pair-Model of magnetoreception can explain these properties by proposing spin-chemical processes in photopigments as underlying mechanism. Applying radio frequency fields, a diagnostic tool for radical pair processes, supports an involvement of a radical pair mechanism in avian magnetoreception: added to the geomagnetic field, they disrupted orientation, presumably by interfering with the receptive processes. Cryptochromes have been suggested as receptor molecules.

Cry1a is found in the eyes of birds, where it is located at the membranes of the disks in the outer segments of the UV-cones in chickens and robins.

Immuno-histochemical studies show that it is activated by the wavelengths of light that allow magnetic compass orientation in birds. Birds can use the geomagnetic field for compass orientation. Behavioral experiments, mostly with migrating passerines, revealed three characteristics of the avian magnetic compass:

(1) it works spontaneously only in a narrow functional window around the intensity of the ambient magnetic field, but can adapt to other intensities,
(2) it is an "inclination compass", not based on the polarity of the magnetic field, but the axial course of the field lines, and
(3) it requires short-wavelength light from UV to 565 nm Green.

The Radical Pair-Model of magnetoreception can explain these properties by proposing spin-chemical processes in photopigments as underlying mechanism. Applying radio frequency fields, a diagnostic tool for radical pair processes, supports an involvement of a radical pair mechanism in avian magnetoreception: added to the geomagnetic field, they disrupted orientation, presumably by interfering with the receptive processes. Cryptochromes have been suggested as receptor molecules.

The precise biophysical origin of animal magnetoreception remains still to be cleared fully. The radical pair mechanism (RPM) hypothesis of magnetoreception was first posited in the late 1970s, following the discovery that electron transfer and related processes can generate a pair of radicals with properties (singlet and triplet spin states) that can be affected by exposure to a magnetic field (MF). Ritz in 2000, first suggested that the blue-light (BL)-sensitive protein CRYPTOCHROME (CRY) might be the elusive magnetoreceptor in magnetically sensitive organisms, based on the fact that the photochemistry of CRY is mediated by the photoexcitation of a bound cofactor, flavin adenine dinucleotide (FAD), and a subsequent electron transfer to FAD from a chain of neighbouring tryptophan residues, generating a radical pair (RP) consisting of a flavin semiquinone (FAD•−) and an oxidised Trp (TrpH•+). Electron transfer has been proposed to be mediated by a triad of Trp residues in CRY, although a fourth Trp residue has also recently been implicated, raising the idea of a Trp-tetrad and/or possible redundancy in the pathway. This radical pair (RP) initially forms with correlated spins that, as previous work on similar systems has indicated, could be influenced by an external magnetic field. Based on a large body of subsequent work the generally accepted mechanism requires an RP, generated by photo-reduction of FAD, to undergo interconversion between the singlet and triplet states. The relative population of each spin state is altered by exposure to an MF. Very importantly, it has been suggested by Bradlaugh and his group that in the canonical model, the reverse reaction in CRY (electron returning to TrpH•+ from FAD•−) can only occur when the RP (Radical Pair) is in the singlet state. Thus, exposure to an MF (magnetic field) is predicted to influence the probability of the reverse reaction occurring, thus modulating the half-life of "active" CRY, correlating with the flavin radical.

We know that birds can only sense magnetic fields if certain wavelengths of light are available — specifically, studies have shown that avian magnetoreception seems dependent on blue light. This seems to confirm that the mechanism is a visual one, based in the cryptochromes, which may be able to detect the fields because of quantum coherence. There exists two basic class of main theories of avian magnetorecption — one theory holds that birds navigate using particles of magnetite located in their head, the other holds that the birds sense the magnetic field as result of photo excitation in the eye whose deexcitation product is affected by the magnetic field. The excitation, which occurs inside the cryptochrome protein in the presence of blue light, produces a radial pair whose singlet and triplet states are essentially degenerate. In the presence of a magnetic field, more radical pairs are pushed into the triplet state, leading to an increased production of triplet de-excitation products. The radical pair thus acts as a quantum chemical compass in the eye — implying that birds can "see" magnetic fields. Cry1a is found in the eyes of birds, where it is located at the membranes of the disks in the outer segments of the UV-cones in chickens and robins. Immuno-histochemical studies show that it is activated by the wavelengths of light that allow magnetic compass orientation in birds.

But, even after that, as usual, the problem remains still unresolved, i.e., whilst the CRY-RPM may provide an attractive explanation for a biological magnetoreceptor, until such a mechanism is shown to directly result in a physiological response, like an electrical response in a receptor cell, all mechanisms remain hypothetical. Thus far, an unequivocal demonstration has proven elusive. To enable biological testing of physical data requires model organisms, facilitating a combination of behavioral, cellular, molecular and genetic manipulation, all together at a time. In this respect, however, the fruit fly, Drosophila melanogaster, offers these advantages, allowing either the entire nervous system or individual neurons to be genetically manipulated and subsequently tested for response to BL (Brain Lobes) with/without an external MF (Magnetic Fields) by electrophysiology. Bradlaugh, together with his group, extensively reviewed this problem, thus establishing this insect to be magnetosensitive through a CRY-dependent mechanism and highlighted the studies, uncovering the mechanistic basis for this "sixth sense".[194] Of course, an often-voiced criticism of using Drosophila, or other insects, to study magnetosensitivity is that they 'do not navigate'. This is certainly true, but does not negate the fact that Drosophila, and other insects, have been unequivocally shown to be able to sense, and to be influenced by, applied Magnetic Fields (MFs).[195] Indeed, **all animals so far studied, from insects through to birds, seemingly share this ability, perhaps, indicating that magnetosensitivity is a primitive sense**. Equally, it seems probable that in animals (including the Monarch butterfly — Danaus plexippus) that do navigate, this sense has been further refined in providing, not only magneto-sensitivity but also, acting as a compass. Thus, regardless of these reservations, Drosophila provides a highly tractable "living test-tube" to explore the mechanistic basis of magneto-sensitivity in a biological system.

Another interesting example of such kind of convincing evidence related to the use of magnetic compass is the migrant monarch butterfly (*Danaus plexippus*) during their spectacular navigational capabilities. Guerra *et al.*,[195] by applying flight simulator technique showed that these migrants indeed posses an inclination magnetic compass which they use in directing their long flight towards equator ward in the fall. During the fall, migration route for eastern North American monarch butterflies (Danaus plexippus) is, from their northern range, to over winter in sites atop the mountains of Michoacán in central Mexico. Monarchs use an antenna-based time-compensated sun compass for the navigation during this long journey in which eye-sensed directional daylight cues,

for example, sun azimuthal position) are integrated in the sun compass found in the mid-brain central complex area and references there in and time compensated by antennal circadian clocks. It is quite interesting and seriously important fact that, as well, during the 'absence' of directional daylight cues that precludes the use of the time-compensated sun compass (for example, overcast sky conditions), migrants have been observed flying in the expected southern migratory direction during the fall migration, suggesting that migrants 'might also use a magnetic compass' to help guide 'directionality'. Again, it has already been observed that the use of inclination compass is light dependent, i.e., it utilizes ultraviolet-A/blue light between 380nm and 420 nm. In this regard, the function of antennae are quite important for the inclination compass as they appear to contain light-sensitive magneto-sensors. Thus, like other kind of migratory species, for migratory monarchs also, the inclination compass may serve as an important orientation mechanism, especially, in absence of directional daylight cues and may also augment time-compensated sun compass orientation for appropriate directionality, throughout the migration. Krylov *et al.*[196] reviewed in detail the orientation and navigation of animals related to different taxa by considering the 'geomagnetic field' in which the mechanisms of magnetoreception in animals have been described. They established the fact that several taxa of fishes perceive the magnetic field via electro-receptors. Again, they also found that some animals are able to sense changes in the magnetic field polarity via sensory cells containing iron compounds. In addition, animals from different taxa are able to perceive the inclination of the magnetic field via a change in singlet and triplets yields of radical-pair reactions under the influence of magnetic field. In such cases, they also observed that a very crucial role is played by the molecules of the cryptochrome which are considered to be excited by 'the shortwave light' capable of producing long-living radical pairs affected by the geomagnetic field.

Lachlan P Lindoy *et al.*;[200] proposed an idea that the stochastic Schrödinger equation (SSE) provides an ideal way to simulate the quantum mechanical spin dynamics of radical pairs. In this, they utilised an well know fact that radical pair recombination reactions are sensitive to extremely weak magnetic fields and can therefore be said to function as molecular magnetoreceptors. Electron spin relaxation effects arising from fluctuations in the spin Hamiltonian are straightforward to include in this approach, and their treatment can be combined with a highly efficient stochastic evaluation of the trace over nuclear spin states, required to compute experimental observables. These features are illustrated in example applications to a flavin-tryptophan radical pair of interest in avian magnetoreception, and to a problem involving spin-selective radical pair recombination along a molecular wire. The classic example is a carotenoid-porphyrin-fullerene (C•+PF•−) radical pair that has been shown to provide a "proof-of-principle" for the operation of a chemical compass. Several important simulations of this radical pair have already been employed by applying semiclassical approximations, which are routinely applicable to its 47 coupled electronic and nuclear spins. However, the above group[202] employed a calculation by applying the exact quantum mechanical spin dynamics, thus presenting a significant challenge. They developed in their approach, a method so that it has been possible in performing numerically converged simulations of the C•+PF•− quantum mechanical spin dynamics, including all coupled spins. A comparison of these kind of quantum mechanical simulations with various semiclassical approximations reveals that, while it is not perfect, the best semiclassical approximation does capture essentially all of the relevant physics in this problem.

5.7.10 *Conclusion*

Birds are capable of true navigation, the ability to return to a known goal from a place they have never visited before. This is demonstrated most spectacularly during the vast migratory journeys made by these animals year after year, often between continents and occasionally global in nature. Though great progress achieved in understanding the mystery of how birds navigate using magnetic cues, it still remains something of a mystery as to how they sense the magnetic field. However, great progress has been achieved in conclusive suggestions that birds sense magnetic values through a light-sensitive molecule called cryptochrome, or through sensory cells containing magnetic iron oxide particles – but definitive concrete evidence for either of these has not yet been provided. Also, behavioural evidence continues to underscore how the Earth's magnetic field is crucial in helping some birds make their epic journeys to breed each year – providing a global positioning system that might just provide birds with a complete navigational map of the world.At present, though the role of magnetic cues for compass orientation has been confirmed in numerous animals, the mechanism of detection is still being debated. Two hypotheses have been proposed, one based on a light dependent mechanism, apparently used by birds and another based on a "compass organelle" containing the iron oxide particles magnetite (Fe3O4). For example, bats have recently been shown to use magnetic cues for compass orientation but the method by which they detect the Earth's magnetic field still remains quite puzzlingly, not yet settled in a concrete fashion.

However, it still remains one of the great unanswered questions in science, despite more than 50 years of dedicated research in this field though the study of true navigation in birds has made significant advances in the previous 20 years, in part thanks to the integration of many disciplines outside its root in behavioral biology, to address questions of neurobiology, molecular aspects, and the physics of sensory systems and environmental cues involved in bird navigation, often involving quantum physics. However, true navigation remains still a difficult front, particularly for those outside or new to the field. Unlike many general texts on migration, which avoid discussion of these issues, this review have tried to present these conflicting findings and assess the state of the field of true navigation during bird migration.

References

[1] G. S. Engel *et al.*, Nature, 446(7137):782–786, 04 2007.
[2] Engel, S. *et al.* Anthropogenic electromagnetic noise disrupts magnetic compass orientation in a migratory bird. Nature 509, 353–356 (2014).
[3] M. Mohseni *et al.*, J. Chem. Phys., 129(17):174106, 2008.
[4] M. B. Plenio and S. F. Huelga, NJP, 10(11):113019, 2008.
[5] M. Sarovar *et al.*, Nature Physics, 6, 462 (2010).
[6] S. Lloyd *et al.*, Nature Physics, 5(3):164–166, 03 2009.
[7] J. Cai *et al.*, Phys. Rev. E 82, 021921 (2010), 0809.4906, 2008.
[8] J. Cai, F. Caruso, and M. Plenio, Physical Review A 85, 040304 (2012).
[9] A. von Middendorff, Die Isepiptesen Russlands, Mem. Acad. Sci. St. Petersbourg VI, Ser. Tome. 8 (1859).
[10] L. E. Foley, R. J. Gegear, S. M. Reppert, Human cryptochrome exhibits light-dependent magnetosensitivity, Nature Communications 2(2010) 1.
[11] R. J. Gegear, L. E. Foley, A. Casselman, S. M. Reppert, Animal cryptochromes mediate magnetoreception by an unconventional photochemical mechanism, Nature 463 (2010) 804.
[12] S. Johnsen and K. J. Lohmann *et al.*, Physics Today, 61(3):29–35, 2008.
[13] T. Ritz, S. Adem, and K. Schulten, Biophysical Journal, 78(2): 707–718, 2000.
[14] N. Keary *et al.*, Frontiers in Zoology, 6:25, 2009.
[15] R. J. Gegear *et al.*, Nature, 454(7207):1014–1018, 08 2008.
[16] T. Yoshii *et al.*, PLoS Biol, 7(4):e1000086, 2009.
[17] M. Ahmad *et al.*, Planta, 225(3):615–624, 02 2007.
[18] T. Ritz *et al.*, Nature, 429(6988):177–180, 05 2004.
[19] C. R. Timmel and K. Henbest, Phil. Trans. Roy. Soc. London A, 362(1825):2573–2589, 2004.
[20] Y. Liu *et al.*, Chem. Comm., 1(2):174–176, 2005.
[21] T. Miura, K. Maeda, and T. Arai, J. Phys. Chem. A, 110(12):4151–4156, 03 2006.
[22] C. T. Rodgers, Pure and Applied Chem., 81(1):19–43, 2009.
[23] C. T. Rodgers and P. J. Hore, PNAS, 106(2):353–360, 01, 2009.

[24] K. Maeda *et al.*, Nature, 453(7193):387–390, 05, 2008.
[25] Mouritsen H, Hore PJ, The magnetic retina: Light-dependent and trigeminal magnetoreception in migratory birds. Curr Opin Neurobiol 22(2):343–352, 2012.
[26] Hamish G Hithcock 2016; The Quantum needle of the avian magnetic compass; PNAS.
[27] Wiltschko W. *et al.* (1972); W. Wiltschko and R. Wiltschko; 2005; Journal of Comparative Physiology A: Neuroethology, Sensory, Neural, and Behavioral Physiology 191, 675 and references there in.
[28] R. Muheim, Photobiology, 465 (2008).
[29] R. Wiltschko, I. Schiffner, P. Fuhrmann, and W. Wiltschko, Current Biology 20, 1534, 2010.
[30] W. Wiltschko, R. Wiltschko, and T. Ritz, Procedia Chemistry 3, 276, 2011.
[31] S. Johnsen and K. Lohmann, Nature Reviews Neuroscience 6, 703, 2005.
[32] K. Able, The condor 97, 592, 1995.
[33] Zachary Brett Walters; (2012); arXiv:1208; 2558v3[physics.bio-Ph].
[34] Schulten, K.; Staerk, H.; Weller, A.; Werner, H.J.; Nickel, B.; 1976; Magnetic field dependence of the geminate recombination of radical ion pairs in polar solvents. Z. Phys. Chem. 1976, 101, 371–378.
[35] K. Schulten, C.E. Swenberg, A. Weller; 1978; Zeitschrift für Physikalische Chemie NF111, 1978.
[36] L. Wu and J. Dickman; 2011; Current Biology.
[37] L. Wu and J. Dickman; 2012; Science.
[38] C. D. Treiber, M. C. Salzer, J. Riegler, N. Edelman, C. Sugar, M. Breuss, P. Pichler, H. Cadiou, M. Saunders, M. Lythgoe, *et al.*; Nature, Vol.484, 19 April, 2012.
[39] W. Beck and W. Wiltschko, Behaviour, 145 (1983).
[40] R. Muheim, J. Bäckman, and S. Ákesson, Journal of Experimental Biology 205, 3845 (2002).
[41] T. Ritz, R. Wiltschko, P. Hore, C. Rodgers, K. Stapput, P. Thalau, C. Timmel, and W. Wiltschko, Biophysical Journal 96, 3451 (2009).
[42] T. Ritz, P. Thalau, J. Phillips, R. Wiltschko, and W. Wiltschko, Nature 429, 177 (2004).
[43] T. Ritz, S. Adem, and K. Schulten, Biophysical Journal 78, 707 (2000).
[44] Stoneham AM, Gauger EM, Porfyrakis K, Benjamin SC, Lovett BW (2012) A new type of radical-pair-based model for magnetoreception. Biophys J 102(5):961–968.
[45] K. Blum, Density matrix theory and applications (Springer, 2012).

[46] E. M. Gauger, E. Rieper, J. J. Morton, S. C. Benjamin, and V. Vedral; 2011; Physical review letters 106, 40503.
[47] Bandyopadhyay JN, Paterek T, Kaszlikowski D (2012) Quantum coherence and sensitivity of avian magnetoreception. Phys Rev Lett 109(11):110502.
[48] Cai J, Guerreschi GG, Briegel HJ (2010) Quantum control and entanglement in a chemical compass. Phys Rev Lett 104(22):220502.
[49] A. Johnson, J. Petta, J. Taylor, A. Yacoby, M. Lukin, C. Marcus, M. Hanson, and A. Gossard, Nature 435, 925 (2005).
[50] J. Petta, A. Johnson, J. Taylor, E. Laird, A. Yacoby, M. Lukin, C. Marcus, M. Hanson, and A. Gossard, Science 309, 2180 (2005).
[51] Deutschlander, M.E., Freake, M.J., Borland, S.C., Phillips, J.B., Madden, R.C., Anderson, L.E., and Wilson, B.W., Learned magnetic compass orientation by the Siberianhamster, Phodopus sungorus, Anim. Behav., 2003, vol. 65, no. 4, pp. 779–786.
[52] Phillips, J.B., Youmans, P.W., Muheim, R., Sloan, K.A., Landler, L., Painter, M.S., and Anderson, C.R., Rapid learning of magnetic compass direction by C57BL/6 mice in a 4-armed 'plus' water maze, PLoS One, 2013, vol. 8, no. 8, p. e73112.
[53] Malewski, S., Begall, S., and Burda, H., Learned and spontaneous magnetosensitive behavior in the Roborovskihamster (Phodopus roborovskii), Ethology, 2018, vol. 28, no. 6, pp. 423–431.
[54] Holland, R., Thorup, K., Vonhof, M.J., Cochran, W.W., and Wikelski, M., Bat orientation using earth's magnetic field: Nature, 2006, vol. 444, no. 7120, p. 702.
[55] Diego-Rasilla, F.J., Luengo, R.M., and Phillips, J.B.; 2013; Use of a light-dependent magnetic compass for y-axis orientation in European common frog (Rana temporaria) tadpoles, J. Comp. Physiol. A, 2013, vol. 199, no. 7, pp. 619–628.
[56] Shakhparonov, V.V. and Ogurtsov, S.V., Marsh frogs, Pelophylax ridibundus, determine migratory direction by magnetic field, J. Comp. Physiol. A, 2017, vol. 203, no. 1, pp. 35–43.
[57] Putman, N.F., Endres C.S., Lohmann, C.M.F., and Lohmann, K.J., Longitude perception and bicoordinate magnetic maps in sea turtles, Curr. Biol., 2011, vol. 21, pp. 463–466.
[58] Putman, N.F., Scanlan, M.M., Billman, E.J., O'Neil, J.P., Couture, R.B., Quinn, T.P., Lohmann, K.J., and Noakes, D.L.G., An inherited magnetic map guides ocean navigation in juvenile Pacific salmon, Curr. Biol., 2014, vol. 24, pp. 446–450.
[59] Naisbett-Jones, L.C., Putman, N.F., Stephenson, J.F., Ladak, S., and Young, K.A., A magnetic map leans juvenile European eels to the Gulf Stream, Curr. Biol., 2017, vol. 27, no. 8, pp. 1236–1240.

[60] Lohmann, K.J., Hester, J.T., and Lohmann, C.M.F., Long-distance navigation in sea turtles, Ethol. Ecol.vol., 1999, vol. 11, no. 1, pp. 1–23.

[61] Kishkinev, D., Chernetsov, N., Heyers, D., and Mouritsen, H., Migratory reed warblers need intact trigeminalnerves to correct for a 1, 000 km eastward displacement, PLoS One, 2013, vol. 8, no. 6, p. e65847.

[62] Kishkinev D., Chernetsov, N., Pakhomov, A., Heyers, D., and Mouritsen, H., Eurasian reed warblers compensate for virtual magnetic displacement, Curr. Biol., 2015, vol. 25, no. 19, pp. R822–R824.

[63] Chernetsov, N., Pakhomov, A., Kobylkov, D., Kishkinev, D., Holland, R.A., and Mouritsen, H., Migratory Eurasianreed warblers can use magnetic declination to solve the longitude problem, Curr. Biol., 2017, vol. 27, no. 17, pp. 2647–2651.

[64] Pakhomov, A., Anashina, A., Heyers, D., Kobylkov, D., Mouritsen, H., and Chernetsov, N., Magnetic map navigation in a migratory songbird requires trigeminal input, Sci. Rep., 2018, vol. 8, p. 11975.

[65] Zapka *et al.* (2009); Zapka, M., Heyers, D., Hein, C.M., Engels, S. *et al.*; 2009; Visual, but not trigeminal, mediation of magnetic compass information in a migratory bird, Nature, 2009, vol. 461, no. 7268, pp. 1274–1277.

[66] Hore, P.J. & Mauritzson, H.; (1993); Hore, P.J. and Mouritsen, H., The radical-pair mechanism of magnetoreception, Annu. Rev. Biophys., 2016, vol. 45, pp. 299–344.

[67] K. Kavokin, Bioelectromagnetics 30, 402 (2009).

[68] Heyers D, Manns M, Luksch H, Güntürkün O, Mouritsen H; (2007); A visual pathway links brain structures active during magnetic compass orientation in migratory birds. PLoS One 2(9):e937.

[69] W. Beck and W. Wiltschko, Behaviour, 145 (1983).

[70] T. Ritz, P. Thalau, J. Phillips, R. Wiltschko, and W. Wiltschko, Nature 429, 177 (2004).

[71] I. Kominis *et al.*, Physical Review-Section E-Statistical Nonlinear and Soft Matter Physics 80, 56115 (2009).

[72] J. Jones and P. Hore, Chemical Physics Letters 488, 90(2010).

[73] I. Kominis, Physical Review E 83, 056118 (2011).

[74] A. Dellis and I. Kominis, Biosystems (2011).

[75] Mauritsen Henrik; 2015; [magnetoreception on Birds and its use for Long-distance migration-Henrik Mauritsen in book: Struke's Avian physiology:Edition sixth:Chap:magnetoreception in Birds and its use for long-Distance Migration, Asssociate Press; Ed: Colin G. Scanes.

[76] Wong SY, Wei Y Mauritsen H., Solovyov IA, Hore PJ; 2021; Journal of Royal Soc. Interface, Nov 10, 18(184)2021–10601.

[77] M. Ahmad *et al.*, Planta, 225(3):615–624, 02 2007.

[78] Haberkorn, R. & Michel-Beyerle, M. On the mechanism of magnetic feld effects in bacterial photosynthesis. Biophys. J. 26, 489–498 (1979).

[79] Steiner, U. E. & Ulrich, T. Magnetic feld effects in chemical kinetics and related phenomena. Chem. Rev. 89, 51–147 (1989).

[80] Salikhov, K. M., Molin, Y. N., Sagdeev, R. Z. & Buchachenko, A. L. Spin polarization and magnetic effects in radical reactions (Elsevier, 1984).

[81] Biskup, T. *et al.* Direct observation of a photoinduced radical pair in a cryptochrome blue-light photoreceptor. Angewandte Chemie International Edition 48, 404–407 (2009).

[82] Yoshii, T., Ahmad, M. & Helfrich-Förster, C. Cryptochrome mediates light-dependent magnetosensitivity of Drosophila's circadian clock. PLoS Biology 7, 813–9 (2009).

[83] Fedele, G. *et al.* Genetic Analysis of Circadian Responses to Low Frequency Electromagnetic Fields in Drosophila melanogaster. PLoS Genetics 10, e1004804 (2014).

[84] Buchachenko, A. L. & Kuznetsov, D. A. Magnetic field affects enzymatic ATP synthesis. J. Am. Chem. Soc. 130, 12868–12869 (2008).

[85] Messiha, H. L., Wongnate, T., Chaiyen, P., Jones, A. R. & Scrutton, N. S. Magnetic field effects as a result of the radical pair mechanism are unlikely in redox enzymes. J. R. Soc. Interface 12, 20141155 (2015).

[86] Bradlaugh A. *et al.*; 2021; Exploiting the fruitfly Drosophila melanoggaster, or indentify the molecular Basis of Cryptochrome(CRY)-dependent magnetosensitivity-Quantum Rep., 2021; 3; 127–136.

[87] Haberkorn, R. Density matrix description of spin-selective radical pair reactions. Mol. Phys. 32, 1491 (1976).

[88] Tiersch, M., Steiner, U. E., Popescu, S. & Briegel, H. J. Open Quantum System Approach to the Modeling of Spin Recombination Reactions. J Phys Chem A. 116(16), 4020–8 (2012).

[89] Kominis, I. K. Te radical-pair mechanism as a paradigm for the emerging science of quantum biology. Mod. Phys. Lett. B 29, 1530013 (2015).

[90] Cintolesi, F., Ritz, T., Kay, C. W. M., Timmel, C. R. & Hore, P. J. Anisotropic recombination of an immobilized photoinduced radical pair in a 50-T magnetic feld: a model avian photomagnetoreceptor. Chemical Physics 294, 385–399 (2003).

[91] Hamdouni, Y. On the partial trace over collective spin degrees of freedom. Physics Letters A 373, 1233–1238 (2009).
[92] I. A. Solovyov and K. Schulten, Biophysical Journal 96(12): 4804–4813, 2009.
[93] O. Efimova and P. J. Hore, Molecular Physics, 107(7):665, 2009.
[94] M. A. Nielsen and I. L. Chuang. Quantum Computation and Quantum Information. Cambridge University Press, October 2000.
[95] J. J. L. Morton *et al.*, J. Chem. Phys., 124(1):014508, 2006.
[96] Ritz T.; Ahmad M. *et al.*; 2010; J.R. Society Interface; 7, p. S135.
[97] W. Wiltschko, K. Stapput, P. Thalau, and R. Wiltschko, Naturwissenschaften 93, 300 (2006).
[98] Betony Adams, Ilya Sinayskiy & Francesco Petruccione; 2018; Scientific Reports; 24 October, 2018.
[99] Perelomov, A. Generalized Coherent States and Teir Applications (Springer-Verlag, 1986).
[100] Wang, X. & Mølmer, K. Pairwise entanglement in symmetric multi-qubit systems. Eur. Phys. J. D 18, 385–391 (2002).
[101] Talau, P., Ritz, T., Stapput, K., Wiltschko, R. & Wiltschko, W. Magnetic compass orientation of migratory birds in the presence of a 1.315 MHz oscillating feld. Naturwissenschafen 92, 86–90 (2004).
[102] Engels, S. *et al.* Anthropogenic electromagnetic noise disrupts magnetic compass orientation in a migratory bird. Nature 509, 353–356 (2014).
[103] Weil, J. & Bolton, J. R. Electron Paramagnetic Resonance: Elementary Teory and Practical Applications (John Wiley and Sons, 2007).
[104] Lau, J. C. S., Wagner-Rundell, N., Rodgers, C. T., Green, N. J. B. & Hore, P. J. Efects of disorder and motion in a radical pair magnetoreceptor. J. R. Soc. Interface 7, S257–S264 (2009).
[105] Liedvogel, M. *et al.* Chemical Magnetoreception: Bird Cryptochrome 1a Is Excited by Blue Light and Forms Long-Lived RadicalPairs, PLoS One 2(10), e1106 (2007).
[106] Pinzon-Rodriguez, A., Bensch, S. & Muheim, R. Expression patterns of cryptochrome genes in avian retina suggest involvement of Cry4 in light-dependent magnetoreception. J. R. Soc. Interface 15, 20180058 (2018).
[107] Günther, A. *et al.* Double-cone localization and seasonal expression pattern suggest a role in magnetoreception for European Robin Cryptochrome 4. Current Biology 28, 1–13 (2018).
[108] Sancar, A. Cryptochrome: the second photoactive pigment in the eye and its role in circadian photoreception. Annu Rev Biochem 69, 31–67 (2000).

[109] Fleissner, G., B. Stahl, G. Fleissner. 2007. A novel concept of Fe-mineral-based magnetoreception: histological and physicochemical data from the upper beak of homing pigeons. Naturwissenschaften. 94:631–642.

[110] Solovyov, I. A., and W. Greiner. 2007. Theoretical analysis of an iron mineral-based magnetoreceptor model in birds. Biophys. J. 93:1493–1509.

[111] Solovyov, I. A., and W. Greiner. 2009. Iron-mineral-based magnetoreceptor in birds: polarity or inclination compass? Eur. Phys. J. D. 51:161–172.

[112] Solovyov, I. A., and W. Greiner. 2009. Micromagnetic insight into a magnetoreceptor in birds: on the existence of magnetic field amplifiers in the beak. Phys. Rev. E. 80, 041919-1–10.

[113] Kirschvink, J. L., M. M. Walker, and C. E. Diebel. 2001. Magnetite-based magnetoreception. Curr. Opin. Neurobiol. 11:462–4.

[114] Falkenberg, G., G. Fleissner, G. Fleissner. 2010. Avian magnetoreception: elaborate iron mineral containing dendrites in the upper beak seem to be a common feature of birds. PLoS One. 5:e9231.

[115] Mouritsen, H., and T. Ritz. 2005. Magnetoreception and its use in bird navigation. Curr. Opin. Neurobiol. 15:406–414.

[116] Schulten, K. 1982. Magnetic field effects in chemistry and biology. In Advances in Solid State Physics [Festkoŕperprobleme], Vol. 22 J. Treusch, editor. Vieweg, Braunschweig/Wiesbaden, Germany.

[117] Canfield, J. M., R. L. Belford, K. Schulten. 1995. A perturbationtreatment of oscillating magnetic fields in the radical pair mechanismusing the Liouville equation. Chem. Phys. 195:59–69.

[118] Hore, P.J.; Mouritsen, H. The Radical-Pair Mechanism of Magnetoreception. Annu. Rev. Biophys. 2016, 45, 299–34.

[119] Vaidya, A.T.; Top, D.; Manahan, C.C.; Tokuda, J.M.; Zhang, S.; Pollack, L.; Young, M.W.; Crane, B.R.; 2013; Flavin reduction activates Drosophila cryptochrome. Proc. Natl. Acad. Sci. USA 2013, 110, 20455–20460.

[120] Mouritsen, H., U. Janssen-Bienhold, R. Weiler. 2004. Cryptochromes and neuronal-activity markers colocalize in the retina of migratory birds during magnetic orientation. Proc. Natl. Acad. Sci. USA. 101:14294–14299.

[121] Möller, A.; Sagasser, S.; Wiltschko W. *et al.* 2004 Retinal cryptochrome in a migratory passerine bird: A possible transducer for the avian magnetic compass; Naturewissenschaften 91; 585–588.

[122] Mouritsen, H., G. Feenders, E. D. Jarvis.; 2005; Night-vision brain area in migratory songbirds. Proc. Natl. Acad. Sci. USA. 102:8339–8344 2005.

[123] Zeugner, A., M. Byrdin, M. Ahmad. 2005. Light-induced electron transfer in Arabidopsis cryptochrome-1 correlates with in vivo function.J. Biol. Chem. 280:19437–19440.
[124] Giovani, B., M. Byrdin, K. Brettel. 2003. Light-induced electron transfer in a cryptochrome blue-light photoreceptor. Nat. Struct. Biol. 10:489–490.
[125] Kottke, T., A. Batschauer, J. Heberle. 2006. Blue-light-induced changes in Arabidopsis cryptochrome 1 probed by FTIR difference spectroscopy. Biochemistry. 45:2472–2479.
[126] O'Day, K. E. 2008. Shedding light on animal cryptochromes. PLoS Biol. 6:1359–1360.
[127] Hogben, H. J., O. Efimova, P. Hore. 2009. Possible involvement ofsuperoxide and dioxygen with cryptochrome in avian magnetoreception: origin of Zeeman resonances observed by in vivo EPR spectroscopy. Chem. Phys. Lett. 490:118–122.
[128] Feenders, G., M. Liedvogel, E. Jarvis. 2008. Molecular mapping of movement-associated areas in the avian brain: a motor theory for vocal learning origin. PLoS ONE. 3:e1768.
[129] Ilia A. Solovyev, Mouritson Henrik, Schulten klaus; 2010; Acuity of cryptochrome and vision based on magnetoreception systems in birds; July 2010, Biophysical Journal 99(1):40-9 DOI: 10.1016/j.bpj.2010.03.053.
[130] Ilia A. Sólovyev; Ritz T., Schulten K. and Hore, P.J.; 2014; A chemical compass for bird navigation; In quantum effects of biology; M. Mohseni *et al.* eds (Cambridge University Press); pp. 218–236.
[131] Sancar, A.: 2003; Structureand function of DNA photolyse and cryptochrome blue-light photoreceptors; Chem.Rev.; 103; 2203–2237.
[132] Kume, K., Zylka, M.J., Sriram, S. *et al.*; 1999; mCRY1 and mCRY2 are essential components of the negative limb of the circadian clock feedback loop. Cell 98, 193–205.
[133] Lau, J.C. Wagner-Rundell, N.; Rodgers, C.T. *et al.*; 2010; Effects of disorderand motion in a radical pair magnetoreceptors; J.R. Soc. Interface; 7(Suppl2); S257–S264.
[134] Niessner C., Denzau, S., S. Gross *et al.* 2011; Avian ultraviolet/violet cones identified as probable magnetoreceptors. PLOs ONE 6; e20091.
[135] Wateri, R.; Yamaguchi, C.; Zamba W., *et al.*; 2012; Light-dependent structural change of chicken retinal Cryptochrom 4; J. Biol. Chem. 287; 42634–42641.
[136] Fusani, L., Bertolucci, C., Frigato, E. *et al.* 2014; Cryptochrome expression in the eye of migratory birds depends on the migratory status; J. Exp. Biol. 217., 918–923.

[137] Bolte, P., Bliebaum, F.; Einwich, A. *et al.* 2016; Localization of the putative magnetoreceptive protein cryptochrome 1b in the retinae of migratory birds and homing pigeons.; PLoS ONE 11; e0147819.

[138] Niessner C., Gross, J.C.; Denzai, S. *et al.*; 2016; Seasonally changing cryptochrome1b expression in the retinal ganglion cells of a migrating passerine bird PloS ONE 11, e0150377.

[139] Qin S., Yin, H. Yang C. *et al.*; 2016; A magnetic protein biocompass; Nat Mater; 15, 217–226.

[140] Y. T. Zhang, G. P. Berman and S. Kais, Phys. Rev. E, 2014, 90, 042707.

[141] Thomas P. Fay, Lachlan P. Lindoy, David E. Manolopoulos and P. J. Ho; 2020; How quantum is radical pair magnetoreception; Faraday Discuss; 221, 77.

[142] Kattnig, I. A. Solovyov and P. J. Hore, Phys. Chem. Chem. Phys., 2016, 18, 12443–12456.

[143] S. F. Huelga and M. B. Plenio, Contemp. Phys., 2013, 54, 181–207.

[144] N. Lambert, Y. N. Chen, Y. C. Cheng, C. M. Li, G. Y. Chen and F. Nori, Nat. Phys., 2013, 9, 10–18.

[145] M. Mohseni, Y. Omar, G. S. Engel and M. B. Plenio, eds., Quantum effects in biology, Cambridge University Press, Cambridge, 2014.

[146] A. Marais, B. Adams, A. K. Ringsmuth, M. Ferretti, J. M. Gruber, R. Hendrikx, M. Schuld, S. L. Smith, I. Sinayskiy, T. P. J. Kruger, F. Petruccione and R. van Grondelle, J. R. Soc. Interface, 2018, 15, 20180640.

[147] Al-Khalili J, McFadden J (2014) Life on the Edge: The Coming of Age of Quantum Biology (Bantam Press, London).

[148] Mauritson H., Nature, 2018, 558, 50–59.

[149] K. Maeda, A. J. Robinson, K. B. Henbest, H. J. Hogben, T. Biskup, M. Ahmad, E. Schleicher, S. Weber, C. R. Timmel and P. J. Hore, Proc. Natl. Acad. Sci. USA, 2012, 109, 4774–4779.

[150] K. Schulten and P. G. Wolynes, J. Chem. Phys., 1978, 68, 3292–3297.

[151] Manolopoulos DE, Hore PJ (2013) An improved semiclassical theory of radical pair recombination reactions. J Chem Phys 139(12):124106.

[152] Lewis AM, Manolopoulos DE, Hore PJ (2014) Asymmetric recombination and electron spin relaxation in the semiclassical theory of radical pair reactions. J Chem Phys 141(4): 044111.

[153] Carrillo A, Cornelio MF, de Oliveira MC (2015) Environment-induced anisotropy and sensitivity of the radical pair mechanism in the avian compass. Phys Rev E Stat Nonlin Soft Matter Phys 92(1):012720.

[154] U. E. Steiner and T. Ulrich, Chem. Rev., 1989, 89, 51–147.

[155] Lachlan P Lindoy *et al.*; 2020; Quantum mechanical spin dynamics of a molecular magnetoreceptor J.Chem.phys.; 2020; Apr 300; 152(16):164107. doi:10.1063/5.0006411.
[156] A. Lewis, Spin Dynamics in Radical Pairs, Springer International Publishing, 2018.
[157] H. J. Hogben, T. Biskup and P. J. Hore, Phys. Rev. Lett., 2012, 109, 220501.
[158] W. H. Miller, J. Chem. Phys., 2012, 136, 210901.
[159] T. P. Fay, L. P. Lindoy and D. E. Manolopoulos, J. Chem. Phys., 2018, 149, 064107.
[160] K. L. Ivanov, M. V. Petrova, N. N. Lukzen and K. Maeda, J. Phys. Chem. A, 2010, 114, 9447–945.
[161] A. A. Lee, J. C. S. Lau, H. J. Hogben, T. Biskup, D. R. Kattnig and P. J. Hore, J. R.Soc., Interface, 2014, 11, 20131063.
[162] A. Schweiger and G. Jeschke, Principles of pulse electron paramagnetic resonance, Oxford University Press, New York, 2001.
[163] Wiltschko W, Munro U, Ford H, Wiltschko R (1993) Red-light disrupts magnetic orientation of migratory birds. Nature 364(6437): 525–527. Red-light disrupts magnetic ori 1993.
[164] Hein CM, *et al.* (2010) Night-migratory garden warblers can orient with their magnetic compass using the left, the right or both eyes. J R Soc Interface 7(Suppl 2):S227–S233.
[165] Hein CM, Engels S, Kishkinev D, Mouritsen H (2011) Robins have a magnetic compass in both eyes. Nature 471(7340):E11–E12, discussion E12–E13.
[166] Lüdemann G, Solov'yov IA, Kubar T, Elstner M (2015) Solvent driving force ensures fast formation of a persistent and well-separated radical pair in plant cryptochrome; J Am Chem Soc 137(3):1147–1156.
[167] Lau JCS, Rodgers CT, Hore PJ (2012) Compass magnetoreception in birds arising from photo-induced radical pairs in rotationally disordered cryptochromes. J R Soc Interface 9(77):3329–3337.
[168] Dellis AT, Kominis IK (2012) The quantum Zeno effect immunizes the avian compass against the deleterious effects of exchange and dipolar interactions. Biosystems 107(3):153–157.
[169] Pauls JA, Zhang Y, Berman GP, Kais S (2013) Quantum coherence and entanglement in the avian compass. Phys Rev E Stat Nonlin Soft Matter Phys 87(6):062704.
[170] Akesson S, Morin J, Muheim R, Ottosson U (2001) Avian orientation at steep angles of inclination: Experiments with migratory white-crowned sparrows at the magnetic North Pole. Proc Biol Sci 268(1479):1907–1913.

[171] Lefeldt N, Dreyer D, Schneider NL, Steenken F, Mouritsen H; (2015); Migratory blackcaps tested in Emlen funnels can orient at 85 degrees but not at 88 degrees magnetic inclination. J Exp Biol 218(Pt 2):206–211.

[172] A. M. Lewis, D. E. Manolopoulos and P. J. Hore, J. Chem. Phys., 2014, 141, 044111.

[173] C. C. Moser, J. L. R. Anderson and P. L. Dutton, Biochim. Biophys. Acta, Bioenerg., 2010, 1797, 1573–1586.

[174] D. R. Kattnig and P. J. Hore, Sci. Rep., 2017, 7, 11640.

[175] D. R. Kattnig, J. Phys. Chem. B, 2017, 121, 10215–10227.

[176] Gill RE, *et al.* (2009) Extreme endurance flights by landbirds crossing the Pacific Ocean: Ecological corridor rather than barrier? Proc Biol Sci 276(1656):447–457.

[177] Lee AA, *et al.* (2014); Alternative radical pairs for cryptochrome-based magnetoreception; J R Soc Interface 11(95):20131063.1.

[178] Rodgers CT (2007) Magnetic field effects in chemical systems. DPhil thesis (Univ of Oxford, Oxford, UK).

[179] Schwarze S, *et al.* (2016) Weak broadband electromagnetic fields are more disruptive to magnetic compass orientation in a night-migratory songbird (Erithacus rubecula).

[180] Timmel CR, Cintolesi F, Brocklehurst B, Hore PJ (2001) Model calculations of magnetic field effects on the recombination reactions of radicals with anisotropic hyperfine interactions. Chem Phys Lett 334(4–6):387–395.

[181] Gauger EM, Benjamin SC (2013) Comment on "Quantum coherence and sensitivity of avian magnetoreception". Phys Rev Lett 110(17):178901.

[182] Lambert N, De Liberato S, Emary C, Nori F (2013) Radical-pair model of magnetoreception with spin-orbit coupling. New J Phys 15(8):083024.

[183] Till U, Timmel CR, Brocklehurst B, & Hore PJ (1998) The influence of very small magnetic fields on radical recombination reactions in the limit of slow recombination. Chem. Phys. Lett. 298(1-3):7-14.

[184] Levy C, *et al.* (2013) Updated structure of Drosophila cryptochrome. Nature 495:E3- E4.

[185] Müller P, Yamamoto J, Martin R, Iwai S, & Brettel K (2015) Discovery and functional analysis of a 4th electron-transferring tryptophan conserved exclusively in animal cryptochromes and (6-4) photolyases. Chem. Commun. 51(85):15502-15505.

[186] Worster, S., H. Mouritsen, and P. J. Hore. 2017. A light-dependent magnetoreception mechanism insensitive to light intensity and polarization. J. R. Soc. Interface. 14:20170405.

[187] Lau JCS (2013); D. Phil. thesis, University of Oxford.

[188] Ball P (2011) Physics of life: The dawn of quantum biology. Nature 474(7351):272–274.

[189] Cochran, W. W., H. Mouritsen, and M. Wikelski. 2004. Migrating songbirds recalibrate their magnetic compass daily from twilight cues. Science. 304:405–408.

[190] Nordmann, G. C., T. Hochstoeger, and D. A. Keays. 2017. Magnetoreception-A sense without a receptor; PLoS; Biol.15: e2003234.

[191] Noboru Ikeya & Jonathan Woodward; 2021; "Cellular autofluorescence is magnetic field sensitive" (PNAS 2021, 118, e2018043118) I.

[192] Ricard Holland *et al.*; 2008; Bats are magnetide to detect earth's magnetic field; Feb 27; PLOS ONE.

[193] Manolopoulos *et al.* 2018; (PNAS 2021, 118, e2018043118) David E Manolopoulos, Christian Kerpal, Sabine Richert, Christiane R Timmel; 2018, July18; Journal of Chem phys; Jul 21; 149(3):034103. doi: 10.1063/1.5038558.

[194] Vacha, M. Magnetoreception of Invertebrates; Oxford University Press: Oxford, UK, 2019.

[195] Patrick A Guerra, Robert J. Gegear & Steven M Reppert; 2014; Nature Communications, 5, Article number 4164 (2014).

[196] Viachesav V. Krylov *et al.*; 2015; Orientational Behavior of Animals with the Geomagnetic field and mechanisms of magnetoreception; 2015; Izvestiya atmospheric and Oceanic Physics; 51(7); 752–765.

[197] Yeagley, H. L., J. Appl. Phys., 18, 1035 (1947).

[198] Yeagley, H. L., J. Appl. Phys., 22, 746 (1951).

[199] Keetan W T; 1971; Magnets Interfere with Pigeon Homing; Proceedings of the National Academy of Sciences Vol. 68, No. 1, pp. 102–106, January 1971.

[200] Lachlan P al Lindoy; 2020; Quantum mechanical spin dynamics of a molecular magnetoreceptor: Lachlan P Lindoy 1, Thomas P Fay 1, David E Manolopoulos [ChemPhys 2020; April 30, 2020; 152[16], 164107.

[201] Lachlan P al Lindoy *et al.* (2021) Spin relaxation in radical pairs from the stochastic Schrödinger equation; J. Chem. Phys. 154, 084121 (2021).

[202] Holland, R.A.; True navigation in birds: From quantum physics to global migration; Journal of Zoology 293(1–15), 2014.

Chapter 6

Coherence in Ion Channel

6.1 Introduction

Nervous systems use electrical signals which propagate through ion channels. These ion channels are specialized proteins and provide a selective conduction pathway, through which appropriate ions are escorted to the cell's outer membrane.[1] Also, the ion channels undergo fast conformational changes in response[2] to metabolic activities which opens or closes the channels as gates. The gating essentially involves changes in voltages across the membrane and ligands. The voltage dependent ion channels have an ability to alter ion permeability of membranes in response to changes in transmembrane potentials. The channels which are Na, K and Ca voltage gated or synaptic channels gated by acetylcholine, glycine or g-aminobutyric acid seemed similar. The magnitude of current across membrane depends on the density of channels, conductance of the open channel and how often the channel spends in the open position or the probability. Hodgkin and Huxley[3] accounted for the voltage sensitivity of Na^+ and K^+ conductance of the squid giant axon by postulating charge movement between kinetically distinct states of hypothetical activating particles. The gating of voltage-dependent ion channels occur in a probabilistic manner and this probabilistic gating is a source of electrical 'channel noise' in neurons. This noise plays a major role in understanding the reliability (repeatability) of neuronal responses to repeated presentations of identical stimuli. This reliability of neuronal responses is closely associated with the computational mechanism available to the brain or more precisely to the predictability of the brain. So the impact of channel noise (which is generated by the random opening and closing of the gating) on the dynamics of single neuron has drawn a large attestation to the community. The movements and dynamics of the ions inside the channel especially in K^+ channel become one of the fascinating area of research towards the quantum effects in biology. Usually, there are two broad approaches to understand the dynamics of the ionic movement. They are namely Brownian dynamics and Molecular dynamics simulation. However, the recent observational results of Mackinnon *et al.*[10] inadequecy of these approaches. In this situation the use of quantum theory seems to be very much promising one. In this chapter we discuss about the dynamics of the ionic movements and the relevance of quantum theory. Before going into the details, the structure and function of the ion channels will be discussed for convenience.

6.2 Structure and Function of Ion Channel

In spite of the detail electrophysiological studies, the atomic structure of voltage gated ion channels still remained in the dark till the discovery of Mckinnon and his collaborators[6, 28, 29] which obtained a crystal structure of a Ca^{2+} gated K^+ ion channel. This is illustrated in Figure 6.1.

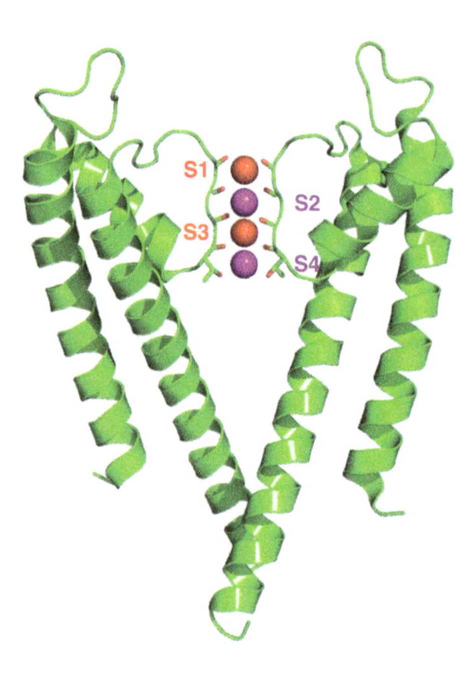

Figure 6.1. Crystallographic structure of the bacterial KcsA potassium channel.

In this figure, only two of the four subunits of the tetramer are displayed for the sake of clarity. The protein is displayed as a green cartoon diagram. In addition backbone carbonyl groups and threonine sidechain protein atoms (oxygen = red, carbon = green) are displayed. Finally potassium ions (occupying the S2 and S4 sites) and the oxygen atoms of water molecules (S1 and S3) are depicted as purple and red spheres respectively.

This provides a mechanism for gating.[7,8] A functional study of KcsA in this context led to a proposal known as the voltage sensor paddle model. Ion channels are membrane spanning proteins with central pores through which ions cross neuronal membranes. The pores through each ion channel flicker between open and closed states, starting and stopping the flow of ions and the electrical current they carry. The moving of potassium ion through pores are depicted in Figure 6.2.

Ion channel pores present a narrow cross section 100 Angstrom and define a path of low dielectric constant across the membrane. When open, the channel pore presents a rather specific ion selectivity filter where the

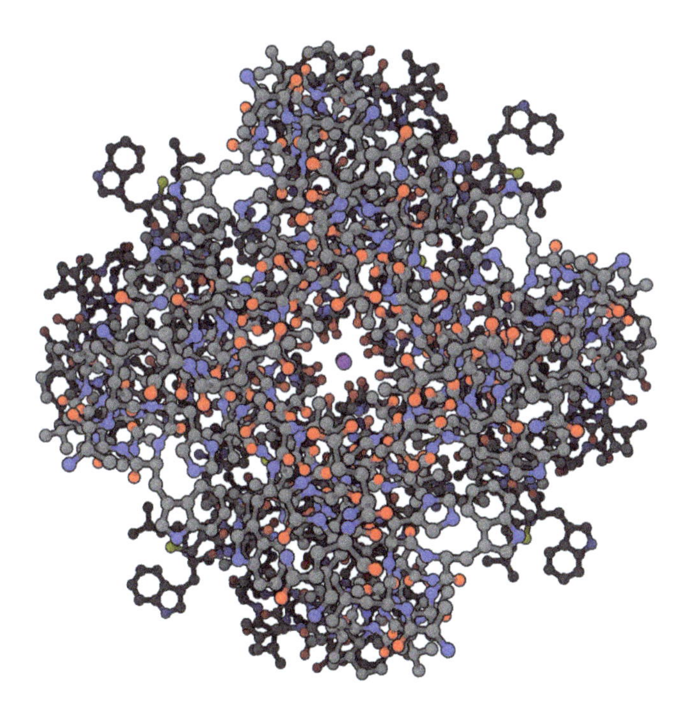

Figure 6.2. Top view of a potassium channel with potassium ions (purple) moving through the pore (in the center).

lines of the electric field tend to be confined to the high dielectric interior of the pore.

The continuity requirement for the orthogonal component of the electric displacement field between the interior of a channel and membrane is given by $\epsilon_w E_w^n = \epsilon_m E_m^n$. Since the lipid membrane has a dielectric constant $\epsilon_m = 2$, while the dielectric constant of water is $\epsilon = 80$, it becomes evident that the orthogonal component of the electric field at the membrane pore boundary must be very close to zero. Indeed, there is only a very slight penetration of the electric field into the interior of the phospholipid membrane. The situation, therefore, is very similar to the expulsion of the magnetic field by a superconductor. As an example, in a channel with a 3 Å; radius and a channel of length $L = 25$ Å, the barrier is about $6k_BT$.

Although it is quite large, it should allow ionic conductivity. This is not too different from such conditions where water filled nano-pores are introduced into silicon oxide films, polymer membranes, etc.

In case of K ion channel, the pore comprises a wide, nonpolar aqueous cavity on the intracellular side, leading up, on the extracellular side to a narrow pore that is 12 inch long and lined exclusively by main chain carbonyl oxygens. Formed by the residues corresponding to the signature sequence TTVGYG, common to all K^+ channels, this region of the pore acts as a selectivity filter by allowing only the passage of nearly dehydrated K^+ ions across the cell membrane. The x-ray crystallographic structure unambiguously demonstrated that the K^+ ions entering the selectivity filter have to lose nearly all their hydration shell and must be directly coordinated by backbone carbonyl oxygens. Specifically, the K^+ ion in the selectivity filter is surrounded by two groups of four oxygen atoms, just as in water. These oxygen atoms are held in place by the protein, and are in the backbone carbonyl oxygen of the selectivity filter loops of the four surrounding filter subunits. In this manner, the filter is constrained in an optimal geometry so that a dehydrated K^+ ion fits with proper coordination, but the Na^+ ions are too small for proper coordination, in accordance with the snug-fit mechanism proposed by Bezanilla and Armstrong.[9] This simple and appealing structural mechanism has been widely adopted to explain the selectivity of the K^+ channel. Indeed, a rigid K^+ pore cannot close down around a Na^+ ion, and so presents a much higher energy than diffusion in water. Indeed, for structural reasons, the selectivity filter cannot constrict sufficiently to bring more than two of the carbonyls within good bonding distance of the Na^+ and as a result, the energy of the Na^+ in the pore is very high compared with its energy in water. This implies a significant structural inability to deform and adapt: the energetic cost upon collapsing to cradle a Na^+ (a structural distortion of about 0.38Å) must give rise to a significant energy penalty (much larger than k_BT assuming the existence of molecular forces opposing a sub-angstrom distortion is tantamount to postulating structural rigidity). Furthermore, the geometry of such a rigid pore must be very precisely suited for K^+ because it would be unable to adapt small perturbations without paying a significant energy price much larger than k_BT. Therefore, precisions in structural rigidity and geometric precision are two underlying microscopic consequences. However, there are fundamental problems with the common view. Proteins, like most biological macromolecular assemblies, are soft materials displaying significant structural flexibility. Despite some uncertainties, the B factors of the KcsA channel indicate that the rms fluctuations of the atoms lining the selectivity filter are on the order of (0.75–1.0)Å. This is in general agreement with numerous independent MD simulations of KcsA. The magnitude of atomic thermal fluctuations is fundamentally related to the intrinsic flexibility of a protein, i.e., how it responds structurally to external perturbations. These considerations suggest that, at room temperature, the flexible fluctuating channel should distort easily to cradle Na^+ with little energetic cost. The flexibility of the pore is further highlighted by the experimental observation that K^+ is needed for the overall stability of the channel structure.

Therefore, even ion channel proteins appear to be inherently too flexible to satisfy the requirement of the traditional snug-fit mechanism. Furthermore, structural flexibility is absolutely essential for ion conduction since in some places the diameter of the pore in the x-ray structure of KcsA is too narrow to allow the passage of a water molecule or a K^+ ion. In the electric circuit equivalent model the channel proteins thereby play the role of field-effect transistors, with a voltage imposed across the cell membrane gating the transfer of ion bound charges through the membrane. Two different aspects characterize channel function: ion selective permeation and gating, i.e., control of access of ions to the permeation pathway. We will base the subsequent concept on potassium channels, employing the crystal structure of the KcsA and KvAP channels at a resolution ranging from 1.9×10^{-10}m to 3.2×10^{-10}m, as revealed by the work of MacKinnon's group.[10]

The channel structure is basically conserved among all potassium channels with some differences relating to gating characteristics rather than ionic selectivity. In the open gate configuration the protein selects the permeation of K^+ ions against other ions in the selectivity filter and can still allow ion permeation rates near the diffusion limit. In the view of Hodgkin-Huxley (HH) type models of membrane potentials, K^+ permeation stabilizes the membrane potential, resetting it from firing threshold values to resting conditions. The atomic level reconstruction of parts of the channel and accompanying Molecular Dynamic (MD) simulations at the 10^{-12} s resolution have changed the picture of ion permeation: the channel protein can transiently stabilize three K^+ states two within the permeation pathway and one within the "water cavity" located towards the intracellular side of the permeation path.[11]

6.3 Dynamics of Ion Movement and Coherence in Ion Channels

The approaches towards dynamics of ion transport in protein membranes (ion channels) are generally considered to be classical and based on molecular dynamics or Brownian dynamics. Recently, one of the present authors showed that[12] the dynamics of K-ion channel can be explained using nonlinear Schrodinger equation which is compatible with the results of MacKinnon's experimental observation. Here, the two K-ions may form an entangled state within selectivity filter during a finite period of time. The temperature within the channel is generally considered to be high enough to destroy the coherence within very short period. Moreover, molecular modes of the protein environment induce dynamical decoherence, which destroys the quantum mechanical superposition of states in a very short period of time. So quantum mechanical approach in these area of research was mostly speculative and far from experimental realization. But in recent times, some experimental demonstration of the presence of quantum coherence in the process of photosynthetic energy transfer[13,14] lead us to reconsider the theoretical approach and understanding. Here we are concerned with the question that whether and under what condition a sustainable quantum superposition is achievable in the process of transfer of ions through biological channels. Now it is quite a practical argument that the quantum state of the traversing ion is strongly coupled with the molecular vibrational modes of the protein environment and hence fast decoherence is almost absolutely unavoidable. But it is to be noted that the traversal time of the ion through the membrane is also quite small and it is the ratio of the time scale of decoherence with this traversal time that plays an important role in understanding the maintenance of coherence. If the decoherence time is larger than the traversal time of the ion, then the quantum superposition of the ionic states is sustainable enough for the traversing entity within the period of ionic transfer. Here we also need to consider the effect of temperature in estimating the decoherence time. With the recent developments of quantum thermodynamics,[22] a new concept of temperature (known as spectral temperature) for micro-states at non-equilibrium condition has been proposed, instead of the usual concept of thermodynamic temperature. Since ion transport through protein membrane is essentially a non-equilibrium phenomena, we propose that the spectral temperature plays important role in understanding the coherence in the channel dynamics.

Considering the master equation of the density operator for a certain quantum system, the decoherence time[15] can be written as

$$\tau_{dec} = \frac{\hbar^2}{2m\gamma K_B T(x - x')^2} \tag{6.1}$$

where γ is the relaxation (dissipation) parameter, T is the thermodynamic temperature and $\Delta x = x - x'$ is the spatial shift of the particle. Now there are numerous definition of traversal time, among which phase time and dwell time are generally accepted by the community.[16] Phase time is equated to dwell time with an additive self-interference term and argued to be the time interval between the energy storage and release in the barrier region.[17–19] So for non-dispersive barrier, where the self-interference term vanishes, the dwell time can be readily interpreted as the life time of energy storage and release in the barrier region. In a previous work,[20] we have calculated the weak value of dwell time as

$$\tau_D = \frac{1}{\gamma} \coth\left(\frac{\gamma \tau_M}{2}\right) \tag{6.2}$$

where τ_M is the measurement time.

The protein membrane can be assumed as an array of molecules with certain modes of vibration. We have mentioned earlier that the quantum states of the traversing entity is interactively coupled to the molecular vibration modes of the protein membrane. The traversing quantum entity loses energy to the vibrational modes, due to the presence of this environmental coupling. So the protein membrane can be interpreted as sort of capacitive system, which can store and release energy with the dwell time as it's lifetime of energy storage. In this case the time interval between the opening and closing of the channel gates can be taken as the measurement time (τ_M). Now we consider a double well potential of the form

$$V(x) = \frac{1}{2}m\omega^2 x^2\left[\left(\frac{x}{a}\right)^2 - A\left(\frac{x}{a}\right) + B\right] \tag{6.3}$$

where A and B are dimensionless constants.

For the particular case of asymmetric double well, $A = 14$ and $B = 45$.

The bistable nature of the potential is useful in practical situation, since the ionic transfer in the channel can be interpreted as tunneling between the two stability regions situated at x_0 and x_2 (see Figure 6.3).

It is also very important to reconsider the concept of temperature in this aspect. Temperature, as we know from the context of thermodynamics, is a property of equilibrium. The ensemble mean of the kinetic energy equals to Boltzman constant times the temperature. So assuming ergodicity, temperature is defined as the time averaged kinetic energy. But the process of ion transfer through channels is a dynamical process interactive to the protein environment and subject to energy exchange with the environment. This is certainly not a situation for thermodynamic equilibrium and hence

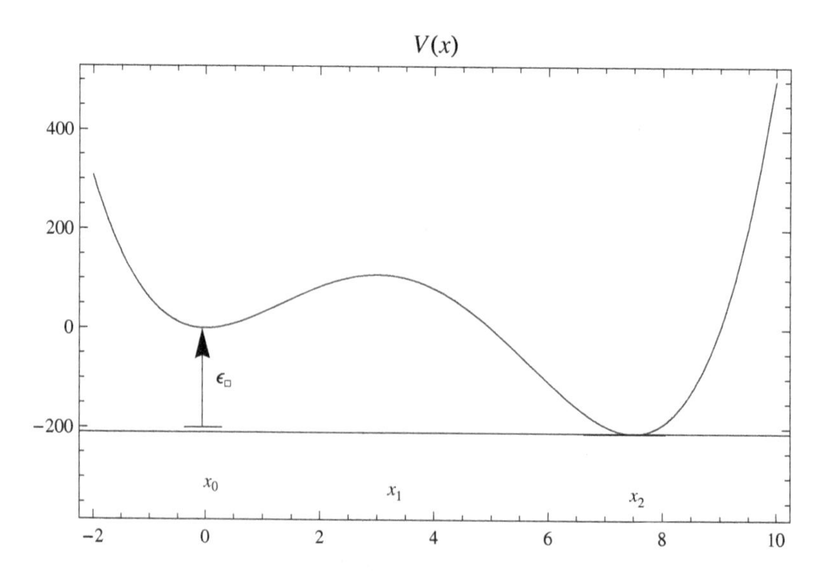

Figure 6.3. V(x) vs. x with parameters $A = 14$ and $B = 45$.

the usual concept of thermodynamic temperature may not be suitable. Here we introduce the concept of spectral temperature, originally formulated by Gemmer *et al.*[22] It is defined as a function of the microstates, to include the non-equilibrium properties of the system. It is formulated as a function of energy occupation probability of the different states of a certain quantum system. This temperature evolves with the evolution of the probability of occurrence. The inverse of the spectral temperature is defined as

$$\frac{1}{K_B T_{spec}} = -\left(1 - \frac{P_0 + P_N}{2}\right)^{-1} \sum_{i=1}^{N} \left(\frac{P_i + P_{i-1}}{2}\right) \times \left[\frac{\ln\left(\frac{P_i}{P_{i-1}}\right) - \ln\left(\frac{\phi_i}{\phi_{i-1}}\right)}{E_i - E_{i-1}}\right] \quad (6.4)$$

where P_i is the probability of finding the particle within an energy compartment with mean energy E_i, having the degree of degeneracy ϕ_i. This definition depends on the energy probability distribution and of course, the spectrum of the concerning system. So it cannot change in time for an isolated system and defined independent of the fact that whether the system is in equilibrium or not. In this definition, the association of quantum probability, which evolves in time, gives temperature a evolving feature representing the dynamical situation between two successive equilibrium. Now here we are approximating the bistable potential as a two-state system having only the ground states of the asymmetric wells. So for our case of non-degenerate two-state system, the expression of the spectral temperature reduces to

$$\frac{1}{K_B T_{spec}} = -\left(1 - \frac{P_0 + P_1}{2}\right)^{-1} \left(\frac{P_0 + P_1}{2}\right) \left[\frac{\ln\left(\frac{P_0}{P_1}\right)}{E_0 - E_1}\right] \quad (6.5)$$

Where P_0 and P_1 are the probabilities corresponding to the ground states of the lower and higher well respectively. Since P_0 is the probability corresponding to the lowest state, it should not decay. Now using the decay probability of the higher state within the time interval equal to the dwell time, we find

$$\frac{1}{K_B T_{spec}} = \frac{1}{E_1 - E_0} \coth\left(\frac{\gamma \tau_M}{2}\right) \coth\left[\frac{1}{2} \coth\left(\frac{\gamma \tau_M}{2}\right)\right] \quad (6.6)$$

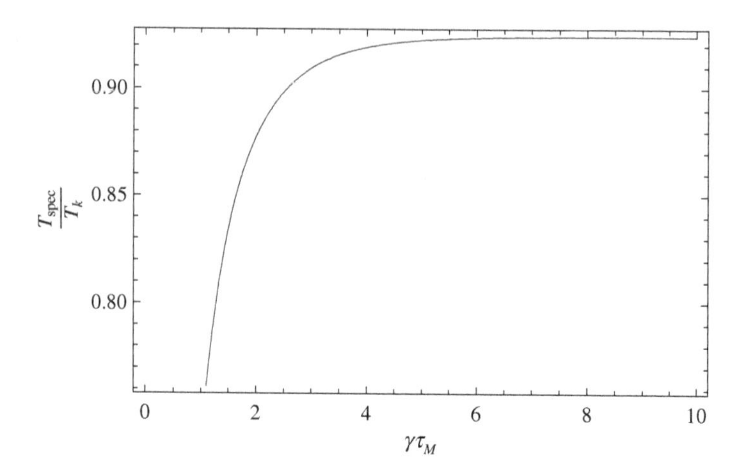

Figure 6.4. T_{spec}/T_k vs $\gamma\tau_M$. Here we keep γ as a constant and basically study the variation with increasing τ_M.

Now if we consider the energy loss by the particle traversing from the higher barrier to the lower one in terms of the usual kinetic temperature T_k as $E_1 - E_0 = \frac{1}{2}K_B T_k$, we get

$$\frac{T_{spec}}{T_k} = 2\tanh\left(\frac{\gamma\tau_M}{2}\right)\tanh\left[\frac{1}{2}\coth\left(\frac{\gamma\tau_M}{2}\right)\right] \tag{6.7}$$

Under the condition $\tau_M \gg \frac{1}{\gamma}$, we find that $T_{spec} \simeq T_k$. i.e. the spectral temperature is almost equal to the usual kinetic temperature, if the time interval of gate opening and closing is very greater than the dissipation time scale. If the gate opening and closing mechanism is slow enough to be considered as a quasi-static process, the spectral temperature remains very close to it's kinetic counterpart. Now it should also be noted that though the process is sort of quasi-static one, it should not be considered as a reversible process, because the coupling to the molecular vibrational modes of the protein environment ensures some generation of dissipative entropy.

In case of a double well potential we get

$$\frac{\tau_{dec}}{\tau_D} = \frac{2\hbar}{w}\sqrt{\frac{2}{m\epsilon_0}}\coth\left[\frac{1}{2}\coth\left(\frac{\gamma\tau_M}{2}\right)\right] \tag{6.8}$$

where

$$\epsilon_0 = E_1 - E_0$$

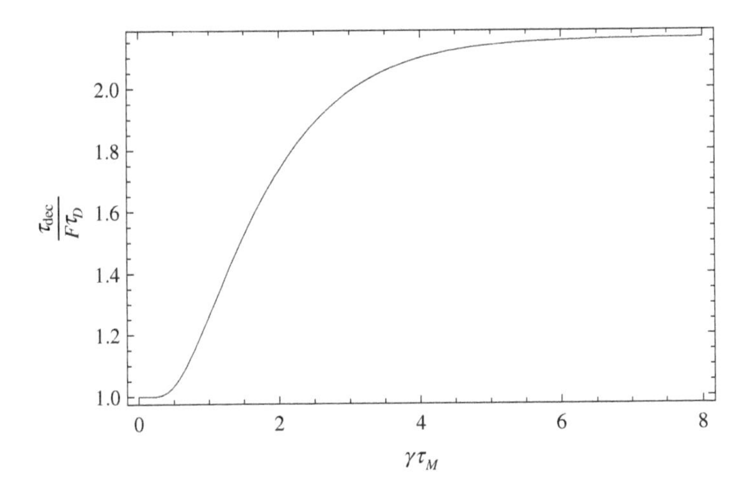

Figure 6.5. τ_{dec}/τ_D vs $\gamma\tau_M$. Here also we keep γ as a constant and basically study the variation with increasing τ_M. Here $F = \frac{2\hbar}{w}\sqrt{\frac{2}{m\epsilon_0}}$. As with increment of τ_M, the process becomes quasi-static, we see that the ratio of the two timescales also reach a stable value.

is the asymmetry energy of the potential and

$$w = \frac{15a}{2}$$

is the separation length between the wells. For the quasi-static condition $\tau_M \gg \frac{1}{\gamma}$

$$\frac{\tau_{dec}}{\tau_D} \simeq \frac{4.5\hbar}{w}\sqrt{\frac{2}{m\epsilon_0}} \tag{6.9}$$

So we find that in the quasi-static region, whether the decoherence time scale is larger than the dwell time depends on the mass of the traversing particle, the length scale of the channel and the asymmetry energy which is basically the energy lost by the particle during the process of traversal. The ratio of the time scales is inversely proportional to all the mentioned parameters. i.e. greater the inertia of the traversing particle, stronger the interaction (hence greater energy loss) and larger the traversal length, faster shall be the decoherence and hence the situation will be more and more classical. But if the parameter can be chosen in such a way in which the process of decoherence is slower than the process of ionic transfer (i.e. $\tau_{dec}/\tau_D > 1$), then quantum superposition will be sustainable enough within the time period of ionic transfer.

Now we turn our attention to the "degree of coherence" for such cases of ionic transfer through protein membranes and try to establish a relation of the afore mentioned ratio of decoherence-dwell time scales with it. Convenient way to discuss coherence in quantitative terms can be done through the introduction of normalized form of correlation functions such as "degree of coherence"[23]

$$g^{(n)}(\xi_1 \ldots \xi_{2n}, t_1 \ldots t_{2n}) = \frac{G^{(n)}(\xi_1 \ldots \xi_{2n}, t_1 \ldots t_{2n})}{\prod_{j=1}^{2n} [G^{(1)}(\xi_j, \xi_j, t_1)]^{1/2}} \tag{6.10}$$

where $\xi = x/a$. For our case of double well system approximated as two-state system is reduced to

$$g^{(1)} = \frac{G_{12}^{(1)}(\tau)}{\sqrt{G_{11}^{(1)}(0) G_{22}^{(1)}(0)}} \tag{6.11}$$

where

$$\begin{aligned} G_{12}^{(1)}(\tau) &= \langle \psi_1(\xi, t) \psi_2(\xi, t + \tau) \rangle \\ &= e^{-\gamma\tau} \int_{-\infty}^{\infty} \psi(\xi - \xi_1) \psi^*(\xi) d\xi \end{aligned} \tag{6.12}$$

and

$$0 \leq g^{(1)} \leq 1 \tag{6.13}$$

For completely coherent situation, the degree of coherence is 1 and for completely decohered case it is 0. Generally the value should lie in between. If we approximate the potential around the left well at $\xi = 0$ as a harmonic potential, we find that the wavefunction can be estimated as

$$\psi(\xi) = \left(\frac{\nu}{\pi}\right)^{1/2} \exp\left[-\frac{1}{2}\nu\xi^2\right] \tag{6.14}$$

where $\nu = \sqrt{B}\frac{m\omega a^2}{\hbar}$. Expressing the asymmetry energy as

$$\epsilon_0 = V(x_0) - V(x_2) \tag{6.15}$$

we find that for a double well potential ($A = 14$ and $B = 45$)

$$\omega = \frac{4}{15a}\sqrt{\frac{2\epsilon_0}{15m}} = \frac{2}{w}\sqrt{\frac{2\epsilon_0}{15m}} \tag{6.16}$$

So for the interval of dwell time given by equation (6.2), we find that the degree of coherence

$$g^{(1)} = \exp\left[-\sqrt{\frac{m\epsilon_0}{2}}\frac{w}{\hbar}\right]\exp\left[-\coth\left(\frac{\gamma\tau_M}{2}\right)\right] \tag{6.17}$$

For the quasi-static condition $\tau_M \gg 1/\gamma$

$$g^{(1)} \approx \exp\left[-\sqrt{\frac{m\epsilon_0}{2}}\frac{w}{\hbar}\right] \tag{6.18}$$

From equation (6.9) and (6.17), we can also establish a relation

$$\frac{\tau_{dec}}{\tau_D} = \frac{4.5}{\ln\left[\frac{1}{g^{(1)}}\right]} \tag{6.19}$$

From this relation we see that for completely coherent and decoherent situation, the decoherence-dwell time ratio is infinity and zero respectively and in general situation it lies in between. So this time-scale ratio gives us a certain measure of coherence. As the value of this ratio gets bigger, the "quantumness" of the traversing entity increases.

Depending on the above analysis, we suggest that the ion selectivity filter may exhibit quantum coherence which can play crucial role in the process of selectivity and conduction of specific ions in biological membranes. For the time scales shorter than that of decoherence time, quantum coherence can be expected to sustain and have vital importance in the dynamics, despite the presence of interactive protein environment. Our analysis shows that for a sort of quasi-static situation, where the gate opening and closing mechanism is slower than the relaxation (dissipation) time scale, the decoherence-dwell time ratio reaches a static value and can also be greater than unity depending on the mass, energy and length parameters. In such situations, coherent phenomena like entanglement can be of vital importance in understanding the mechanism of selectivity and transport. In case of $K+$ filter, there exists two energetically almost degenerate binding states, commonly referred as $(1, 3)$ and $(2, 4)$ states.[24–26] Presence of quantum superposition may lead us to explain the transport phenomena in terms of quantum mechanical tunneling between these two states. The interplay between quantum coherence and environmental noise induced dephasing may also be of fundamental importance. Progress in the atomic spectroscopy of membrane proteins indicate that protein membrane organization may carry a certain coding potency, implying quantum entanglement within ion channels.[24–27] Increasing number of

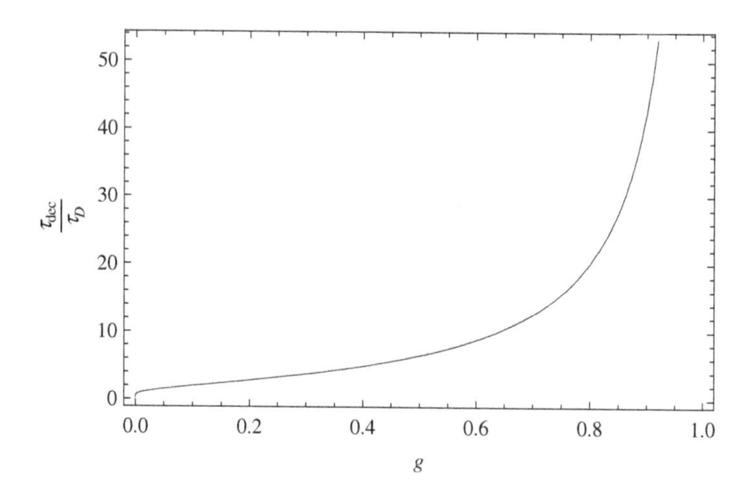

Figure 6.6. τ_{dec}/τ_D vs. $g^{(1)}$.

studies are indicating towards the probabilistic nature of ion channel gating mechanism. In the light of these researches, we conclude that it may be relevant to build certain model of quantum information system driven by the entangled ionic states in the voltage gated selectivity filters, which can provide necessary inferences into the biological ion channel dynamics.

6.4 Discussions

The above discussions on possible existence of quantum effects on microscopic structures in the brain give rise to many challenges. For example, Salari *et al.*[30] studied the possible quantum effects on the infrastructures in the neurons and its role in cognitive processing. Usually in the mainstream cognitive science the quantum effects at the neuronal level underlying cognition and consciousness is not considered. However, the results of Salari *et al.* clealry show that the decoherence time is of the order of picoseconds considering the superposition states of potassium ions. This decoherence time is not long enough for the cognitive processing.

The selectivity of ion channel is still an open question in biology. Recently, a novel attempt has been taken by several authors[31] to understand the issue of selectivity. They proposed whether quantum interference of ions through ion channels can be used to understand the selectivity. Their results indicate that the quantum interference of ions seems unlikely due to environmental decoherence.

References

[1] Mitra, I., & Roy, S. (2008). Neurons, cooperativity and the role of noise in brain. NeuroQuantology, 6(2).
[2] Hille, B. (1992). Ionic channels of excitable membranes. 1992. Sunderland, Massachusetts: Sinauer Associates Inc.
[3] Hodgkin AL, Huxley AF. A quantitative description of membrane currents and its application to conduction and excitation in nerve. J Physiol (Lond) 1952;117:500–544.
[4] Doyle, D. A., Cabral, J. M., Pfuetzner, R. A., Kuo, A., Gulbis, J. M., Cohen, S. L., Chait, B.T. & MacKinnon, R. (1998). The structure of the potassium channel: molecular basis of K+ conduction and selectivity. Science, 280(5360), 69–77.
[5] Zhou, Y., Morais-Cabral, J. H., Kaufman, A., & MacKinnon, R. (2001). Chemistry of ion coordination and hydration revealed by a K+ channel–Fab complex at 2.0 Å resolution. Nature, 414(6859), 43–48.
[6] Jiang, Y., Lee, A., Chen, J., Cadene, M., Chait, B. T., & MacKinnon, R. (2002). Crystal structure and mechanism of a calcium-gated potassium channel. Nature, 417(6888), 515–522.
[7] Cha, A., Snyder, G. E., Selvin, P. R., & Bezanilla, F. (1999). Atomic scale movement of the voltage-sensing region in a potassium channel measured via spectroscopy. Nature, 402(6763), 809–813.
[8] Durell, S. R., & Guy, H. R. (1992). Atomic scale structure and functional models of voltage-gated potassium channels. Biophysical Journal, 62(1), 238–250.
[9] Bezanilla, F., & Stefani, E. (1994). Voltage-dependent gating of ionic channels. Annual review of biophysics and biomolecular structure, 23(1), 819–846.
[10] Lockless, S. W., Zhou, M., & MacKinnon, R. (2007). Structural and thermodynamic properties of selective ion binding in a K+ channel. PLoS Biol, 5(5), e121.
[11] Roy, S., Mitra, I., & Llinas, R. (2008). Non-Markovian noise mediated through anomalous diffusion within ion channels. Physical Review E, 78(4), 041920.
[12] Roy Sisir and Llinás Rodolfo (2009). C.R. Biologies **332**, 517 (2009).
[13] G.S. Engel *et al.*; Nature **446**, 782 (2007).
[14] Mercer I.P. *et al.* (2009); Phys. Rev. Lett. **102**, 057402.
[15] W.H. Zurek; arXiv:gr-qc/9402011v1 3 Feb 1994.
[16] E.H. Hauge and J.A. Stoveng; Rev. Mod. Phys. **61**, 917 (1989).
[17] H.G. Winful; Opt. Express **10**, 1491 (2002).
[18] H.G. Winful; Phys. Rev. E **68**, 016615 (2003).
[19] H.G. Winful; Phys. Rep. **436**, 1 (2006).

[20] S. Bhattacharya and S. Roy; Phys. Rev. A **85**, 062119 (2012).
[21] Mohsen Razavy; *Quantum Theory of Tunneling* (2003), World Scientific Publishing, p. 255.
[22] J. Gemmer, M. Michel and G. Mahler; Quantum Thermodynamics, Vol. LNP657, Springer, Heidelberg, Berlin, 2004.
[23] R.J. Glauber; Quantum Theory of Optical Coherence, WILEY-VCH Verlag GmbH and Co. KGaA, 2007.
[24] G. Bernroider and S. Roy; SPIE 2005; **5841**:205.
[25] J. Summhammer, V. Salari and G. Bernroider; Journal of Integrative Nanoscience **11**, 123 (2012).
[26] V. Salari, J. Tuszynski, M. Rahnama and G. Bernroider; JPCS **306**, 012075 (2011).
[27] Hans J. Briegel and Sandu Popescu; arXiv:0806.4552v2 (2009).
[28] D. Doyle, R. MacKinnon *et al.*; Science **280**, 69 (1998).
[29] Y. Zhou, A. Morais-Cabral, A. Kaufman, R. MacKinnon; Nature **414**, 43 (2001).
[30] V. Salari, N. Moradi, M. Sajadi, F. Fazileh, and F. Shahbazi (2015) Quantum decoherence time scales for ionic superposition states in ion channels; Physical Review E 91, 032704.
[31] Vahid Salari, Hamidreza Naeij & Afshin Shafiee (2017) Quantum Interference and Selectivity through Biological Ion Channels; Scientific Reports — 7:41625 — DOI: 10.1038/srep41625.

Chapter 7

Conclusion

This book is about the application of quantum theory to aspects of biology that cannot be accurately described by the classical laws of physics. It means quantum theory has to be applied to understand those processes. Quantum physics is a successful theory so far as it describes the nature at the inanimate microscopic level. Biology and quantum physics have long been considered as unrelated disciplines. In recent decades the idea of quantum coherence, quantum entanglement and other nonclassical effects attracted large attention to clarify some of the fundamental issues towards systems of increasing complexity. On the otherhand, more and more refined explanations of macroscopic phenomena for living organisms are put forward based on the improved understanding of molecular structures and mechanisms. These developments in animate and inanimate objects help us to study some of the phenomena related to living organisms using the concepts of quantum theory. In this book we have discussed some phenomena in biology which need quantum principles for their explanation.

At the primary level, the dynamics in living organisms are associated with transfer of charge and energy. The charges which involve are electrons, proton and ions. Charge transfer and excitation energy transfer in photosynthesis are considered to be most established areas of quantum biology. The study of enzyme catalysis based on the coupling of electrons and protons attracts large attention in more recent investigations.

At this present stage of scientific developments, taking into account of different light harvesting systems, photosynthesis has been transformed to be a fascinating challenge. In that regard, we have already mentioned the works of Förster as unique, pioneering, and, if not, revolutionary in this field of research. The theoretical framework developed by Förster and others describes how electronic excitation migrates in the photosynthetic apparatus of plants, algae, and bacteria from light absorbing pigments to so-called reaction centers where light energy is utilized for the eventual conversion into chemical energy. The role of quantum coherence in photosynthesis attracted a large attention to the community. Following the discussions for the Quantum coherence in photosynthetic light harvesting, we experience different meaning for different models, applied for the explanation of the coherence phenomena, meant for different circumstances. The most important conclusion that we can draw from the vast amount of works over recent years on coherence and coherent dynamics in light harvesting is that both the suggestions by experiments and predictions by theory have been accumulating evidence that the mechanisms for energy transfer in light-harvesting involve quantum coherence. Careful analysis is needed to differentiate between classical and quantum coherence in photosynthesis.

Quantum effects in biology are not just a quirk of plants and other organisms that do the peculiar job of turning sunlight into fuel. They may also provide an answer to a scientific puzzle that has been around since the 19th Century: how migratory birds know which way to fly. Bird navigation is a complex enterprise, requiring birds to make repeated and varying orientation decisions based on directional and positional information. Though great progress achieved in understanding the mystery of how birds navigate using magnetic cues, it still remains something of a mystery as to how they sense the magnetic field. The integration of many disciplines outside its root in behavioral biology is needed to address questions of neurobiology, molecular aspects, and the physics of sensory systems and environmental cues involved in bird navigation, often involving quantum physics.

Sensing the smell, i.e., Olfaction process, is accepted as one of the most ancient and again at the same time, one of the most intriguing characteristics of living organisms, maintaining a typical contrast nature when compared with the associated environment. It is the oldest and most fundamental aspect of chemical sensing which are being applied by almost all kind of lifeforms in interpreting their surroundings. The process of smelling is caused by certain kind of small molecules, neutral, volatile in nature, known as odorant. This process has certain typically interesting and fascinating characteristics, thus attracting the science community greatly, delivering a great number of unique theories each of which tried to define and explain henceforth the mechanism behind such process. The lock and key paradigm was one of the earliest attempts to rationalize remarkably selective responses to different molecules especially for large molecules. However, for small molecules, the underlying idea that shape is the sole critical factor fails apart quite badly. Our sense of smell allows us to discriminate between small molecules in very low concentrations via scent molecules interacting with receptors in the nose. Presently, the biomolecular processes of olfaction are not yet fully understood, and some evidence suggests that a mechanism based solely on the size and shape of odorant molecules is inadequate.

The possible existence of quantum effects on microscopic structures in the brain give rise to many challenges. For example, some works have been done on the possible quantum effects on the infrastructures in the neurons and its role in cognitive processing. Usually in the mainstream cognitive science the quantum effects at the neuronal level underlying cognition and consciousness is not considered. However, some recent investigations on possible applicability of quantum theory in the functioning of ion channel (K-ion channel) open up new vistas in quantum biology.

A large number of people do believe quantum mechanics provides a framework into which all physical theories must fit. In other words, this is known as the trivial application of quantum theory to understand the physical universe. In 1932, Pascual Jordan published the first book on quantum biology. Since its publication the explanation of many phenomena in the world of living organisms clearly show the inadequacy of classical physics. Recent experimental findings clearly suggest that the classical probability theory is still not successful to explain modalities in human cognition, especially in connection to decision making. The major problem seems to be the presence of epistemic uncertainty and its effects on cognition at any time point. Moreover, the stochasticity in the model arises due to the unknown path or trajectory (the definite state of mind at each point in time) a person follows. The consideration of "black box model of human mental functions" produces much ambiguity. A generalized version of probability theory, borrowing the idea from the quantum paradigm, may be a plausible approach. Quantum theory enables a person to be in an indefinite state (superposition state) in the context of neurobiology, especially in relation to central nervous system and allows all these states to be potentially (of course, with prior probability amplitudes) expressible at each moment. Thus, a superposition state seems to provide a better representation of the conflict, ambiguity or uncertainty that a person experiences at each moment. However, the framework of quantum probability considering the superposition of mental states is an abstract framework devoid of material content like the concept of elementary particles and various fundamental constants in nature. This framework can be applied to any branch of science dealing with decision making, such as biology and the social sciences. Very few attempts have been made so far in the context of neuroscience and higher order cognitive activities.[1]

For centuries, the various mental states and decision making have been widely discussed in various Indian traditions. In Buddhist traditions, the various mental states and decision making have been widely discussed in Abhidamma and one of the states of citta or mind has been considered as "upekkha sahagata" or "citta with neutral feeling" i.e. like state of citta or mind with "neither pleasure nor pain". The similarityof these states can be drawn with the intermediate states associated with the superposition of complementary states in quantum theory. This kind of mind can be analyzed in relation to the state of mind known as "equanimity" in other school of India philosophy like Bhagavad Gita. It is described as the state where the human being will be able to make the mind steady and balanced in all possible conditions of life and as well, will be able to think and make a decision, being in an absolute tranquil and equiposed state of mind. Essentially, it refers to a state of mind that can not be swayed by biases abd preferences.[2] It is worth mentioning, similar to the potentiality interpretation of quantum theory,[3] we can think of such a state of mind, where the mind is capable of of decision making even while remaining in a state, suerposed of two complementary aspects. As explained earlier, the process of decision making is a measurement process that occurs at certain hierarchial level of neuronal architecture present in central nerovous system. So even when this state of mind can be considered as a superposed state of complementary aspects, potentially always remains for the actualization of one of the complementary states. Then it raises the possibility of describing the state of mind called "equanimity" using framework of quantum logic where a decision or judgement can be made without any bias or preference.

In this book we emphasized the non-trivial application of quantum theory in the sense that it is not possible to explain some phenomena in the world of living organism using the laws of classical physics.

References

[1] Sisir Roy, Decision Making and Modelling in Cognitive Science, Springer, 2016.
[2] Anuruddha B. and Bodhi B.; A comptehensive mannual of Abhidhamma, 2000.
[3] The Family of Propensity Interpretations of Probability; Lecture at University of Bologna,, Archives of Scientific Philosophy, Special collections Department, University of Pittsburg.

Index

G

H

I

J

K

L

M

N

O

P

www.ingramcontent.com/pod-product-compliance
Lightning Source LLC
Chambersburg PA
CBHW071801130925
32297CB00004B/17

9789813109827